8/1/90

ON THE GLASSY SEA

ON THE GLASSY SEA

an astronomer's journey

Tom Gehrels

AMERICAN INSTITUTE OF PHYSICS NEW YORK

The author has assigned his royalties from this book to the Spacewatch Fund of the Lunar and Planetary Laboratory, University of Arizona, Tucson, Arizona 85721

Printed in the United States of America.

Library of Congress Cataloging-in-Publication Data

Gehrels, Tom, 1925–
On the glassy sea.

Bibliography: p.
Includes index.
1. Gehrels, Tom, 1925– 2 Astronomers—Netherlands—Biography.
I. Title.
QB36.G44A3 1988 520′.92′4 [B] 88-34949
ISBN 0-88318-598-9

CONTENTS

Of Passengers and Pilots

Research and Reality

INDIA AND BEYOND

People and Science

Experiments in Health

Integration

Preface

The children had asked that I write down my tall tales that range from the Netherlands through the United States to India. The writing started when daughter Jo-Ann needed to interview someone for a term paper; the teacher liked the stories, and so, perhaps, will you.

This is therefore a book of stories, to be read for entertainment, perhaps out loud, skipping the parts that do not interest you. I will tell of peace in the 1930s, war in the forties, student life in the fifties, the changes of the sixties, space flight in the seventies, and India in the eighties. I will describe the life of an astronomer, an observer who works with telescopes. My branch of astronomy is surveying, searching for new worlds and comets and asteroids using the telescopes mostly in an old-fashioned manner with the observer being present underneath the nightly sky. An astronomical observatory dome is a good place to become aware that the observable realm is at least 10^{79} times larger than I am, that is 10,000,000,000,000,000,000,000,000,000,000,000,000,000,000,000,000,000,000,000,000,000,000,000,000,000,000,000,000 times. No one can help but be in awe at that vastness of our universe. John Milton describes it in *Paradise Lost*[1] :

> Witness this new-made world, another heaven
> from heaven gate not far, founded in view
> on the clear hyaline, the glassy sea;
> of amplitude almost immense, with stars
> numerous, and every star perhaps a world
> of destined habitation. . . .

This book has an undertone of trying to understand our origins. As a child I could not believe what we were taught about creation so I decided to find out for myself as a scientist. It is therefore also a detective story, in search of God.

In writing this book I came upon a few other questions, such as: Did von Braun know about the Dora-Nordhausen concentration camps? Who won the infamous "England Spiel" between the Germans and British? What happened in the cockpits at Tenerife to cause the worst disaster involving aircraft? What all can we see from an airplane window? Could Astronaut Virgil Grissom have drowned after his landing in the water in 1961? How did we experience the early history of planetary sciences and NASA's space program? Was the discovery of the rings of Uranus a case of serendipity? What is the proper balance between a strong national defense with a concern for nuclear proliferation? What can yoga do for our health? How old are the atoms within us; do they all come from the Big Bang? I have searched for answers to these questions. You may judge how successfully.

In the peaceful years before World War II it was fun to bike through the Netherlands. Later I was fortunate to be sent as a paratrooper on various assignments, also to the Far East. After the war I was trained as an astronomer and went on to observe with various telescopes, four of which were on balloons or spacecraft. I was involved with the Pioneer space program; and since its story had to be told abroad, a lot of travel ensued with exposure to dissidence and environmental issues. My full-time professional work is in planetary sciences, and some of it is done in India. The gentle influence of India seems to have brought me a clarity and radiance for further thought and action.

What to call this book? Because of the search for our origins, the title might have been *Where Are You From, Tom*? I am overly sensitive to that question because of my foreign accent, which seems to be getting worse with each trip abroad. Going into a drug store, for instance, even to get a small item, I may be asked that question, and I always wonder whether it is out of curiosity or to put me in my place. In any case, replying with "Gila Bend," a small Arizona town, usually works wonders in changing the topic and getting my purchase fast. Other titles were considered such as *Observing the Universe*, after Freeman Dyson's sensitive book with a similar title.[2] However, the glassy sea encompassed it all.

I thank Shirley Marinus for her patient assistance, Tom von Foerster for his perceptive editing, and many friends and colleagues for checking my descriptions and recollections; disagreements are in the Notes, where other remarks and references to books may also be given (a superscript in the text indicates that there is such a Note).[3] Correspondence from readers will be appreciated, so that I can correct mistakes or misimpressions in a revised edition. I have tried to write what actually happened. Our high-school teacher Mr. G. R. Bax of Hilversum taught us to mistrust history: It is usually written by the winner, selected for the better events, and embellished in the retelling with the wisdom of hindsight. During my checking I found that the core of a recollection is usually correct while we may not agree on the details. One case demonstrated most clearly that there is no Absolute Truth. I sometimes have a nightmare of having made a fatal mistake as a commander in the field by not listening to a more experienced sergeant. However, in checking with the sergeant, forty years later, it turned out to be his nightmare too, as he believed that he had not followed my advice! Now we both sleep better as we learned to let it be, whatever way it occurred.

Tom Gehrels
Space Sciences
The University of Arizona
Tucson, AZ 85721

Now then, Neil and Ellen, George and Jennifer, Jo-Ann and whoever else wants to come along, imagine us on a Time-Reversing Flying Machine that can send us back more than half a century and set us down near the center of the Netherlands, in a town called Baarn. There, among the clouds, do you see the church and steeple?. . . . Down we land and tiptoe in. . . . It is Sunday morning. . . . They are singing. . . . Up the stairs, to a gallery in the back of the church. . . . Just smile at the Postma family, he is the Principal of our school, the Father of Sep and Doortje. . . . Now look. . . . Up front with the Elders. . . . The little boy with the blond hair. . . . That's him. . . . Now we can start. . . .

EUROPE AND THE FAR EAST

The Netherlands in the Thirties

1. Does a Steeple Point to Heaven?

A small boy in a Dutch Reformed Church, watching and wondering. The inside of the church was whitewashed and simple without ornaments. We knew that the Catholic statues and paintings and ceremony had been removed from the churches in the sixteenth century, when the Netherlands became an independent, Protestant nation, its churches following the teachings of Martin Luther or John Calvin.

The Calvinist church in our town of Baarn held about 600 people, and it was usually full. Our family belonged to that church almost literally. The doctrine ruled everything in our lives, and the indoctrination was thorough. On Sundays we would walk to this nearby church with its steeple among the trees and the flowers. The steeple was important, Father said, to guide us from afar by pointing to heaven. So heaven must be beyond the clouds or the stars which was all I could see in that direction.

We could afford to come not so early, as our seats were reserved, and it was nice to see how many people would greet Father and Mother. We guessed that this was not only because our parents were kind people, respected for their opinions and leadership, but that there were financial reasons also. Many years later I was to find out that they had provided major funding for the dominant steeple that can be seen halfway between Haarlem and Amsterdam. However, one should not speak about such monetary affairs. "It is God Who provides, and He takes care of His chosen people."

Father was an Elder, ranking above the Deacons, whose duty was to collect dimes and quarters with a black velvet bag at the end of a long pole during the singing of the hymns. The Elders would not have to get up from their special benches placed to the right of the Minister's elevated pulpit, which looked like a throne. Whenever one of the Elders was absent I was allowed to walk up with Father and sit with him, which was

The steeple at Halfweg (halfway) between Haarlem and Amsterdam.

a special honor. It must have been a gentle sight, the big man in his fifties in a formal dark suit and next to him the little blond tyke. But within that dreamy head flowed faraway flight and rebellion with some arrogance—as the others around did not seem to be much awake or conscious.

Church lasted from 10 o'clock in the morning till almost noon; another session was held from 5:00 to 6:30 PM. First came a formal Opening and Blessing, some singing and announcements, readings from the Scriptures, and collections by the Deacons. The sermon was most important, much more so than in the Catholic Church, for we thought of our doctrine as being the most logical, being based on reason with only one act of faith, namely to believe that the Bible is God's Word. Once this was accepted, everything would be explained by the Minister in his Sermon, a carefully worked out lecture of about one hour, divided into three parts. (The threes in the Table of Contents of this book show my everlasting indoctrination.) The sermon was based on a short Bible text, and other passages were quoted for documentation. The greatest of these lectures might be published or sent out on the radio. Famous preachers prided themselves on working through all of the teachings of Calvin.

The principles and their application to our daily lives were further explained to us children in a catechism class on a weekday evening. On Saturday evenings we participated in a church club which was subdivided into age groups, for boys and girls separately. It was not unpleasant, because we did more than Bible studies—excursions and camping trips, for example. It was somewhat similar to Baden-Powell's Boy Scouts, but quite separate: they did so much on Sundays that we were not allowed to join them. We could not even ride a bicycle on Sundays.

On Sunday afternoons we would walk into the forests around Baarn and maybe spy on a young couple, which was dangerous because if we were detected the fellow might come chasing us. Another entertainment was to dare who would lie closest to the tracks where fast steam trains would come thundering by at frightening speeds. We had a small club to do such things together; friendships were informal and easily made. Cars were rare; ours was used on Sundays to take invalids or the feeble to church and we boys sometimes stowed away on the back bumper! On weekdays we would bike to the "School with the Bible"—that was its dedication, in large letters above the entrance. Movies and dancing were never permitted. We were living guarded lives, busily occupied within the triangle of church, school, and family. Our young minds were thoroughly brainwashed and gullibility was the result: The safe and known church environment with friendly people around, sitting close to Father or Mother in an atmosphere that was not too stark, provided a secure foundation, even for a rebellious young chap.

The sermons were a backdrop for daydreams through the clouds and out into the night, sailing inside a submarine or flying from an aircraft carrier. In the 1930s there were wartime exploits in Spain and East Africa with stories of faraway places and travels on ships and airplanes. A famous occasion was an international race of aircraft from London to Melbourne in which our national pride, KLM Royal Dutch Airlines, did so well that we all seemed to fly with that crew to become national heroes, while in the process we learned geography along the Persian Gulf and into the Far East. Perhaps these daydreams helped steer me on my way.

The protective atmosphere made us mature slowly. Any mention of sex or crime was avoided, while thrift and hard work came across as the respected virtues of Calvinists. That did include success in business and encouragement of businessmen through capitalistic principles. There was not much talk about money in our family, but all the same it seemed a matter of pride to reach a degree of individual enterprise and financial independence. All this security and protectiveness, however, left in later life a certain frustration and awkwardness in how to deal with the real world.

These were the standards in the 1930s of a small church community in which the most highly educated people were the doctor, the teacher, and the preacher. The old doctrines and procedures are not followed so strictly in the Dutch Reformed Church any more, as its members will proudly point out. However, exactly the same was pointed out to us children about earlier practices in the church. In Geneva, for instance, Calvin had ruled the community so strictly that we would now call it sheer fanaticism, and it was not long before my time that a pregnant maiden would be brought before our congregation to admit her sin. Changes were made, but they were more or less forced from the outside, as without modification the church would look ridiculous and members would be lost. Even so, the church always seemed at least twenty years behind what I found in books, some of them rather forbidden or at least frowned upon. My parents, however, were remarkably tolerant and allowed me to buy books almost without restriction. Father seemed less extreme or dogmatic than the other Elders, probably because in his own youth his parents had been of different denominations. When one of his own sons had to get married, his future daughter-in-law was certainly not brought before the congregation. But then, he was a pillar of the church and of the local bank as well.

Just because the Church prided itself on being so logical, I could not help but do some reasoning of my own and that is where my problems started. My doubts did not come easily, because of the indoctrination and the pride of church tradition in our community. For instance, on Sunday mornings after church we would assemble at home for a cozy coffee get-together. We sometimes had a distinguished guest, perhaps the preacher of the day, who might be a Professor, and such a learned man always had answers and explanations for any and every question. His university was called "The Free University," which seemed a bit of a misnomer because its teachings were bound to the Bible. Anyhow, an extensive lunch was served, which, like all meals, started with a prayer and ended with Bible reading and a longer prayer said by the guest. These prayers sometimes seemed a little self-centered with a lot of asking for good weather which was, of course, important in the rainy Netherlands. The Bible readings seemed long, but we would snicker a little when Father tried to skip over the begets and other explicit stories considered improper for the ears of young boys and of our not-much-older maid, who could not eat with us but was called in for the Bible and prayers.

Eventually I collected a fabric of doubt, while still being so influenced that a quick denial seemed out of the question and scientific investigation a necessity for my own future. I remember the doubts as follows. Literal acceptance of the Bible as God's Word seemed strange. Eve was made from Adam's rib and it was easy to suspect that this was a way to keep the women in an inferior place. How funny to choose a rib as a reproductive organ! We could make fun of that even within the family, as long as we did not doubt it, for God could do anything. It seemed incestuous that the people must have resulted from mating within the family of Adam and Eve: Only they and their children would have started all of the human race. But incest is now forbidden; it was hard to see the logic in this. Again, it seemed clear that keeping certain racial groups in their place was the point in the story of the sons of Noah, one of whom, together with all of his descendants were doomed to be subservient to the others, a favorite theory of the Dutch Reformed Church in South Africa.

The church insisted that God is good and wise. But the basic doctrine did not appear loving or just, because the Dutch Reformed Church believes in original sin, that people are born as sinners. Their salvation would come in a rather indirect way. God sent His Son to be killed, in a cruel fashion, nailed to a cross, in order to take care of the sins that He put into people to begin with. If they were fortunate enough and predestined to have heard of this salvation mechanism, and if they believed it, then they were saved and allowed into heaven. But if they were not saved, they could not get into heaven. My horse would not go to heaven—even though she seemed to have a noble soul, a little naughty but not sinful. People who had never heard of the salvation, as the Papuas in New Guinea, could not go to heaven either, and would burn in hell forever. To a young chap that seemed unfair and terrible. Burn in hell? Have you ever read Dante's descriptions of hell? Its images were implanted in our minds as young children.

Heaven also was a puzzling place. Where was it? And what were we going to do there for such a long time—that is, forever? Well, we were going to sing. Hallelujah! And I would look around the congregation and wonder how these people could believe that or look forward to it. So few of them liked to sing. When the Church changed to a more rhythmic and lively form of singing the Psalms, there were members who did not agree and would sit during the singing with their mouths shut in a demonstrative fashion. These people were going to sing forever? Was that their greatest ambition? In heaven we would get to see God, but that would not be a surprising vision as we were made in His image. God looked like us! Presumably he was a white male, not black, of course, nor oriental. That God had made everything seemed a bit of logical evasion, for Who had made God?

The more we learned science in high school the greater the conflicts became, even though the church modified its teachings as rapidly as it was forced to. Evolution was always dismissed. There was an insistence on a "missing link," an evolutionary step between monkeys and people that would never be found. Earlier we had learned to work out the age of the Earth. If the Bible is taken literally as God's Truth, then one can count forward from the Creation to the time of Christ. In the Old Testament there is an uninterrupted sequence of begets: Adam begat Eve and they had two sons, etc., etc., all the way to Jesus Christ. (There had been a little reluctance in explaining begets, because children were not supposed to know about that, but this could somehow be glossed over.) An average age was assumed for a person to get to the begetting. It was also known that Christ walked on the Earth near the beginning of our own calendar, the year zero. Thus it was estimated that the world was about 6000 years old. That is a lot of years in the minds of simple people, hard to imagine, really, and that age determination was therefore not challenged for a long time.[1] But it was hard to fit in with what we were learning about biology and the Earth.

Some of these doubts seem so obvious now, and they had been treated long before especially in French literature, but I had to work it out for myself, against tradition and a fear of hell.

But while the church held on to the past, there was continuous haggling over points so small that congregations kept breaking up into smaller new churches. Even when the Netherlands were united against the German occupation in 1943, our church kept busy with a split over a point of dogma so fine that even my father, the Elder, could not explain it. That was a last blow for me, and at the age of 18 I broke with the church. It was my First Liberation.

This is not to say that I became an atheist at that time. The influence of having been raised in that Calvinist environment was so strong that some of it will never leave. And to some extent it is good. A great culture is based on the biblical script. Honest Christians, as my parents were, were leading their lives according to high ethical standards. In any case, the doctrine brought questions to the fore that made me wish to explore for myself. If it is not by Creation, then what is the origin of life, the Earth, the solar system, and the universe? Is there a godliness anywhere?

2. The Dutch Nation and Some Birth Statistics

Before the sixteenth century, the people of de Nederlanden, the Low Countries or Nether Lands, lived in a feudal society beholden to local rulers, or in fortified cities such as Haarlem and Leiden. By historical accident, involving a number of marriages and bequests, they came under the ultimate control of the Hapsburg king of Spain. The country was a watery landscape dominated by three great rivers: the Maas coming north from France to about the middle of the Netherlands, where it turns westward toward the North Sea; the Rijn (Rhine) coming north from Switzerland and turning west, paralleling the Maas to join its delta, and the Waal, which flows through the Netherlands between the Maas and Rhine.[1] There was an unending succession of lakes, swamps and estuaries along the North Sea from the northern part that is called Friesland toward the southwest that is called Zeeland, "Sea-land." The stretches of land in between were never far from the water, and the Netherlanders became experts in sailing the waters or in reclaiming the land for agriculture. Their familiarity with the sea—and the strategic location at the mouth of the Rhine—made them successful traders and prosperous merchants.

The nation has a fortunate geographical location: enough in a corner to maintain a separate entity (more so than neighboring Belgium); enough on the crossroads of large countries to be exposed to their ideas and travelers (more so than Denmark, farther toward the east).

In the early sixteenth century, the Reformation of Luther and Calvin brought about a religious renewal and a social and political upheaval. The Protestant emphasis on the individual must have appealed to the individualists on the Dutch waters and to the shopkeepers and businessmen in their cities. The currents of renewal from Germany and Switzerland added religious fervor to the political rebellion against increasing taxation by the Catholic Spanish overlords. The war lasted from 1568 to

1648—one of the longest wars in history, although the fighting was not continuous, and there was a truce from 1609 to 1621—but it holds within it the birth of a nation.

An early leader was William of Nassau, Prince of Orange, known as Willem de Zwijger, William the Silent, a forefather of the present royal family, but he was assassinated in 1584. The defeat of the Spanish Armada by the British Navy, helped by a storm, in 1588 was a major milestone in the war. This is not dusty history, but surprisingly alive. When Princess Irene married a Catholic Spaniard in the 1960s, the country rose up in protest as though the Eighty-Years' War was still being fought!

The war, which resulted in the independence of the Netherlands[2] in 1648 also made the country powerful and rich, producing a flowering of the arts and sciences now referred to as a Golden Age. It is a great example of the enterprise that adverse conditions can bring about. These people who for generations had lived near and on the water now sailed out with their small vessels to faraway places such as South Africa, Ceylon, Java, China, and Australia. Dutchmen went west as well, to what is now the United States, to the Caribbean and to South America. The exploration of the world brought a return of riches and spices and countless observations of faraway places. The Dutch also became cruel experts in the slave trade.

Surprisingly in contrast with Calvinist adherence to Bible and church, there was a tolerance for other opinions which has been maintained as a Dutch tradition until the present day. There was a major conference, the Synod of Dort (1618–19), to discuss and settle differences among the leaders. During following centuries, streams of refugees trekked to the Netherlands: the Protestant Huguenots came from Catholic France, the Pilgrims from England (for a while), and many Jews from Germany. They brought variation, with new philosophies and business practices. Presiding over liberty became a tradition; *Praesidium Libertatis* is the motto for Leiden's University as we shall see in Chap. 9. The philosophers Baruch Spinoza, René Descartes, and John Locke worked at various times in the Netherlands. The diplomat Constantijn Huygens stimulated the arts. Frans Hals, Rembrandt van Rijn, Jan Steen, Jan Vermeer, and Albert Cuyp were famous painters of this period. Huygens's son Christiaan (1629–1695) was a versatile scientist, the forerunner of a tradition in natural sciences in the Netherlands.

The swamps have gradually been drained and the floods pumped away. So much in fact that environmentalists are becoming concerned over dwindling wetlands, but these are changes of the last few decades; during my youth, the ecology was still anchored in its more original forms.

I was born on a farm in a "polder," a reclaimed area pumped dry after isolating it with dams and dikes. There are many such polders in the western parts of the Netherlands. This one is called the Haarlemmermeer, Haarlem's lake—the word *meer* is originally from the Latin *mare* for sea. In this case, it was an inland body of water between Haarlem, Amsterdam and Leiden, menacing all three of them when it overflowed in every storm. It was therefore pumped dry by 1853 and used at first to grow wheat. The farmers, my father, Jacob Fredrik Gehrels (1878–1944), among them, were a proud people who lived an independent life, not beholden to anyone, employing laborers in the field and servants in the house. My mother, Wilhellemijnsje van Baren (1883–1945), added to our family pride as she had been raised as an only daughter of a wealthy gentleman farmer. She was a cheerful person who loved to sing. It was she who brought the riches, but it was he who handled them. That was the tradition following Biblical script: the Father leads the family. But he did it gently, as he was the nicest of men and beloved by all. They lived as Christians who practiced what was preached and were willing to help others within the sphere of the church.

Yet it was fortunate for me that we left the farm when I was only five years old, in 1930. The polders were not healthy places, with open sewers into the canals, many rats and a high incidence of cancer. At every major intersection a café could be found for solace in drink. The farmers were isolated and not used to traveling. Studying in city schools or universities was generally considered unnecessary, dismissed as mere "boekenwijsheid," wisdom from books without practice. Nowadays, the Haarlemmermeer is bustling with activity other than farming. Amsterdam Airport lies in it, and on the approach from the northwest one sees, from the right-hand side of the plane, a single red roof, the farm where I was born. Until World War II, however, it was a place with a uniformly agricultural population. There was regular work in the spring for the sowing of wheat and beets and the planting of potatoes. The months of May, June, and July brought a gentle task of weeding while the crops grew. In August, however, one entered a critical period of harvesting the grains, and, until modern times, it required great physical effort of the farm laborers. The owners shared the long workdays in their supervisory role and were under stress as they worried about the weather affecting the harvest. There are many rainy days in the Netherlands and the grains may spoil if not harvested on time. August was their busiest time of the year, with long workdays since at that latitude of 52 degrees there is daylight from about 4 AM till 9 PM The grain harvest was followed by collecting potatoes and sugar beets, an activity that might stretch into October, depending on the weather. Some steady work of plowing the fields would run into November. During the winter months, starting in November and ending in March, the work would generally entail more regular hours during shorter days for processing of stored harvests and for repair and maintenance of equipment.

Of that situation I once made a small statistical study because a high-school teacher had declared that at times of hard work more boys are conceived than girls. The Haarlemmermeer was ideal to test this because the people worked periodically hard in summer, and not so hard in winter. The Municipal Office of the Haarlemmermeer allowed me to collect birthdates of 5590 children in 1392 families. They were tabulated in three groups of farm owners, farm laborers, and a control population in the small towns. Plots of the number of birthdates, minus nine months, showed a statistically meaningful drop in August in the conception rate for the farm owners. The percentage of boys was, however, always near 51.5%, the worldwide average; there was no correlation with hard work. One hears similar rumors that more boys are conceived during times of war; that must be superstition also, easy to verify in a similar study.

There was, however, a peculiarity for the control population in the towns of the Haarlemmermeer. These people had, throughout the year, a somewhat lower percentage of sons, namely 49.1%. They also had smaller families, 3.1 children per family compared to 4.2 for the farmers and 4.5 for the farm laborers. Apparently they would only continue to produce children if there had been no son. There must have been a concern for the future if additional boys were born, since only one could inherit the business.

By tradition, a father would hand over the business or farm to the oldest son when he comes of age (there are parts in the eastern Netherlands where it is to the youngest son). In our family, the first son, Cor, did not want to become a farmer, but the second and third sons did, and Father provided the latter also with a good wheat farm. My father therefore left the Haarlemmermeer at the age of 52. Following doctor's advice for my mother, who was not in good health, we moved rather far away for those days, to Baarn, a little north of the city of Utrecht. It is a beautifully wooded area, in a drier part of the Netherlands. There were two more sons, of whom I was the younger and, according to united family opinion, quite spoiled. These two youngest boys came along to Baarn, and thereby into an environment more conducive to studies.

3. Ships, Bikes, and Horses

While Christmas is soberly religious in Western Europe, December 5 is the Feast of Saint Nick. He lives in Spain, but around the time of his birthday he comes to the Netherlands on a ship. He must like horses because he has a beautiful white one and rides it proudly down the streets. His servant, a Spanish Moor, walks; Black Peter is his name and children are afraid of him. Kindly old Saint Nick brings presents for those who were good, but Peter, Black Peter, may thrash the children with a bundle of switches.

Our three married brothers would come with their wives and children and we had a roomful when the Saint arrived at our doorstep. That was early in the evening, in the dark and windy Dutch winter—but cozy inside. It was the right moment for singing:

Zie ginds komt de stoomboot
uit Spanje weer aan.
Hij brengt ons Sint Nikolaas
ik zie Hem al staan.

See, there comes the steamboat
from Spain again ashore.
It brings us Saint Nicholas
I see him standing there.

Zijn knecht staat te lachen
en roept ons reeds toe:
"*Wie zoet is krijgt lekkers,*
wie stout is de roe."

His servant stands laughing
and shouts out to us:
"The good ones get candy;
the bad ones the switch."

The door swung open and Peter threw in a hail of gingerbread and peppermints. We scrambled to fetch them all over the floor, but not for long, as Saint Nick was entering! The Saint was a stately figure, a big man in a beautiful red robe and a bishop's miter. He carried a staff and walked slowly and truly seemed a saint. But little Black Peter was the one to watch as he darted about and would swing his switches. The little children believed that these two really came from Spain. Nothing of the tale was doubted, not even how a popish priest was to receive such a welcome in a Protestant home. The older children had a favorite game of guessing who these two might be whom our parents had hired for the job. They probably were teachers, but it was hard to penetrate the disguises.

What a performance it was! The Saint seated himself and called everyone by name for a short sermon on what was right and what went wrong during the past year. The naughty ones had to be teased and dropped over Black Peter's knees. That was awkward, but the rest of the Feast was just fine. After a while the Saint and his servant departed, because all homes in all towns had to be visited, and then we entered the Room of Presents that had been carefully locked during the past week for the laying out of gifts by our parents. And great presents they were, for young and old. Our parents must have spent hundreds of guilders every year. They were the only ones of us who could afford it, but we were also expected to buy small presents for everyone, and it was the custom to make a poem or rhyme to tease the recipient.[1]

For the people who were not poor it was a good life in the Netherlands during the 1930s. There were portents of war in the air, but there still lay a haze of beauty and peace over the landscape.

The Dutch climate is conducive to creating contrasts. The rays of the Sun may come and go in between the fast-moving clouds. The high humidity gives rise to mists and hazes and often a drizzly rain, and it softens the hues of the landscape. There is a lot of rain and clouds and darkness, but when the Sun comes shining through one notices it more than in the steady skies of Arizona. The trees show the seasons, losing their leaves in the fall, looking stark and barren in the winter, and greening again in the spring. And then there are flowers everywhere, tulips and daffodils, and roses for export, and even the dry heather shines in a flamboyant purple.

The fields and flowers and vistas were close to us on our bikes and ships and horses. There was a lot to see. We followed bike paths through heather and woods, by brooks and canals and lakes and rivers, past sunlit meadows, with cows and haywagons everywhere and endless views; in other places, gloomy forests with the fragrance of trees and mushrooms invited us to make camp and lie down right there. The tent would come out and with it the challenge to set it up dry and comfortable, to cook a meal with wood that always seemed damp, to sleep soundly, and to undo the operation the following morning.

It was a quiet life, without freeways or runways or television. Radios and record players were still expensive. Our music was mostly classical, but we heard some jazz, and we had many books to read for school assignments. Children now would think it a dull life, but we did get out for camping and trekking. It was safe for young ones to go out and explore, hiking, biking, boating or whatever, even some hitchhiking, which could be quite a challenge along the quiet Dutch roads. After the age of about thirteen I was allowed to take off for six days, but home on

Sunday, and that was meant with Calvinist rigor, from 12:00 midnight to 12:00 midnight.

In six days an aggressive chap can cover a lot of territory on a three-speed bike. The extreme was 250 kilometers (155 miles) in a single day, at the end of which one was likely to be reprimanded by the Father of the youth hostel for not observing nature enough. Well, he may have missed the point, for nature and activity that were being enjoyed lay deep inside the stormy restlessness that can be so strong in teenagers.

It was interesting to observe the people, too. Within 10 kilometers or so there could be drastic changes in custom and costume. More challenging than youth-hosteling was trying to stay overnight with farmers. In the evening I would have to approach a farmer to ask if I could sleep in his haystack or barn. That salesmanship failed only a few times, and more often it succeeded with my staying in the house instead of the barn, sometimes becoming friends for life, and a few times having to handle a not-so-subtle attempt at being coupled to one of the daughters. One farmer came home from the fields, took off his socks and wooden shoes, and stepped into a large trough that his wife had prepared for him to walk around in for kneading the dough of the weekly bread.

I preferred to make these travels alone or with my horse. There would be other groups around of twos or threes or fours, but with not much mixing of girls and boys. Dangerous drugs were unknown, but during our all-night parties there was drinking, but also eating, games, and some dancing—which may have controlled the drinking. There seemed to be less sex in those days, or perhaps it was just me and my Calvinist upbringing. In the schools there was of course a lot of flirting and falling in love, but not much sex if any at all. Calvinism taught us a fear of passion, equating it with disease and early parenthood, and contrasting it with purity for those saving themselves for marriage. The lesson stuck with me, awkwardly, even through soldiering and a lot of tempting circumstance. The church had spread an aura of sacredness over sexual relations. There was a belief that a certain man and a certain woman had been ordained to belong together, in a unique and everlasting combination. Divorce was not heard of in our circles, and, indeed, it was generally rare in the Netherlands of those days.

My first love had soft lips, could be nice and naughty, and was full of a tireless spirit. Her name was Nublina, of the Oldenburg race: She was of medium size with a beautiful brown, short-haired coat, black mane and a long tail, with a few white markings on the forehead and near the feet. Even in the late afternoon after school we could cover a lot of territory in the wooded environs of Baarn and Soestdijk Palace. There were more riding lanes than roads, and though there would be other riders occasionally, mostly we were on our own. We trotted primarily, galloped and jumped a bit, and I sometimes walked long distances beside her. I had total care of her, except during visits with the stallion, because I was not supposed to see that. So there was a great deal of brushing and combing and trimming and cleaning of hoofs and soaping and polishing the sad-

Nublina and friend.

dle and gear. I learned a oneness with the horse, and to have a gentle but thorough concern and control. The first thing to be learned was a good seat and grip so as to stay on and never fall off. On Saturday afternoons we would meet in a riding club and learn to maneuver together, obey commands exactly and ride figures precisely. There would be occasional competition in "Concours Hippique" to show off our expertise; these were strenuous efforts for horse and rider alike. On the long way home we would sing out loud, which had an amazing ability to revive the spirit and bearing of horses and people. The great riders of the era became my heroes, especially as they progressed toward the Olympics, or if they were national figures such as Prince Bernhard. I remember a stately girl from Denmark, Lillian Witmak, a fantastic horsewoman and oh, how it hurts to fall in love!

One day there was the Great Experiment. Would Nublina go into the house—my parents were away—and come up the stairs to my room as billy goat Frits liked to do? Well, yes, she would, but chickened out half-way up those stairs. Nublina must have weighed more than half a ton, so that Mother's pride of marble floor plates in the hallway became cracked forever. This is one of the tall tales my children would not believe until a pilgrimage was made from Tucson to Baarn, allowing them to observe for themselves the Marmoreal Cracks.

All this was, however, not conducive to schoolwork, and Nublina was responsible for my flunking a year in high school. Surprising that our parents tolerated so much. What wisdom and patience, especially in that thrifty Calvinist environment!

Around the lakes and the sailing there would be great parties during the long summer evenings in youth hostels and especially in the famous sailing town Sneek in Friesland (Snits in the Frisian language). That is a special area of the Netherlands—country up north with almost as much water as there is land. Frysk was spoken; it sounds halfway between Dutch and English.[2] One of the youth hostels was named after a beacon near the edge of the lake nearby, and its name was therefore "It Beaken" (in Dutch it would have been "Het Baken"). The hostel was a member of the international youth hostel organization, which had a touch of a youth movement—like the boy or girl scouts: We were kept busy, controlled, and educated with sessions of song and dance and games and lectures in the evening. Father and Mother Piersma were in charge of It Beaken; in Frysk they were *Heit* and *Mem* Piersma (the *ma* ending is practically proof that a family is Frysk); they had the tact to handle teenagers, and they taught us to sail and cook, wash dishes and fold blankets. Up early in the morning, sails readied, and soon five or six small Frisian ships with Heit in command would go out on the endless lakes, sailing some competition while everyone learned safety and sportsmanship, keeping course, and to observe the ways of the wind. Heit could sail nearly straight into the wind, passing us all, on the side of a canal where reeds would catch a nearly head-on wind and thereby change the angle of wind direction to make it come from the side. Tradition forbade the use of the word *boat*, it was proudly referred to as a ship, no matter how small—and female, of course.

Later I had a ship of my own, and she was large enough to go on the biggest of lakes, the IJsselmeer, which had originally been an inland sea. I would set out from Amsterdam in the evening and sail through the night on a compass course, reaching the opposite shore of Friesland before daybreak, maneuvering into harbors and around obstacles with the use of harbor lights, or horns when the fog was too thick. The Netherlands being flat, one could make week-long trips going from lakes to canals to rivers to estuaries and out onto the North Sea if one dared, which I never did.

I still sing, in my telescope dome at night, the hollering song for announcing to the bridge man in the distance that the ship had to go through; he could collect a nickel for opening the bridge by holding out a small wooden shoe at the end of a fishing pole.

Brio! *Brio*!
De brêge sit ticht
en it skip moat der troch.
Doch brêgeman dyn plicht!
In skip foar in stûr
det is net djoer.
Brio!

Bridge! Bridge!
The bridge sits tight
and the ship must go through.
Do Bridgeman your job!
A ship for a nickel
that is not dear.
Bridge!

With great satisfaction we would watch the line of cars waiting for the open bridge, and we would speculate how the weighty citizens felt about having to pay us this homage.

We were, in fact, rather spoiled. For instance, once a week our bicycles would be collected by the Shop, washed and polished, and returned to the house. Well-to-do children lived idyllic lives—self-centered and protected, without knowing poverty or Depression, and blissfully unaware of the gruesomeness that was on its way. It was said sometimes that another big war was coming. We were learning international history, and we learned about earlier wars, about the great trek to Moscow by

Napoleon in 1812 and the German surge into the trenches of northern France in 1914. But frankly, as young people, we were hankering for action and hoping that we would not be left out of whatever the next war would bring. One could not say that to older people, of course, for they had been impressed with the horrors of gas, bullets, and bayonets in France and Flanders.

Our parents took us to the memorial arch near Ypres in Flanders with its countless names of the dead on its walls, and to Douaumont near Verdun with its lugubrious vaults holding the bones of hundreds of thousands of unknown soldiers. The French had been the first to try gas early in 1917—tear gas, which is non-toxic and still is sometimes used by police. In October 1917 near Ypres, the Germans used chlorine gas that killed 5000 French soldiers; the remaining 10,000 French troops in the area fled in panic, but the Germans were so surprised by the results that they did not follow up to break through the French lines. Later in that war, gas was again occasionally used. Many of the survivors were maimed for life by "gas lungs," destroyed lung cells. My father feared that the next war would bring all-out gassing of soldiers and civilians on both sides, much as we now fear a global atomic holocaust and nuclear winter if World War III were to come. We children felt, however, that time would bring its own adventures and solutions, and I cannot remember any fear for the future that we ever had or even discussed.

War and Resistance

4. Nazi Occupation

Hell came to Earth in Poland on September 1, 1939.

Because a German invasion might eventually come into Holland, the Dutch prepared to stop it by planning to flood a wide strip of land near the central section of the country, adding to the obstacles presented by the Rhine, the Waal, the Maas, and the IJssel (which branches off from the Rhine in a northerly direction near Arnhem). My parents did not wish to be on the German side of the floodings, separated from their two sons and farms in the west, so we moved to Heemstede, south of Haarlem where I continued high school.

The Dutch were rather pro-German in 1939. Nazi propaganda stressed that the Peace Treaty of Versailles in 1918 had been unfair to the Germans who then had a rough time economically during the 1920s. The fact that the Germans had set an example in 1917, through their own Peace Treaty with the Russians being excessively vengeful, was not widely known and, of course, not discussed by the Germans.

It was hard to judge whether the Germans had reasons for World War II other than their territorial ambitions. One needed to understand the causes of World War I, but these were complex; Nehru, in his *Glimpses of World History*,[1] shows a viewpoint that contains a warning even today. He wrote to his daughter Indira:

> I have tried to examine with you some of the causes of this war: how the greed of capitalistic industrial countries, the rivalries of imperialist Powers, clashed, and made conflict inevitable. How the leaders of industry in each of these countries wanted more and more opportunities and areas to exploit; how financiers wanted to make more money; how the makers of armaments wanted bigger profits. So these people plunged into the war, and, at their bidding, and that of elderly politicians representing them and their class, the youth of

the nations rushed at each other's throats. The vast majority of these young men, and the common people of all the countries concerned, knew nothing of these causes which had led to the war.

In any case, Germany's propaganda minister Joseph Goebbels was convincing the Dutch of the awfulness of the Versailles Treaty, leaving poor Germany without colonies—in those days in Europe nothing was thought wrong with having colonies. Goebbels, the famous explorer Sven Hedin, and other celebrities told great tales of a new Germany. And, indeed, as we drove on their new *autobahnen*, the first freeways in the world, there was much to impress us as tourists from a small country. Germany had overcome runaway inflation in the 1920s and it did not seem to be affected by the Great Depression, while in the Netherlands food was being destroyed in order to keep up prices while poor people went hungry. The youth movements of the *Hitler Jugend* and *Bund Deutscher Mädel* appeared idealistic; we found it curiously intriguing that young Nazis were encouraged to mate in order to produce a purer superrace. We were too naive to see the deviousness of racism and segregation.

In my first *Lyceum*, or high school, in Hilversum near Baarn, there was a growing membership in an organization that carefully concealed and even denied any connection with Germany. It was called *De Jeugdstorm*, the Youth Storm; it was for both boys and girls and had sports and camping and outings. I wanted to join. At this point, my father explained Nazi principles and it became clear that I wanted no part of that. It was, however, fortunate that we moved out of that budding Nazi nest, as through friendships and perhaps association with some strong youth leader I might have been influenced permanently. I had read Adolf Hitler's *Mein Kampf* and been impressed with it, apparently too young to discern its underlying lust for power. It is even stranger now to recall how I would listen to Hitler's tirades and his screaming, "Wir werden England ausradieren!" ("We will eradicate England!"), and cheer along with his intoxicated audience. But that is the way it was.

In the early morning of May 10, 1940, the Germans invaded Belgium, The Netherlands, and Luxembourg on their way to France. We stood on our roof, and what a sight it was! Bird-shaped Heinkel-111 bombers, sleek Messerschmidt fighters, and lumbering Junkers transports were in the air around us, with little in the way of Dutch fighter planes or antiaircraft guns to stop them. Four days later they bombed Rotterdam, wiping out part of the city; the Queen and her ministers escaped to London and the army capitulated. The parade of German forces into the Netherlands within the following week was impressive to the eyes of a 15-year-old. The Heinkels we would see again later in large numbers evening after evening, as they flew from Amsterdam's airport, so heavily laden with bombs for England that after 7 miles, flying full throttle, they would still come low over our rooftops.

There was no thought as yet of resistance in underground organizations. There was a tendency to consider the war lost, finished, and to try to work with Germany toward a new future. It was remembered how even the British had made accommodations with the Nazis in the years before 1939, how Prime Minister Chamberlain had made a pilgrimage for peace to Munich. There were distinguished Americans too who advocated cooperation with the Nazis, such as Charles Lindbergh and Ambassador Joseph Kennedy in London. If the Germans had been more clever than arrogant, they could have made good use of the situation.

The atmosphere changed gradually. It became clear what had happened in Rotterdam: a murderous act of bombing civilians. The Germans generally showed a peculiar trait of treating people according to their status, with heel-clicking for superiors and cruelty toward the oppressed. It made them a despicable breed to us. These convictions took about six months to come about; for many people it was a radical change of mind.

We witnessed the gradually increasing persecution of the Jews. First they were required to wear a yellow Star of David on their clothes, below the heart. It was followed by their elimination from parks and theaters, and professorships too. Later they were collected, to work in camps it was said, but that this would lead to their final elimination we never knew, and found hard to believe even as we heard of it after the war. While it is true that within our circles there seemed to be no segregation, and we were not supposed to even think "Mr. Cohen is a Jew," it is also true that most of us were not concerned enough to hide them in our Calvinist homes. Perhaps deeper down there was the feeling that these were not the chosen people, not predestined to go to Heaven, but I cannot remember having heard such things said out loud.

Leesha Rose, in *The Tulips are Red,*[2] describes the Jewish struggle from meekness to resistance. She begins before the war, subtly setting a stage of a happy life and a young optimism for the future within a close family of two children and loving parents. During the occupation, however, her parents and her brother were taken away; by chance she was not at home when the Nazis came. She worked as a nurse in a Jewish hospital, still quite docile, obediently walking toward the German trucks when they came to empty the place. So well she describes how something clicked in her head, a sudden realization that a choice had to be made between life and death. So she walked calmly away as if she had been a Dutch bystander and not a Jew at all, which she must have carried off with conviction because the police and soldiers did not stop her.

Surprisingly, she went back to work when patients again came to that hospital for help. This time she was loaded by the Nazis onto a truck and then into a train. An officer on the station's platform called with a bullhorn a few names of nurses to come back out of the train and one of these she knew not to be there; she pretended to be that person, knowing that if she were found out she might be killed.

Back again into that hospital. When the Germans collected Jews for the third time, a fantastic escape occurred that must be read in her book for sheer suspense. But this time she was contacted by the Dutch Underground and worked with them for years, mostly in Heemstede, our town. It is an absorbing story of heroism and suffering. She gives compliments to the assistance she received from the non-Jewish Dutch, but this seems overly generous as we could have done so much more; I have a feeling of guilt and neglect, but perhaps the Germans outsmarted us by letting the evil come gradually.

One of the first requisitions the Nazis wanted was brass to be made into bullets. Brass ornaments and utensils were to be brought to a central collection place. We would not do that, of course, and buried them, even though the penalty would be severe if the Germans found out. One had to be careful of traitors, and a few people meekly turned in at least some brass, and their radios later. Cars and trucks were requisitioned. Bicycle tires were not available anymore. We had to be indoors most of the night; during those hours we could not walk or bike through the town without a pass. Sometimes there were manhunts when the Germans would close off a block of town and comb the homes for Jews or for young men to serve as forced laborers in Germany. Food began to be scarce. My family of course had relatives on farms where we could always get food, and it became quite an art to smuggle 80 kilograms of wheat on a wobbly bike past control posts manned by Dutch agents under German orders. Like a slowly turning vise, the war more and more affected the life in the occupied countries.

Our Lyceum building was taken for lodging of German troops, so we went on a half-day schedule in a Catholic girls' school. That, of course, brought out the rascalry in us. The nuns were afraid of our being attracted to each other and would check our desks for messages and appointments for dates. They never suspected the little inkwells in our desks. Did you know you can write in green on a slip of white paper and hide it in black ink? So Treesje and I arranged to take off—and that meant the whole day—and we dared to have fun and go sailing together.

Our day was dominated by the school, with a program of instruction that would not let up on its heavy load. But our real interest was in involvement with illegal activities that seemed so much more interesting and important.

Terribly upsetting events occurred frequently. For instance, we heard the story of an Underground group that had held an instruction period in hand grenades. The instructor accidentally let the safety pin-and-handle mechanism go free. In 15 seconds the powerful fragmentation device would explode. . . . The room was full of people. They'd be safe if he'd toss it out the window, but there also were people. . . . So the man slams his hand with the grenade into his stomach and double bends himself over it into a corner of the room—just in time. . . . Would I have had such courage?

In our school, Principal van der Elst was a wise educator, coaching us to keep the outer forms, our daily routines, going as normally as possible, acting calm, pretending to be calm, and continuing with the work at hand. This training helped us greatly through later upsets and disasters.

We kept up a routine that appeared normal enough. Bike to school, always late—but not too late, as punishments were severe. Students stayed mostly in a home room, with the teachers walking between classes. Long hours of homework every day and weekends. Not much entertainment—maybe a tennis club or other sport once a week. There was not much to do if you refused to go to Nazi movies and activities. I came to detest my given name, Anton, because the Dutch traitor in charge of the civilian government was Anton Mussert. So it became Ton or Tom, or other names as well, for illegal tasks, with falsified documents if necessary. There still is irrefutable proof, with stamp and photograph, that I am Arie Graafland.

We originated a high school debating society that spread quickly over our city and nation. It was an idealistic forum to promote a future for our society with a new unity and without Germans. It was vague, but not more so than what Her Majesty Queen Wilhelmina was talking about in her radio speeches from England. She envisioned a structure, simpler and more united than before, when fine points of difference in politics and religions had caused so many subdivisions. That high school organization never was betrayed to the Nazis. We also collected support for the citizens of Nijmegen. That city and Arnhem, Enschede, and Deventer had been bombed through regretful error by Americans on February 22, 1944.[3] It was a shocking disaster for which we collected quantities of money and clothing nationally to help the people who had lost everything. Our treasurer disappeared with a large sum, to our great surprise. We had no experience in this sort of thing, and all we could do was hope that the money had gone into even more illegal underground activity, as it was not feasible to pursue the case through police or other investigative channels.

There was the task of typing and distributing underground newspapers. The one I seem to remember was *Trouw*, "Faithfulness," which after the war became the national Protestant newspaper and it was then clear to see that in our environment that source had prevailed. The copies had to be pecked out on a typewriter, with carbon sheets, hitting the keys so hard that the tenth copy was still readable. We charted the defense positions around the harbor of IJmuiden, with careful detail as to the disposition of troops that manned the guns and of fields of fire from each position. One tried not to remember from whom the request came because there was a danger of getting caught and having the Sicherheits Polizei succeed in making us confess.

Hatred crept into our lives. In demonstrating our disdain for the new authorities, it was a challenge to see how much we could get away with. Near us lived a Dutch Nazi family and, even though they were decent people, we experimented with unfriendly acts toward them, such as making a face. That was more than one dared to do to Germans, as their punishment would come instantly, but even to them one could show contempt. The least one could do was to sing:

Die Gedanken sind frei,	Our thoughts are free,
kein Mensch kann sie raten.	no person can guess them.
Sie fliehen vorbei	Past us they flee
wie nächtliche Schatten.	like shadows at night.
Kein Mensch kann sie wissen,	No person can know them;
kein Jäger erschiessen.	no hunter can shoot them.
Es bleibet dabei:	And thus it will be:
die Gedanken sind frei.	Our thoughts are free.

While singing it out loud, I once walked along the wall of the prison of Haaren when my brother was in it. Nothing stirred. Who heard it?

Strain and disgust got worse, a somberness setting in like a stagnant chilly mist. Yet, these were good times too. There was a unity as never before. Life had become simple: The goals were survival and finding food. We were certain of a clear solution to all these problems: the elimination of Hitler and the troopers that seemed so dumbly beholden to him. When the Germans invaded the Soviet Union on June 22, 1941, we celebrated, since we knew from our classes in history that Hitler would lose that war: When winter came, his troops would freeze, just as Napoleon's did, and that is what happened. When the Japanese attacked Pearl Harbor on December 7, 1941, we had another celebration, because now America was forced into the wars, which were certainly going to be won.

We learned international geography as we followed the campaigns in detail from garbled voices in London that we received with a radio hidden between ceiling and top floor. To minimize interference by German transmitters that were loudly blurping away on British frequencies[4] we made the antenna directional. The antenna wire was wound tight on a frame, and with that plane pointed toward England one could catch the station, provided the German transmitter was not on the same line. Higher pitched voices came over best and we would listen entranced to the fiery speeches of Her Majesty the Queen. She must have suffered hard times, especially when Prime Minister de Geer, whom she had taken to England, returned to Holland as a disheartened private citizen. That tough old lady would never give up. We later heard a story that Winston Churchill once said: "I'm afraid of no man, but only of Queen Wilhelmina of the Netherlands." Her requests to him were for arms and ammunition, and later for food when her people were starving. She became a legend, together with her son-in-law Prince Bernhard, who was with her in England, and her only daughter Princess Juliana, who was in Canada with little children. They were all involved with the troubles in the Netherlands and with the sailors, the airmen, the Dutch who escaped, the secret agents, and the invasion forces preparing to return.

The astronomer Bart Bok later told how the Queen came to inspect a newly formed outfit, which wanted to show off by having a band. She came too soon, however, and there was only one simple piece of music they could play, and play it they did—over and over:

> The old grey mare
> she ain't what she used to be,
> ain't what she used to be,
> ain't what she used to be.

According to Bok, Her Majesty—who was quite experienced—never let on and had a fine time. The new band director must have been embarrassed to learn the words later, for she was dearly beloved.

Back in the Netherlands I had a daily routine that seems childish now, but it shows the desperate dedication descending upon us. Above my bed there was a small picture of Her Majesty, and every evening before turning in I would stand at attention before Her. I also would train for tougher times ahead with cold showers and running and jujitsu and boxing and studying French even with private lessons, in preparation for an escape through Belgium, France, and Spain. Anything illegal, whatever we could do against the Nazis, became more important than schoolwork. I was going to have to repeat another year, but Principal van der Elst interfered personally, as he trusted I was working for good Resistance projects; he arranged a conditional passing to the next grade, the condition being a few special tasks and an examination at the end of summer. I did not even open any of the required books that summer and felt that my flunking was a sacrifice for the Cause.

We gradually came to understand the term "total war" as a total commitment to do or die in a desperate situation with only one way out, namely the elimination of the evil oppressors. We knew it was going to be a long haul; it already seemed endless in Europe, and after that there would be a rough round in the Far East. I prepared for that future by taking private lessons from KLM Captain Abspoel, who was writing a textbook on navigation for airline pilots, and who was glad to practice on this student—especially as he brought sacks of wheat. My wish to become a scientist eventually was still there, but the immediate urgency was to get out of the country and to fight as a flyer, a naval air pilot. I was hankering to fly among the clouds. The white clouds in a blue sky were a magnificent sight for me, lying flat on my back in a meadow. It contrasted the foreign oppression and lack of freedom, of being tied to our small country, not even able to get near its seashore. Then to dream of flight and freedom, watching those billowing clouds rising up strongly, higher and higher, and moving away forever!

5. Humanity in a Concentration Camp

My oldest brother was born in 1906, and he had been abroad to faraway places; of all the family he saw international relations most clearly and he had no doubts regarding the intentions of Adolf Hitler and Benito Mussolini. He was a lively person with clearly expressed principles, which he lived by and could die for if necessary. His full name was Cornelis Adolf Gehrels, Cor for short, and he must have had some embarrassment over his middle name being the same as that of the man whose deeds he loathed. Cor was a rebel, the hero of the family. As a youngster, he had no interest in farming nor in formal schooling; his fascination was with short-wave radio, the fad of the 1920s. It was illegal to have an amateur transmitter in Holland in the 1920s—licenses were not as yet being issued—but that merely made it more exciting. Later, he went off to sea as a wireless operator, not on regular sailings but in "de wilde vaart," "the wild sailings" of merchant ships that could be sent anywhere. Back home on furlough, we would see him on a fast motorbike with his blond hair in the wind and a twinkle in his blue eyes while he kept inventing new ventures and any scheme to catch the most popular girl in the Haarlemmermeer, Grietje, his sole love.

Grietje's first matrimonial ruling was to return him to the shore. So he landed a radio-engineering position with the electronics firm of Philips in Eindhoven.[1] Even then, she did not see all that much of him, because when he was not at work he was wirelessing away in the attic with a transmitter and receiver array that he built himself—call letters PA0QQ. Still, they had eight children. Grietje fondly recalls how he would see to it that she had her full night's rest, and he'd crawl in with her for a while. As soon as she was asleep he would hurry back to his radio, usually till 3 AM; but he would never be late at Philips, for he loved that work too.

Cornelis Adolf Gehrels (1906–1945)

"Why so many children?" I asked her, and she answered that they loved to have them, that he was the one who took care of them if there was a crying or some problem during the night, and that they would have had more children if the Germans had not taken him away.

He was full of life and concerned with many issues; he could not keep a dime in his pocket, literally, when there were poor people around. He would not have any butter or anything on his bread during the war years, asking his wife to give anything that they could spare to the poor and hungry. When she wanted to join him in the meager diet, he refused that because of her pregnancies.

Cor had a tendency to go all out for whatever he believed in, and to make that into his Magnificent Obsession. His fulfillment began when the German bombs came down on Rotterdam. He hated those brutes right from the start, and he saw clearly that it was going to be Total War. One of his colleagues had been called up for reserve duty in the Navy and was now in England. A group of Philips personnel collected gifts and financial support for the family of this man and of others who were in the military abroad. Cor was the leader—so much so that after the war it was referred to as "Organization Gehrels." In addition to food and clothing, an amount of about 78,000 guilders, equivalent to at least a million dollars now, was collected and distributed. He had found other involvements, perhaps too many in view of the safety of his own large family. It was done quietly. The rule in those days was not to tell wife or parents or friends more than strictly necessary about what you were doing and whom you were doing it with. Not only because traitors might be hidden in unexpected places, but also because the Germans were thorough in interrogation. Anyway, he distributed illegal newspapers, collected intelligence that was sent to England, helped people in hiding, took Jews into his home, and worked for an escape route for downed Allied flyers. A longer-range project was to build a powerful transmitter to provide public information during the aftermath of the war. It started broadcasting from Eindhoven soon after the liberation in the autumn of 1944, under the name "Herrijzend Nederland" (rejuvenating Netherlands).

But there was a betrayal. Just how and by whom is not clear—two Dutch *Marechaussee* policemen are suspect, but this was never pursued. In any case, late at night on July 25, 1943, the Nazi police dashed into several homes in the Netherlands, and also into theirs. The interrogations lasted four days. Cor did not know which activity he had been arrested for, so it was important to give no information at all, lest he betray anyone. The Germans could torture even without leaving visible markings. They refined it by letting him see his sick father and his wife and baby in the distance. Just what they might do to them could be imagined. Cor did not betray. He was flung into jail.

And now the real story starts: of the human spirit maintained in prison and concentration camps, with secret seminars for fellow prisoners on science, politics, the future, or whatever would lift them up from their frightful surroundings. At first there still was some contact with his family. Grietje was not easily scared by Germans and she succeeded a few times to bluff a visit with him in the Haaren prison, even taking one or two of the children along. It was typical for them that in these odd circumstances, having only a short time with a guard standing nearby, they did not talk much about themselves; they talked about others, and they managed to smuggle a few cigars in and a farewell message out for another prisoner, who had been condemned to death.

On or before September 4, 1944, Cor was loaded onto a train from the concentration camp of Vught, near Haaren, to Oranienburg north of Berlin.[2] After the war we heard reports from that dreadful train ride. The trip lasted several days. Cor had a screwdriver and worked on a small opening, but the escape was not executed; there was something about a principle—not wanting to cause punishment for others. The story is vague here and will remain so, unless someone who was in that wagon could tell us what happened. It was hot; painful thirst. Inside that cattle wagon the men were packed tightly together. When the Germans threw in a bucket of water, the crazed prisoners fell upon it, such that it toppled over and no one got anything. Somehow, Cor got them under control. The next time, the water was distributed by the thimbleful, but everyone got some.

I got a report from the man who was next to Cor during the nights of his last three months. His name was Henk van Hoeve. He was seen in the Dutch version of the film about Anne Frank, where he played himself, the supplier of vegetables to the Franks' hiding place, now a museum in Amsterdam. van Hoeve talked about the human spirit. How Cor encouraged them, until the ghastly end. I protested that he was stressing this because he was talking to Cor's brother. He denied that. He had never seen a more extreme case of holding up an ideal: You could not lower yourself to the level of the Nazis, or the war would be lost for you. All that had been done and suffered would then have been in vain. So now, reader, sit tight and bear with me. Not just for the sake of his memory, but for an example of what the human spirit can endure.

From Oranienburg they were made to march, pressed against each other, with gravel in their mostly wooden shoes to the Heinkel factories near Berlin. Quoting van Hoeve: "The guards, mostly from the Ukraine, behaved so beastly. In our miseries I came to admire Cor, always encouraging others."

Back to Oranienburg. Off to Grosz-Rosen near Breslau (the present Wroclaw). That camp had perhaps 40,000 people, some Jewish, mostly prisoners held for political reasons. The guards were German criminals, released from their own prisons for this purpose. They used heavy sticks with sharp corners, hitting hard, or they kicked with their boots. Without reason, at any time. Roll calls at 5 AM and again in the evening, standing at attention. For hours. Anyone could be pulled out of the ranks to be hit. One tried to be inconspicuous. Work all day, under dreadful conditions. Hardly any food. Wounds from the beatings. Frozen feet. It was cold that winter. There were days of 20 degrees below freezing, with the wind blowing. The only clothes one had were taken from the dead. Once there was an uprising, but it was quashed with wrath. van Hoeve said that the aim was to kill 5000 people per week.

It was the ultimate test of the human spirit. The circumstances are too ghastly to describe. Standing at roll calls till collapse. Long daily marches to and from work. Hunger. Weakness. Cold. Mud. Snow. Wind. Betrayal. Despair. Shouts. Blows. Sickness. Misery. Hangings. This was their life during those long months of the cold winter of 1944–45.

Cor was still doing his best to lift his fellows up. At night or on Sundays they would have periods without guards. Seminars were no longer possible because of overcrowding in the barracks—fifteen hundred men in shacks built to hold one hundred. But at least some discussion could still be held.

One had to watch one's words, for there were traitors, who did that for food. Life became a matter of survival. One would have to steal—a morsel or bread crust—from others. I have heard that from three people who were there—it was the only way to survive, and these were nearly the only survivors. But Cor would not do that. Again, his reason was that then everything would be in vain and life not worth living. van Hoeve had a special recollection. Once or twice Cor brought him a cigarette. That does not mean much to us, but there. . . . It was a miracle. van Hoeve kept referring to the Human Spirit, the strength to go on with a

The museum at the Dora–Nordhausen concentration camp.

smile. I asked him if religion had come into their discussions. Cor was raised with our Calvinism, although in Holland he did not seem to subscribe to it. van Hoeve said that religion had never been discussed.[3] Cor apparently did not take a final and easy refuge back into the fold.

When the Russians came closer, the prisoners were moved west. It was a death transport. In some of the coldest weather, in open coal cars. For five days and four nights; no food; no water. They were headed for Nordhausen, in the Harz Mountains, to work on von Braun's V-2 rockets that were fired on London and Antwerp. van Hoeve remembered that on April 2, 1945, the day after Easter Sunday, Cor was split off from him because he now was a *muselman*, that is, a person hopelessly emaciated, who had been given up. He had dysentery, he was wounded. Cor was put into nearby Camp Dora, in a barracks for sick prisoners. He perished sometime in April, only days before liberation. Or was it March, as others reported?

Viktor Frankl gives a description of life in the camps in his *From Death Camp to Existentialism*.[4] For instance:

> On the average, only those prisoners could keep alive who, after years of trekking from camp to camp, had lost all scruples in their fight for existence; they were prepared to use every means, honest and otherwise, even brutal force, theft, and betrayal of their friends, in order to save themselves. We who have come back, by the aid of many lucky chances or miracles—whatever one may choose to call them—we know: the best of us did not return.

Back in the Netherlands, Grietje kept up the action. "Organization Gehrels" had to be kept going because families depended on it, and it was done with the help of others and also of their teenage daughter Coco, an exceptional child who had been with her father enough to know where some of the money came from and to whom it was distributed. In March 1945 Coco biked on or near a bridge halfway between Haarlem and Amsterdam. These were the days when Allied pilots would come in fighter planes and shoot or bomb everything in sight. Some of them did not seem to know the difference between Dutch and Nazi, between civilian and military. Garbage trucks, for instance, were their favorite, even though they were being pulled by horses at that time. Orders were orders. The bridge was bombed and Coco died.

Grietje was not known to be down or to cry, even when ghastly blows hit her. Their house was bombed by chance, as happened in those days. Her mother was killed and Grietje herself severely wounded. After the war ended came the agony of waiting. Where was Cor? Would he come back? How would she tell him about Coco, their love child?

Eight years later a widower with seven children proposed. So she became the lively mother for both families. A car crash in which she was severely injured left her second husband an invalid; he died some years later. When I visit her occasionally now—she is into her eighties—she will give an enthusiastic embrace, nearly lifting me off the floor. Thirteen children and their spouses and offspring have her near the center of their lives, with visits, consultations, and phone calls from abroad or wherever they are. Asteroid numbered 2049 in the International Catalog of Minor Planets is "Grietje." It might just as well have been named "The Human Spirit."

6. Scrutiny and Liberation

I asked van Hoeve whether he and Cor had discussed the cause of so much German cruelty. Yes, he said, and they had come up with an answer: upbringing. They had become experienced observers of the Nazis, before their arrest, during interrogations, and in the various camps. van Hoeve and Cor concluded that in Germany the upbringing at home and in the schools was too strict. Schools that are too disciplinary cause a reaction in the end—it might be as a relief from a day's submission, by pestering the dog or a younger sister or by killing a bird, and it might have lifelong effects of a similar nature. Along these lines they learned to understand the despicable trait of heel clicking for superiors while shouting or worse at people below. I wonder if this trait is yet instilled into Germans: At a scientific meeting in Heidelberg not long ago, a young scientist seemed to us too docile when his Director interrupted his talk with a question; a few minutes later he seemed too hard on his young student–projectionist who showed a slide upside down.

There was a touch of too much discipline also in the Netherlands in the 1930s. We disliked school, and we rebelled. Such rebellions were common and the punishments were frequent. I was a rebellious child and dreamt of freedom and running away already at an early age. On the first day of school, I had to be carried in, kicking and screaming, by four older students. A punishment in the second grade was to make a difficult drawing of a rooster, in class after hours. For long times we had to sit with arms folded behind the back. When the Principal entered our classroom, we had to stand up. Students would be punished by having to stand in the corner or in the hallway; this happened even in high school. The beauty of understanding nature or arts or algebra were not emphasized, but rather the importance of passing the grades. There was a lot of homework. Much time was spent learning languages: Because Dutch is spoken by only few people in the world, English, French, and German were required, and for the best students Latin and Greek as well. If one failed one class, and some 15% did, one had to repeat the whole year with new and younger classmates. Nonetheless, a high mark was not something to be proud of; too many would earn one the title of *Brave Hendrik*—"prig" or "grind." We dreamt of burning the place down.

The war shattered old traditions and overturned old attitudes. As children we had been urged to follow in the footsteps of Father and Mother. But what if they were Nazi traitors? The Germans showed a slavish following of commands: orders were orders, and orders might excuse even committing a crime. In reaction to our oppression, we learned to suspect authority, challenge commands, and mistrust laws and governments. We learned not to believe tradition, state, church, family, or Bible without scrutiny. Even now, I cannot help but question rules and laws.

To understand this one has to appreciate how thoroughly the Nazis had penetrated our daily lives. It was not just a military occupation, but a regime that produced devious laws seemingly similar to those of normal times. The ruling Dutch government in the Hague consisted of Nazi puppets. They were supported by only some 7% of the people; 7% is about right for extremists in any nation. Their laws had to be obeyed. A policeman would be unlawful if he did not arrest us when we were out at night without a pass, or if he found a radio in our home. If we were Jewish, he had to put us on a train. It was the law.

There was too much fanaticism, and too many people followed a flag or a banner that would lead them to evil or death. Politics became too serious; there was no room for humor or skepticism. I wonder with some bemusement when I hear Americans make their Pledge of Allegiance to the Flag; the war has left within me a peculiar view of national flags. On the one hand, the flag-wavers during World War II were usually the first ones to run and hide; it was the quieter ones who suffered and died for their principles. On the other hand, there was the oath to Hitler that the German officers took so seriously.

Perhaps as a reaction to attitudes forced upon us during the war, there came an urge to be only oneself, not to be labeled as a Jew or pro-German or anti-German or whatever. I still react with resentment when I hear a person labeled, for example, as upper, middle, or lower class; sociologists defend such terms as describing reality, while they are really promoting discrimination.

It seems that in the United States our children are taught to be impressed with forms of lawfulness, rather than being encouraged to be moral and to think for themselves, to always scrutinize the laws and rules.[1] Please note that I am not promoting the lazy freedoms of grabbing all one can. I would be in favor of sending these very same children, without exemptions, for a year into tough military duty. It is difficult to find the proper balance between freedom for oneself and respect for the rights of others, but this balancing will be ever more important as society becomes ever more intricate. People will have to become involved, "engagé," in politics and environmental concerns, just as a good fraction of the Dutch population were forced to become during the war.

I had to face a conflict of ethics with practice in 1944 when coming out of another strict school, Great Britain's interrogation center called "Patriotic School." To go to England had been my dream for years, with an obsession to fly in the Fleet Air Arm. Escaping the Netherlands was difficult after 1940, because the Germans had barriers for checking identities and only with special passes was one allowed to cross or even come close to any borders. One of the routes involved crossing the Dutch–Belgian border, the Belgian–French border, a demarcation line running east–west through northern France, the border with Vichy France, and the Pyrenees Mountains into neutral Spain, and even the Spanish were known to be harsh. But we also knew of people who had succeeded, and it seemed a more direct route than by going first to Sweden or Switzerland. Crossing the North Sea was out of the question after 1940 because of German control of the beaches where they were later building "Atlantic Wall" defensive works. I prepared for the escape via France in 1942, but failed even to get out of the Netherlands. A good opportunity finally came in 1944.

After breaking out from Normandy in July of 1944, the Americans under General Patton stormed east and the British under Field Marshal Montgomery northward. Brussels was liberated on September 3, 1944, and Antwerp, with its extensive harbor works intact, the next day. The book and film *A Bridge Too Far*[2] describe what happened next as an Allied failure, a lack of progress on the following days. That makes it a haunting tale for the Dutch, because it leaves the impression that the war was unnecessarily prolonged and that the suffering of the dreadful starving winter of 1944–45 could have been avoided. Such conclusions, however, seem exaggerated. September 4 was the day of reorganization of the German Command in the West; troop movements had already been ordered and were being executed ahead of the British along the Albert Canal in northern Belgium—a line of defense recognized by strategists—its bridges were blown and it was being reinforced by paratroops coming by train from Germany. Perhaps on the two days of September 5 and 6 an all-out air-supported Allied advance could have freed Antwerp's estuary. The ground troops were, however, exhausted after their blitz from France, and a two-day rest was called. In any case, the big rivers of the Netherlands were still far away, especially in view of the stretched supply lines and the intervening German troops. I saw examples of that: The grand Moerdijk bridge was held by Germans on September 5; an orderly column of German troops and horses moved near the Belgian border a few days later; a modern motorized group resisted the Allies' crossing the Meuse–Escaut Canal; the Hermann Goering Division fought a stiff fight near Hechtel. A strategic decision by Montgomery, endorsed by Eisenhower, was made to move from Brussels in a northeasterly direction first, clearing Antwerp's estuary later; it was not a blunder by a local commander in the field, as *A Bridge Too Far* seems to imply. A move in a northerly direction toward Rotterdam was not planned. The goal was Germany.

September 5 was *Dolle Dinsdag* in the Netherlands, "Wild Tuesday." How fast were the allies coming? Where next after Brussels and Antwerp? On that crazy Tuesday the rumors built up to a frenzy, while radio news from England seemed to confirm such optimism, and it was even said that Allied tanks had been seen in the south of the Netherlands and were crossing the great rivers. Out came the flags, a hesitant jubilation started in places with jeering at Jerries and traitors trekking east toward Germany, and people gathered on the highways in order not to miss seeing the allies coming in glory.

I did not want to be merely liberated. Dolle Dinsdag seemed just the right day to bike to England. Germans were still stopping everyone on the Moerdijk bridge across the western Maas, but not at a ferry farther east. The Belgian border seemed deserted enough to cross. But where were the Allies?

The young chap on his bike, coming from the direction of Rotterdam, did not encounter any advancing allies. That became clear while biking from west to east on sandy paths along the familiar Belgian border. It is fortunate I did not try to go directly toward Antwerp or Brussels, for I probably would have been caught by German patrols along the Albert Canal who might not have been kind to apparent messengers. I was probing with local people, "where are the Allies?" and stayed overnight close to the border with a farmer who had many children and fly-covered food, which we all ate lustily. With fantastic luck the place where I moved south was exactly where the evening before the British had come north across the Maas–Schelde Kanaal (Meuse–Escaut canal).[3] That happened along the highway toward Eindhoven, the road for their advance northward about a week later. I got onto that highway, but did not dare use it. Too quiet. . . . Frightening. . . . Quickly off the road. . . . Into some cover on the east side. . . . No sound. . . . No birds. . . . Can't stay here. . . . No going back. . . . Exhilarating! I am now committed to this!. . . Walk the bike. . . . Steady. . . . Just a biker-hiker. . . . Young; appearing not more than 15 years old; that had worked before. . . . Keep going. . . . Right into a gun-sight. . . . British!

It had been perfectly safe, after all, because no one was going to give his position away by shooting. The British were friendly but not in a mood to celebrate my Second Liberation. They advised me to keep low and stay in the area. They were spearheading it, and the situation was confused. Ah, but the Fleet Air Arm was in England! So, off on the bike. Across the bridge over the canal. Straight south to Hechtel. The proper turn would have been toward the west. But the weather was nice and liberations are exciting. There were a few dead along the way and even some trees across the road, but I was not paying attention, just trekking along with the little flags on that bike fluttering in the wind. When you

Brussels, September 14, 1944.

are singing you don't notice that it is quiet again. That the birds have flown away again.

"*Wer da*?" "Who's there?"—This from the Hermann Goering Division, a fanatical parachutist outfit. So, 15 years old again, understand no German. Just a hiker; ain't seen no British. They were not stupid and talked about how to get rid of the little errand runner. They could not shoot—have to keep things quiet. Bayonet? It was a frightening discussion. But I was taken through holes in the walls from house to house toward the center of the small town. It became a long night, locked up in a bar room, but interesting. There were some British prisoners, "Tommies," and we could whisper. Where to pee? Into stored beer bottles? Would the Jerries notice the difference? Deeper down in a cellar there was a headquarters where heel-clicking of messengers was heard.

Early next morning, Montgomery advanced to take Hechtel. As he always did, with his artillery plastering the place before sending in his soldiers. In a panic we rushed into the cellar, practically standing on top of the officer who was wounded and not much in command anymore. The bombardment rolled forward and the British tanks and infantry came and fought the Germans hand to hand, cutting and shooting, kicking and screaming. No one paid attention to me standing there, watching it totally fascinated, too scared and untrained to know what to do and where to find cover away from that burning house. When it quieted down I found myself talking to German wounded, all of us feeling better and beginning to cheer that we were still alive. That is how a British officer took me. He wouldn't respond to my joy: That little chap, coming out of a German Headquarters, was suspect. What should have been my Third Liberation got a little delayed. To get away from British custody, however, was not difficult because they were busy. I was dumped with nearby Dutch troops of the famous Princess Irene Brigade, and these did not bother to be suspicious but invited me for lunch to tell them about the Netherlands.

The bike was lost in the meantime, but that was good since hitchhiking is faster and much more interesting. So Brussels was next, which was a blast, with great celebrations of its liberation still going on. Strangely though, I was taken into custody again, this time by the Underground and this was serious. It was not clear what they saw wrong in me. Only later did I realize that in Belgium there was a suspicion of young German spies. A chap by the name of Karl Arno Punzeler, a 16-year-old, was caught spying on US troops in Belgium. Actually, he may have been caught later that winter, but that I looked like him may be seen on page 43 of *The Secret War* by Francis Russell.[4] These Belgian Resistance leaders seemed to be liquidating people, then and there, right in the heart of Brussels! Just when it began to look frightening, a British Major came through the prisoners' hall as if on an inspection, trying to find out what was going on. This was my chance and I rushed toward him saying that I had just come from Holland, had the defense works of IJmuiden Harbor in my head, and could give him other information from up north where they would be going soon. Indeed, he took me with him, but that was not a liberation, for the next night I had to spend locked in a closet without relief.

After a thorough interrogation the British handed me over to the Dutch, to an officer of the Headquarters of Prince Bernhard, the dashing Prince, Commander of the Dutch Interior Forces. That was exciting! But his interrogation was followed by advice to lie low; soon I could go back home. Home? I was on my way to England! No, England was forbidden. It had been closed even for diplomatic traffic since before the invasion in Normandy on June 6. That was done to guard strategic secrets and was still in effect. England was still closed.

All right, Spain again. At first it was delightful travel with military traffic of jeeps and lorries and trailers and messengers in endless array. The roads were lined with them, it wasn't moving fast, and the drivers were pleased to pick up the young bloke, share their chow with him and let him sleep with them underneath their trucks. The greatest unforgettable ride was on a tank carrier, a huge armored tractor pulling a trailer, which was empty so that it could fetch another tank to be brought up north. I sat for hours and seemingly days with the crew on top of this giant, with one of them driving and the others having plenty of time for swearing and laughing and telling tales to impress this young greenhorn who seemed bent on rushing to England to go at the throats of the Japanese. Japan! That was far away—better step on it to get him there faster! We were barreling along the route of the Red Ball Express, a one-way track marked out over French roads to speed up and better supply the armies from what was then still a preliminary harbor system near the beaches of Normandy.

Somewhere I met an American pilot who offered to fly me to England, but a Dutch Recruiting Officer overheard the conversation and warned me against the scheme. Apparently these fliers were taking guests, preferably female, over to England for the night, bringing them back again in the morning—as they had to, or the guest would be imprisoned by the British. Everyone entering England was put through Patriotic School, to weed out German plants and other suspicious characters. The Dutch recruiter offered to put me on a guarded shipment from Cherbourg directly into Patriotic School where I then would have to fend for myself to get a clearance. What luck to have met this man, for otherwise there is no predicting what might have happened in France, Spain, or elsewhere.

The docks of Cherbourg were in a mess, as the Jerries had done a thorough job of blowing them up and dumping the remains into the waters. One of the first repaired was a pier for ships with railroad tracks to bring engines and cars to the Continent. I was accommodated on one of those ferries, together with other dubious Dutch types that had been working for the Germans in France or had gotten there somehow or other. In Cherbourg I had stayed in a red-light house without realizing it—I actually thought these were nice girls—and my shipmates were a bunch of homosexuals, but such worldly insights occurred to me only later since our church's education had not prepared me for that. Not knowing what was wanted saved the day for me, and we were good friends.

Next came the hard part, Patriotic School at Camden Park. Without proper identification or anyone knowing me in England, how would the interrogators know that I was reliable and a prime candidate for their Fleet Air Arm? I do not know the answer. Day after day I had to tell my story, particularly of crossing the front line, in the greatest detail of roads and troops and guns and trees and bushes. There were long interviews, friendly but strict, and days and weeks of waiting. But not alone, as it was easy to make friends. Unfortunately I have lost track of the Polish refugee who wrote the following poem; his name is hard to read: L. Pawlowski?

A mon ami Tom Gehrels, 5–10–44
Nostalgie
Je te comprends mon doux ami
lorsque tu me parles de ton pays,
que tu as quitté un jour, car il
a connu l'occupation à son tour.

L'occupation de ce peuple haïs
qui je voyais tout peinis
et surtout se croyait invincible.
Chaque pays pour lui était une cible.

Tu me disais souvent dans le noir
"mon doux pays est si joli
que sitôt un étranger voit ses jolis
soirs, je jure d'y écouler toute
sa vie."

Bientôt tu la reverras ta rue,
ta maison, tes jolis moulins à vent,
car les boches sont battus
et ton pays reprendra son rang.

Je te comprends mon ami car moi
aussi je regrette les pittoresques
paysages de mon pays ou s'est écoulé le
quart de ma vie et que j'ai aimé
de mon plus jeune age.

To my friend, Tom Gehrels
Nostalgia
I understand you my dear friend
when you tell me about your
country that you left behind one day,
as it also came to know
occupation.

The occupation by these hated
people whom I saw in a painful
way and who believed
themselves invincible. Each
country was for them a target.

You told me often in the evening
"my beloved country is so pretty
that once a stranger sees its
gentle evenings,
I promise he will live out his life
there."

Soon you will see your street
again,your home, your pretty
windmills, because the krauts are
beaten and your country will
take its rank again.

I understand you, my friend,
because I also long for the
picturesque landscapes of
my country where a quarter of my
life was spent and that I have
loved from my youngest age.

Suddenly it was over. I felt like I was the most often liberated person on Earth! Next, however, came a frustrating session with a Recruiting Officer. Fleet Air Arm? No, but Royal Air Force Bomber Command had openings. This was the moment of ethics clashing with practice. Of deciding for oneself, and going against the current if necessary. On the one hand, I hated the Germans bitterly and the only solution, it seemed, was to eradicate the race; bombing their cities and factories might do just that. On the other hand, the bombing of Rotterdam in 1940 had left a ghastly impression. Allied bombers missed Haarlem's train depot by hundreds of meters on April 16, 1943, and I had been on rescue. I had seen Nijmegen too: The horror in the rubbled homes of innocent citizens. So I had my doubts about Bomber Command. The officer was angered by all this contrariness. In imperfect English, coming from a greenhorn foreigner! For such a brat, he had just the right solution. A bit terminal maybe. But I surely wouldn't mind doing something dangerous, hm? No, that was all right. No further questions.

The next Recruiting Officer was Lieutenant Piet de Jong, from the Office of Special Tasks (Bureau voor Bijzondere Opdrachten, or BBO) in cooperation with the British Special Operations Executive (SOE). Their mission was to parachute Dutchmen back into the Netherlands, operating in civilian clothes to make contact with the Underground and keep in touch with London through a small radio set. Exciting! Lieutenant de Jong tried to warn me that it was dangerous too, even mentioning rumors of betrayal and accidents.

As far as dangers were concerned, soldiers have a strange feeling that nothing bad happens to themselves, even if others may die or be wounded. Furthermore, the years of occupation had brought about not only a hatred for Adolf Hitler and his marching boots, but also a sense of connivance and subversion. No Dutch family was left unscathed; we were all in a struggle for life, and we were therefore prepared for just about any action or sacrifice. I was not the only one to carry the following poem and to sometimes say to myself the last two sentences.

A Soldier—His Prayer

This anonymous poem was blown by the wind into a slit trench at El Agheila during a heavy bombardment.

Stay with me God. The night is dark,
the night is cold: my little spark
of courage dies. The night is long;
be with me, God, and make me strong.

I love a game. I love a fight.
I hate the dark; I love the light.
I love my child; I love my wife.
I am no coward. I love Life.

Life with its change of mood and shade.
I want to live. I'm not afraid,
but me and mine are hard to part;
oh, unknown God, lift up my heart.

You stilled the waters at Dunkirk
and saved Your Servants. All your work
is wonderful, dear God. You strode
before us down that dreadful road.

We were alone, and hope had fled;
we loved our country and our dead,
and could not shame them; so we stayed
the course, and were not much afraid.

Dear God, that nightmare road! And then
that sea! We got there—we were men.
My eyes were blind, my feet were torn;
my soul sang like a bird at dawn!

I knew that death is but a door.
I knew what we were fighting for:
peace for the kids, our brothers freed,
a kinder world, a cleaner breed.

I'm but the son my mother bore,
a simple man, and nothing more.
But—God of strength and gentleness,
be pleased to make me nothing less.

Help me, O God, when Death is near
to mock the haggard face of fear,
that when I fall—if fall I must—
my soul may triumph in the Dust.

Parachutes and Independence

7. Jumps to the Underground

It was said that Mahatma Gandhi had stayed here, in this mansion in the country halfway between Oxford and Kidlington, but the house seemed too much an abode for the wealthy, with its large rooms and open fireplaces, and without another home in sight over the rolling meadows with a brook rippling through them. It was a fine place for getting in shape, shooting guns, throwing grenades, moving around in the dark, trying camouflage and practicing ambushes, and testing dash and determination.

The instructors were a varied lot. It was said that one of them, now in a captain's uniform, had been a burglar, released from prison for the purpose. Another one delighted in making us use fresh cow dung for face paint. The training was not just in the fields, but also inside the mansion, to open doors and sneak around without making a sound. A test assignment was to go into a house in town to steal something without geting caught. My burglary succeeded, even though a door slipped out of my hand with what seemed a very loud bang. In retrospect it seems we were probably set up not to get caught—the inhabitants warned in advance—to give us confidence as well as experience. The training was intense: several crafts to be mastered within about four months. I was to be an "organizer," to be sent to the Netherlands accompanied by a wireless operator trained in sending and receiving of Morse code. For all of us there was an emphasis on physical training as well as fieldcraft, that is, reading maps and moving around obstacles with a lot of crawling, running, and learning to hide. We were trainees with the rank of sergeant.

At the end, my evaluation mentioned a still strong influence of German propaganda. Indeed, compared to the Germans the British seemed so phlegmatic. Yet I was in awe of them, for they had stood alone in 1940 when everything appeared lost and everyone discouraged. Had Hitler won, the ten million people of the Netherlands might have been dumped into a salt mine in Poland—there had indeed been such a plan! Winston Churchill and his people had saved my life.

We had all sorts of weapons to fire and clean, some of them equipped with silencers. Silent killing was a special subject on the curriculum. British paratroops were also trained to kill with a judo chop, an all-out swing with the edge of a stretched hand to the side of an enemy's neck. British and American paratroopers were therefore not billeted in the same town, as the inevitable rivalries could have fatal consequences. The Americans had more pocket money, and so had more favor with the girls. They also had different techniques of parachuting. Limeys had only one parachute, yanks had a main chute and a spare emergency chute: "don't have the guts to jump without one," and the fists and chops would start to fly!

Sharpshooting, with a telescopic sight mounted on the rifle, was my favorite. We learned the insides of a German tank, how their troops and equipment were organized, and how to recognize their units. We were also taught some theory of logical thinking and decision-making. A frustrating choice between two nearly equivalent possibilities was to be worked out on a sheet of paper by listing the advantages of the one on the left and the other on the right: "Ways open to us," against "Ways open to the enemy." With the choices thus clearly displayed, a decision was easy to make, and once it was made there was to be no further hesitation. If time was lacking to work it out on paper, the same procedure was to be followed in the head. We were taught to switch off our emotions and senses, to imagine somewhere in the head a toggle switch that turns them off. One can practice it simply by taking a cold shower—the showers were frigid that winter—and we gained a good feeling for the usefulness of the concept. The instructors also simulated German interrogation, having me stand on a table with pants down, while they gathered around for some blows and degradation. I was their youngest trainee, but it did not make much of an impression. We did not really know at that time how cruel the SS and Sicherheits Polizei could be. Anyway, why worry beforehand?

We went out to a simulated landing zone in order to practice the S-phone, a walkie-talkie shortwave radio to tell a plane where and how the delivery was to be made. We told the plane crew, somewhere out there in the distance, that we had a good-looking female agent with us so they'd better come in low! They did not believe it at first, for there were only few girls in this business, and they wanted to speak to her first. Jan van der Weyden then put on such a good imitation that they came in, out of nowhere, precisely homing in on our S-phone, with their Hudson so low that we had to duck, while van der Weyden showed them his hairy legs.

Some of the training came from the tough and famous British commandos and parachutists. The jumping was done at Ringway, the present international airport of Manchester. The pretraining already was a cure for my acrophobia, as we jumped off high slides and buildings.

Drinking and singing were part of the training too; the melody is of the song "The Red River Valley"[1]:

> Come sit down by my side in a D.C.
> Do not hasten to bid me adieu.
> Just remember the brave parachutist
> and the job he is willing to do.
> When the red light comes on, we are ready
> for the Sergeant to shout 'Number one!'
> Then we all huddle up close together,
> and we jump through the door one by one.
>
> When we're coming in for a landing,
> just remember the Sergeant's advice:
> 'Keep your feet and your knees close together
> and you'll find Mother Earth very nice.'
>
> So stand by your glass and be ready
> and remember the men of the sky.
> Here's a toast to the men dead already,
> and a toast to the next man to die.

There were other traditions. A paratrooper jumped or parachuted and he landed—he was never dropped, delivered, or thrown out. The right foot was used to enter the plane; if one paratrooper used the left foot by mistake, then all would come out and enter again.

Our first jump was from a drafty old bomber; I seem to remember it was a Stirling. At take-off we were sitting by an open hole in the bottom of the hull. A mighty roar of the engines, the runway going by faster and faster, deeper and deeper down sank the beautiful countryside on a sunny day. It was my first flight in an airplane, but this was no joy ride. Those engines pulled us up quickly. The red light came on, at which I swung my feet into the hole. Green light, and I was out, barely hearing the "Go! Go!" of the instructor, out straight as prescribed. A jerk of the chute. . . and then the peace of floating freely and the loveliest view all around of the English landscape, so fresh and green in hedge and meadow. The first thing I saw straight down was the red cross of an ambulance. An instructor boomed with a bullhorn "Let up number one! Let up number one!" In my panic I heard it as Dutch, not English. "Let op" means "Pay attention," and what the hell does the bastard think I'm doing? I'm paying damn good attention! What he was really trying to tell me was to let the front lines of the parachute up a bit so as to come down more gently without sliding forward too fast. So it was a hard landing, but with a nice rolling motion, just as we had practiced. Back on the feet and run with the wind. Paw the parachute in on one side to get the wind out. Collapse the chute, fold it, pack it, report to the instructor. Straight at attention. Get hell. . . . "Yes sir." Dismissed.

The most interesting jump was at night from a tethered balloon, veered up to 300 feet with four of us sitting around a hole in the bottom, petrified. It was agonizingly slow, on purpose, to make us more aware of how the parachute was packed and would function if we did everything just right. The "static line" came out of the backpack and was hooked to a supposedly strong part of the gondola, and when we jumped we could actually feel the line come zigzagging out from the pack and then the parachute, zigzagging too. Finally, a string between the static line and the top of the parachute broke, and that was proper. The chute had then come out completely; it spread by catching the air and let us float down gently—or at least relatively gently, for the plopping down on the ground was still comparable to a free jump from some eight feet or so. A clever jumper could soften the impact with a slightly forward component of the motion, by pulling a bit on the two front lines of the four main ones that went up to the chute. The parachute thereby tilted slightly to act like a sail. The resulting forward motion would let the parachutist break the fall by rolling on the ground. But this was a delicate trick that had to be done just right or broken toes and bones were a result.

In parachuting all is well if everything is done just right at the right time, but punishment is immediate and severe if not. It was a training in doing things with determination, doing them once and correctly. There were plenty of accidents, and the ambulances were busy with sprained ankles, broken backs, and worse.[2]

But let us return to the roaring fireplace in Gandhi's mansion.

The students were few, but from various backgrounds. One never knew if their names had anything to do with reality, and we held back on telling about families and where we were from, as that type of information would soon be a dangerous trust. Peter Tazelaar, a real name, came through for a few days, probably more to teach us than for him to learn. He was one of the early agents working with Erik Hazelhoff Roelfzema (see his *Soldier of Orange*[3]) in the early organization "Contact Holland." When Queen and Government arrived in London in May of 1940, there had been no preparation for contact with people and resistance in the Netherlands. Hazelhoff was asked to establish that. Between September 1941 and May 1942 there were several landings on the beaches by a few agents such as Tazelaar and Hazelhoff. The rough rides on torpedo boats, the audacity of walking on the beach, in evening clothes and acting intoxicated, the tense moments of meeting up with Jerries, the successes in contacting the Underground, and the inefficiencies in London's offices, are all described in a lively fashion by Hazelhoff; his book makes fascinating reading and the film made of it has been seen by millions of viewers.

Later in 1942 one could no longer land on the Dutch beaches, as the Germans had become aware of these operations. A new organization was started in London to parachute agents into the Netherlands; the "England Spiel," which I will describe below, started as a deadly game of wits between British and Germans that lasted until late 1943. After a reorganization, the operation was replaced by a program in which dozens of agents were sent over by parachute in relative safety. These were the sergeants trained in the large house near Oxford.

Just before our mission we had a few days in London. There were interesting tales about the ladies of Picadilly, but I was too Calvinist and spent the time instead exploring the great city and sitting through some twenty showings of the film "The Fighting Lady," who was the American aircraft carrier *Yorktown*, and I went several times to see Richard Tauber conducting *Gay Rosalinda*, a version of *Die Fledermaus*. London was a busy place, the center of our world. The city was still under blackout in early 1945, but the German bombers and V-1 buzz bombs did not fly over anymore. Occasionally there were loud explosions, with a rumbling sound afterwards. Not much explanation was issued, but we began to understand that these were V-2 rockets, which came down from up high at supersonic speeds.

For us, there were preparations, some briefings and issuing of equipment, and promotions to Second Lieutenant, with a swearing in by General J. W. van Oorschot. The lady in charge of arrangements was Sybil Bond, who was a veteran in these matters. Even so, a leather jacket I was to wear seemed conspicuous to me, and it was of British manufacture. The most astonishing item was a potassium cyanide pill in a rubber capsule, for quick death by crushing it between the teeth. This would surely be a most direct give-away of an agent if it were found in his luggage. Someone said that a safe place to store it was by swallowing it, and collecting it again at the other end. Funny to imagine that, like swallowing a live toad—but one would have to wonder how good the rubber was. Anyhow, the pill came along.

A spooky ride to the airport in the dark. Everyone tense. I remember on such occasions a bit of schizophrenia: an indifferent Observer Half of me watching how the Real Half behaves, with the two halves sometimes talking to each other and the Observer Half snickering a bit and having an interesting time. A large Royal Air Force plane with four engines was used, but the hole in the bottom was round and smaller than the one of the Stirling. It probably was a Halifax. The pilots were sympathetic and allowed me to fly with them in the cockpit and this was fascinating. Their job was dangerous, what with guns and radars on the ground and experienced German night-fighters overhead. The plane flew between the devil and the deep blue sea, literally. They were safest against radar low near the ground—these guys were really flying "on the deck"—but it was dark and hard to know where "the deck" was, with all its obstacles of ships, dunes, dikes, homes, trees, water towers, and, indeed, guns and radars. Sure as hell, they had to pull up sharply for a ship, which they did with great delight, "Tally-Ho!" not knowing whether it was a friend or foe. The crew of the ship, which looked like a small freighter, must have been surprised too, but it happened too fast for an exchange of signals or hardware.

The landing zone was at the very east end of the Netherlands, not far from Enschede, so the two of us had time to get ready. The radio operator was Second Lieutenant Wouter Pleysier, code named "Rumble." By the time the red light came on we were ready for "action stations," sitting straight and trim, feet in the hole, hands on the edge, ready to push off. Green light, instantly out. . . . Feel the jerk of the chute. . . . Just enough time to release the kit bag on a rope, and smack in the heather—My, that Halifax had been low!. . . Hit the chute's quick release, parachute off, gun in hand. . . . Wait to see who comes.

Ah, but it was professional perfection. This ground crew had done it many times, talking on the S-phone, lining up in a prescribed figure on the ground, for instance the letter "L," and pointing their flashlights straight at the pilots. With the Halifax low and us coming out fast, we were right on target in the middle of our reception committee, and they were good Dutch. The times of the Sicherheits Polizei doing this were far behind, the England Spiel finished. What a marvelous reception! It was I, Second Lieutenant van Kampen from the Royal Netherlands Army, code named "Grunt," with a uniform in his kit bag, who was the greenhorn here because these Resistance veterans had already received agents and tons of supplies from previous flights. Our plane had also brought containers loaded with guns and ammunition. There was a splendid spirit of conspiracy and a willingness to do or suffer anything necessary to make home and country free again. Soon we were providing communication between them and London to prepare the Underground for action, should British or Canadian commanders request this in an offensive which was to come soon. It was possible to move around on bicycles, in single file with a large distance between us: the Commander of the Resistance group first, Pleysier and me next, and Ank, our courier who carried the most dangerous items, at the end.

Biking along a highway I made one of my non-thinking mistakes. We four were the only ones on the road. The Allies had everyone else in hiding. Anytime, out of the blue, a sharpshooting Spitfire or Mustang fighter plane could appear. However, here came a German staff car at great speed on that highway. And what should I do but look up as if there were a plane in the sky. Tires screeched, the Germans jumped out. . . . To hide in the ditch—What fun! This was a favorite schoolboy's dare with the Germans. But now I was no longer a schoolboy. Biked on, laughing. The Germans had stopped right by Ank who had our equipment in her bicycle bags and who must have been petrified.

Ank was a beautiful blonde, a very special person.[4] I fell in love with her, but we were trained and had been reminded to work as professionals and not as human beings. Personal feelings were not to be discussed. I am not sure the feelings were mutual, anyway, because Ank had a broken heart. She had loved my predecessor, Majoor Henk Brinkgreve, "Major Hank," who had saved her life. They had been working in a small group with some maps and equipment in a farm house when Germans showed up. They made the Dutch put up their hands. Henk's instructions had been that Ank should, if possible, never be caught because she knew all addresses in their Resistance group and one never knew how long one could keep such information when under torturous interrogation. So she moved toward the door. When one of the Germans raised his gun to shoot her, Henk punched him in the face such that the soldier fell backwards. There was firing then and a bullet hit Henk in the head. In that terrifying confusion, Ank and another Resistance person got out, onto their bicycles, and raced zigzagging away while Germans were shooting. Henk died.

One day in a home near the center of Groenlo, a small town close to the German border, Pleysier was transmitting to London while I stood guard at the window. Tension had been building because a great offensive across the Rhine was coming toward Groenlo, and the Germans had gathered as many troops as they could round up among their old and young. Their intelligence units were active to prevent the Underground from spying and preparing and transmitting to London, just as we were doing. The street below was quiet, and all seemed well. We saw nothing. We had done this before, but you never knew—capture would be deadly, that was clear. Still nothing. So we sent a long message. Suddenly a truck came round the corner. "Stop the radio! Hide!" The vehicle coming by slowly was a bread truck or some silly thing that had no business being there—just the type the Germans were using to hide equipment and personnel to beacon in on radio signals. Ours were off just in time. If the Germans had been able to drive by the house during our transmission they might even have learned from the sudden change in DF, the direction finding, that we were in a front room. As it was, they must have heard the signals stop, but the city block was too large for searching it or burning it down. We did not use that location again. Only after the war did we learn how seriously the Germans took this—what they did to people they caught. If they had used a pedestrian with a small DF instrument instead of that vehicle, as they sometimes did, we would probably have been caught, because I was not trained much in the possibilities of DF-ing. We were under British influence, under the impression that these were games one played with Jerry.

I staked out a German headquarters in a forest for a possible raid by the Underground. Two German soldiers saw me hiding in a culvert, but they just laughed—I still looked 15 years old, and scared—and it was a relief to get away with it, but also an annoyance that those Krauts did not recognize the demeanor of a great Paratroop Lieutenant.

Suddenly, it was over. The Canadians and British came through fast, too fast for us to stay ahead of them in German-held territory, and we were ordered back to London. They had successfully crossed the Rhine in a massive boat and parachute operation. Compared to that, the German forces were not strong anymore. Farther north in the Netherlands and east in Germany, there still would be fighting.

We made our way back to London and were soon parachuting again, radio operator Second Lieutenant Ferdinand Stuvel and I, near Ter Aar,

National Air & Space Museum

Manna from heaven, May 1945.

east of Leiden. The western part of the Netherlands, north of the Maas and west of the IJssel, were still firmly in German hands. The Dutch population was in terrible shape, with people in the cities dying of starvation. Bombings and shootings, boredom and frustration, stench and confinement, hunger and lice, no electricity or fuel, and trips in vain to farmers for food. How long could it last?[5]

Stuvel and I were again well received by experienced people of the Resistance and located in the Hague. I stayed with a girl courier, Anne Marie Aarts, who was proud she could make a meal of lettuce. Our weapons were hidden behind books in the library of the Hague's international Peace Palace, "Vredespaleis." On a bike without tires I went to Amsterdam to report to "Peter," the famous agent and commander of western Holland's Underground, Peter Borghouts. His command was a fighting force, trained and equipped to do anything the Allies would ask for. In a way, these men and women must have been hoping for a military offensive as an opportunity to use their training and preparations, to act out their hatred and frustration, and to show how much they were willing to sacrifice for the Cause. A disastrous slaughter of millions would have been caused, however, by a fighting advance into this tightly packed community of starving creatures, occupied by Germans who were fanatically dedicated to Hitler without scruples or common sense. This too the Underground knew. Fortunately for all, no military attack on this island of misery was necessary.

So there was fantastic delight, with jubilation in the streets beyond description, when suddenly over the rooftops came the B-24s, the B-25s, the Halifaxes, the Lancasters and other aircraft with open bomb-bay doors. German guns and radar were still armed and alert, and it was a risky proposition for pilots and their gunners, but though they also were ready, there was no firing. Instead, there were thousands of loose and flapping sacks dropped to cushion the impact for the cargo on assigned fields: for the bread and the grain and the cans and the goods, and there was enough to give at least some food to everyone, falling like manna from heaven.

Stuvel and I went back to London with a motorbike taken from the Germans, carefully finding a route through newly cleared mine fields and over half-destroyed bridges to an airfield in the long-liberated south. It occurred to us only later that we had not learned to ride a motorbike or steer a car; that had somehow been left out of the training. In London I was asked if I could drive a truck and said without hesitation "Yes sir," whereupon we took a large lorry through the heart of London. It must have gone well, because the orders came to proceed, to load it with a strange assortment of leftover luggage. At the Tilbury docks I was to drive the truck onto a Landing Craft for Vehicles for sailing to Ostend. "Remember they drive on the right in Europe." I was to go to the Netherlands equipped with a little notebook with some fifty addresses. It was urgent and confidential. A special mission. The truck carried the belongings of agents killed in the England Spiel.

To understand the England Spiel we have to go back to 1942. When Hazelhoff's landings on the Dutch beaches became no longer feasible, a new organization was set up in the Special Operations Executive, a British unit for subversive warfare. The SOE was Churchill's brainchild to "set Europe ablaze," in the hope of creating confusion and possibly even some insurrection. The governments in exile in London participated by supplying the agents who were to be parachuted into their country. Huub Lauwers was one of the first to jump into the Netherlands, but he was caught by German DF-ing of his transmissions.[6] Lauwers had been instructed, as all agents were, that he could come back on the air to London if he was forced to by severe interrogation. For just such an eventuality, however, there was a crucial security check for each transmission. I do not remember ours, except that it seemed clever and unbreakable, involving a prescribed mistake that would be omitted if we were caught instead of a positive signal that would have been easier for the Germans to prevent. In the case of Lauwers, as long as he was safe he would make a spelling mistake in the sixteenth letter of a message or anywhere in a multiple of sixteen letters.

Lauwers felt that it was all right to comply with German persuasion—he was not tortured—and he sent whatever they wanted, even with details of when and where the following agent would be received in the Netherlands. This was the "England Spiel," the England Game, for the Germans to play the radio link to fool the British. The question is to what extent the England Spiel was useful in fooling the Germans.

Lauwers was horrified when London reminded him about using his safety check; he even sent a few extra warnings such as the word "caught" in some of his following telegrams. But other agents arrived at the time and place he had wired to London. They were captured by the Germans, and some of the radio operators also transmitted, but nearly all of these obeyed their instructions regarding security checks. Others were caught in various ways until there were several radio sets sending and receiving messages between the Netherlands and England. A few agents refused to work with the Germans, no matter what was said or done to them. But in the end, tons of equipment were received by the Germans, at least fifty flying personnel of the Royal Air Force were lost in pinpoint night-fighter attacks that destroyed twelve ferrying planes, and at least twenty-eight members of the Dutch Resistance and fifty-four agents eventually lost their lives. Two agents escaped from the prison at Haaren, while Lauwers and two others, one of them a female agent, survived German concentration camps. All others were murdered, 39 of them on the steps of the "Parachutist Wall" of Mauthausen camp in Austria on September 6 or 7, 1944. They were put to work in a stone quarry, forced to run barefoot up the 186 steps while carrying heavy granite blocks. They were killed by SS guards.[7] The quarry of Mauthausen now is a memorial, and there is a special plaque there for these martyrs.

There are four theories on how and why the England Spiel happened, of which I dismiss the first two readily. It has been said that the game was a sinister scheme by Russian interests to eliminate certain influential groups from the future of Western Europe. There were, in fact, three people in fairly high circles of British Intelligence who later defected to the Soviet Union: Guy Burgess, Donald Maclean, and Kim Philby. But this theory seems too far fetched. It would not have been in the interest of the Soviet Union to diminish Dutch Resistance.

A second explanation is that there were traitors in London, possibly in touch with German spies there. Queen Wilhelmina's closest advisor, François van't Sant, was mentioned. Within our group of agents we heard about that sometime in 1944, and there was some discussion of liquidating van't Sant. It is now clear that he was not a traitor. Far from it: van't Sant had a distinguished record of personal service to Queen Wilhelmina, much of which came to light in the 1970s to explain how the Queen had come to trust him as a confidential advisor. Furthermore, British society was so spy- and traitor-conscious and so successful in spotting them (witness, for instance, the careful scrutiny in Patriotic School), that it seems unlikely that spies or traitors could have operated for any length of time in a sensitive area like the SOE, unless they were allowed to do so in a double-cross play by the British.[8]

The debate that is still going on in the Netherlands is between two other possibilities, namely whether it was carelessness within the SOE and Dutch bureaus, or whether the situation was permitted by the British. To understand the uncertainty on this topic in the Netherlands one has to take into account its secrecy. Everyone connected with the SOE had sworn not to talk about that work until 25 years after the end of the war, that is, until 1970. The expectation was that within this period we would have World War III with the Russians, just as there had been less than 25 years between the first and second World Wars.

There always had to be a cover story, during the war and afterwards. Mine was that I was attached to the British 6th Airborne Division, our first jump was near Bocholt, which is not far from Enschede, just before the Rhine crossing. When I jumped into western Holland, my cover was that it was an emergency landing from an action by Allied paratroops, presumably in eastern Netherlands or northwest Germany just before the end of the war, but our plane had been fired at or shot down on the approach over Holland. This was my tale for 25 years. I did not tell it often, because these were times of looking forward, not back into the past, but even now it is difficult to get off it. I still catch myself claiming with pride to have been attached to that famous British Division. Those lies make an odd contrast to a scientist's search for truth.

The files with more closely guarded secrets in England will be opened only 100 years after the event, but even by the year 2045 the complete truth may still not be learned because many of the records have been destroyed. In any case, a shroud of secrecy lay over the England Spiel at least until 1970.

An hour-long BBC television program made in 1984 starts by saying that the England Spiel was a victory for the Germans. The most knowledgeable person regarding the England Spiel in that pessimistic program, which concluded that carelessness in London gave the victory to the Germans, is the decoding officer Leo Marks. On the program, Marks appears to be in a quandary: He speaks of his oath of secrecy, yet he feels that the truth should be told. He speaks deliberately: "the Dutch nightmare. . . has a very unexpected answer. . . ." I have since heard from Marks that he has written a book to explain what happened in the code room, but the British Foreign Office has not yet granted permission for him to publish the book or to tell me what the solution to the nightmare might be.

The England Spiel can have played only a small role compared to the games that were played with captured German spies, the breaking of German radio codes, and the 1944 feint of General Patton's "Army Group" in Eastern England to make the Germans expect a second invasion far to the north of Normandy. However, the England Spiel must have contributed to the overall impression successfully created in German minds, all the way up to Hitler's, that the low lands north of Normandy were an important area of military activity. In German eyes, the Allies seemed to be preparing during the years 1942 and 1943 for bigger things that could come any time. There was a landing at Dieppe in 1942. Even before the Normandy plans—SOE started operating long before—Churchill must have aimed at holding the largest number of German forces in Western Europe. The fear in England was that Stalin's forces would collapse so that the full might of the Germans would come west again. In 1941 and 1942 it did appear as if the Russians were close to collapse because of the rapid advance by the Germans, and perhaps also because of the poor performance of the Soviets in the Finnish war of 1939–40. A major invasion was therefore not dared before 1944, but to the Western Allies it seemed important to relieve their Russian allies as much as possible, and Stalin urged them do so. Indeed, the Germans kept more than a million troops in Western Europe that might have swung the balance at Leningrad or Stalingrad.[9] They and the Abwehr and Sicherheits Polizei were necessary, in German eyes, to prevent Churchill and his agents and invasion forces from "setting Europe ablaze."

The England Spiel was a part of the Allied fighting and ultimate victory. The crucial feint was to keep the Germans guessing where and when the invasions would take place. Especially during the first few days of the Normandy landings it was essential to keep the German reserves away. This succeeded, as is clearly documented in the history of World War II. If the German reserves had come to Normandy promptly, the invasion might have failed. Even if only a few more German divisions had been there, at the dangerous time of the first landings, it might have cost tens of thousands of lives of Allied attackers. For two weeks, however, powerful German reserves were held in northern France because Hitler believed another landing was coming north of Normandy.

One should take into account that this was a dangerous business in any case; few of those who jumped before 1944 would have survived. Even after reorganization in early 1944, the Dutch part of SOE, BBO, had 32% mortality: 18 of its 56 agents died. This number is an average over the last year of the war. Many of those who jumped before the fall of 1944 were caught and killed. They did, however, assist with guerilla fighting near Apeldoorn and with a united Underground army ready for action where necessary. Let the following story of Sep Postma, Gerrit Reisiger, Maarten Cieremans, and Bob Vree be an example of what all SOE agents were willing to do and die for.

Seerp Postma, Sep for short, was from Baarn, the town where I had lived; he was a friend of my brother's. Sep escaped from the Netherlands in 1943 and reached England via France and Spain, a difficult trip, because the Spaniards were pro-German and could be cruel whenever they caught escapees like Postma. He had prepared his trip carefully by enlisting as a sailor on ships that were in service of German commerce. One of those ships brought him to Germany, where he managed to visit his brother in a forced-labor camp. On another ship he went to France, and his credentials were proper enough to get him through all of France. Crossing the Pyrenees was difficult. He had prepared for the interrogations in Spain with an invitation from an uncle who lived in the United States. But instead of going to the US, he went directly to England to fight the Germans and Japanese. He was driven by a sense of justice; already in 1939 he had wanted to help the underdog Finns in their war with the Soviets. He was devoutly religious and willing to sacrifice, but he had a light touch, an interest in people and journalism. He once agreed with his sister that even in the event of a disaster, they would emphasize and remember whatever good and cheerful there had been in their lives. Whatever had to be done would be done. The family is of

Frisian background and they live by a Frysk saying: "Doch dyn plicht en lit de ljue rabje," literally, "Do your duty and let the laity brag."

After he reached England, Sep was cleared in Patriotic School and recruited by Lieutenant Piet de Jong to join the BBO and SOE. He was trained and sent back to the Netherlands by parachute with radio operator Gerrit Reisiger on August 7, 1944, with the assignment to improve the contact between London and the Resistance. He was to unite sometimes competing organizations that ranged from fighting groups to services primarily interested in controlling the country after the war. He united several resistance teams successfully, traveling to various parts of the Netherlands on a bicycle. With Maarten Cieremans, another agent, he organized and trained fighting teams near Apeldoorn, where there was great activity and some guerilla fighting. In September large numbers of British paratroops had been defeated at Arnhem, just south of Apeldoorn, and some of these were still being helped back to their units across the Rhine. The following is a translation I made of parts of a report written by Cieremans after the war.[10]

> With our transmitter we arranged weapons supplies. We trained local teams in the use of arms and explosives, helped the formation of active fighting units, and took part in several actions. On 22 October 1944 our headquarters in the forest near Apeldoorn was attacked by surprise. So many were killed or arrested that we could not consider an immediate continuation of what had been started successfully in that area. Postma, Reisiger, and I escaped through the woods and heather country, taking along our transmission codes, frequency crystals, and small arms. Another shooting occurred; it was with Dutch-speaking SS who had surrounded us. We escaped and could resume our work in Utrecht a few days later. Our request for another transmitter was immediately taken care of by London and our direct radio contact was thereby resumed.
>
> On November 22 a final meeting was to be held in order to unite various resistance groups that had been operating separately until that time. Postma and I were to attend because our assignment from London was to expedite such unification. That meeting took place in the building of the "Kamer van Koophandel" (Chamber of Commerce) in Utrecht, but precisely at 2 o'clock, the time of our getting together, an SS detachment stormed in. They were from Apeldoorn, possibly the same we had fought with before. We learned later that the meeting was known to the SS because a girl courier who carried information regarding the meeting had been arrested. Postma and about twelve others were caught. With "Uncle John" Kool, who used to be a policeman in Jutphaas, I managed to hide during 24 hours on top of a cabinet in the attic of that building. The ones who had been arrested were terribly beaten with gunbutts and rubber hose: the number of raincoats found in the vestibule did not agree with the number of people caught. We were,

> however, not betrayed and, despite several searches, not discovered either.
>
> The building was still occupied when we finally escaped in full daylight with a ladder that we used as a feeble bridge between our roof gutter and the one next door. Kool went to Jutphaas and I, after some lonely wanderings, made contact with others who were saved because they had been a few minutes late for the meeting. To continue the unification of resistance in the center of the Netherlands demanded great care. None of the prisoners, however, gave up any information. Soon after their capture, about ten of them[11] were brutally murdered with bayonets and displayed in the streets of Apeldoorn with notes: "terrorist." My colleague Postma shared this lot.

Sep Postma was the sort of person who could feel what the following poem, by Leo Marks, expresses.

> The life that I have
> is all that I have,
> and the life that I have
> is yours.
> The love that I have
> of the life that I have
> is yours and yours and yours.
> A sleep I shall have,
> a rest I shall have,
> yet death will be but a pause,
> for the peace of my years
> in the long green grass
> will be yours and yours and yours.

Maarten Cieremans continued until the end of the war, organizing resistance groups, dozens of weapons drops, and direct contact with the Allied ground forces that were liberating the southern and eastern parts of the Netherlands. He worked under oppressive and exceedingly difficult circumstances, continually menaced by a gradually more cruel and desperate enemy. Friends and fellow agents were caught, but Cieremans managed to survive and remain effective. He emerged on 8 May 1945 at the age of 23 as a starved and lonely figure who wanted to forget what had happened during that dreadful year. Reisiger had been caught and had vanished into a German concentration camp, not to return. Cieremans had found a local operator, Bob Vree, who had been trained as a radio ham and had worked in the Dutch telegraph system. This kindly person was also tough and versatile; he moved from place to place for his frequent and extensive transmissions to London, never staying long enough for the Germans to catch him. In the end, these people—Postma, Reisiger, Cieremans, Vree and their associates, with the help of SOE and the Royal Air Force—had succeeded in welding together in the central parts of the Netherlands an Underground army of about 5000 men and women. They performed many tasks for the Allies and were ready to undertake a major uprising. Fortunately that was not necessary. The war ended in May 1945.

Soon after the end of hostilities, my truck went from home to home. I delivered whatever belongings the agents had left in London, but most of them had been in England only a short time and their possessions were usually not more than the contents of a battered suitcase. The addresses they had left behind were mostly of parents; the Postmas in Baarn were among them. Some parents had not even been told that their son was dead. It was deeply sad. So little was known about what had happened. What I did know at the time, and of our suspicions of treason, I could not tell. These were days of contrasts: of joy over food and liberty, and of sorrow over loved ones who would not return.

And what of the future? What about the war against Japan? With a letter of recommendation from Prince Bernhard, Second Lieutenant Henk Geysen and I made an application to the American equivalent of the Fleet Air Arm. In anticipation, we smuggled a billy goat onto a plane to London. Surely the aircraft carrier would need a mascot! Prime Minister Professor Dr. P. S. Gerbrandy was also on the plane and must have wondered where the goat came from, but he played with him joyfully. Just when the American plans began to look successful, however, there came the summons to come to the Far East right away for the same work we had done in Europe, parachuting in small groups. That was an honorable call to duty, and promptly obeyed.

8. Jungles and Papuas

Some twenty agents floated towards a new venture in their Dakota DC-3, or, in military notation, a C-47. The pilots were also enjoying themselves, flying not overlong stretches in the daytime to stay over in places that promised activity during the night. Malta offered a swim in crystal clear water. The next day we flew low over North Africa, bringing back names of battlefields from the Rommel campaign: El Agheila, Tobruk, El Alamein. The pyramids and the Sphinx were impressive that night, as were the out-of-bounds districts of Cairo. Then, on into the Far East, with hops and stops along the Persian Gulf and a stay in Karachi, with our final destination the Ratmalana Airport of Colombo, Ceylon (now Sri Lanka).

India and Ceylon were still under the British raj, or reign. Admiral Mountbatten was gathering forces for his great offensive, planned for September 1945, to match General MacArthur's drive towards Japan.[1] Ours was "Force 136," similar to the Special Operations Executive in Europe.

Our first business was a familiarization with the tropical terrain. Walking a compass course at night was a struggle between fright and laziness. When one came upon a clump of growth, the lazy way was to go straight through, if possible, to stay on course and not deviate from the straight-line path. But in the dark jungle there seemed to be a big snake rattling in every bush—at least that is the way it sounded to a newcomer from Europe. There were leeches too, which required a special mind-set of paying no attention to them instead of pulling them off, as their little sucker heads would stay in the skin and cause an itchy infection for weeks. If you could just let them be, they would suck themselves full and then drop off; that did not cause infection and was even said to be healthy—blood refreshing or something. After they dropped off, bloated full, the wound would bleed some more, and our bloody uniforms would have to be washed green again after each exercise. The birds and the beasts in the jungle of Ceylon and the large bats hanging in trees along the rivers, it was all new to us, strange and intriguing.

It was remarkable how little talk there was of what had happened before. Here were some sixty veterans of the most peculiar actions, but I do not remember that any tales were told. For a while I was in a hut together with Captain G. van Borssum Buisman who had a wide and deep scar across his back. I had heard something about an escape by jumping from a moving train transport to a concentration camp. But we discussed not this, nor anything else of bygone days. We were under an oath of silence, but there also was an urge to forget the nightmare past.

When we heard of the Atom Bomb, I condemned it forthwith. The first bombing of open cities such as Warsaw, Rotterdam, and London had given the Allies justification for carpet bombing of German cities, with fire storms raging through Hamburg and Dresden: "The Germans did it first." Anyone could now feel justified to use the Bomb by saying: "The Americans did it first." It was only later, much later, that I could see how Hiroshima and Nagasaki may have prevented a nuclear World War III.

Hindsight appears to bring questions as well as many answers. Why did I not quit soldiering in 1945 and become an astronomer? For one, there was a transportation problem from Ceylon after the end of the war, with priority given to senior combatants to go home first, but I could have hitchhiked to Europe one way or another; one of us, Puck van Lienden, hitchhiked around the world on planes and ships, and came back to Ceylon within about a month. I was asked to lead a small convoy of Dutch vehicles via the Burma Road to the Embassy in Chungking. A little later, however, that plan was scrapped. It could have changed my life drastically, for I would have been expected to stay on in Chungking in Foreign Service employ. At least 80% of life seems affected by chance occurrences and only a little by free will. I had a wish to be trained as a proper paratroop officer, rather than merely as a secret agent. And I was tempted to stay because of a budding attachment to these Eastern societies. Finally, there was a persistent recruiting by the Dutch; we later learned it was for the fight to retain the Dutch East Indies (now Indonesia) as a colony. We had no knowledge of a struggle for independence in Indonesia; such information must have been effectively suppressed. We had heard nothing in the 1930s of the concentration camp in New Guinea for Sukarno, Mohammed Hatta and other leaders. Now, in 1945, we were told that some Japanese were not giving up, were embarrassed to go home and therefore continued to fight in the interiors of New Guinea and with terrorists elsewhere.

All of my old SOE jumping buddies went back to the Netherlands, and I was left with new friends, who mostly were professional soldiers. We were now waiting to be shipped to the Dutch Indies. For lack of transport and other problems, we had time to explore Ceylon: beautiful beaches, strange trees, jungles, elephants, people in villages with foreign habits and traditions, priests and temples, customs, and religions. I was temporarily free of military duty, free to hitchhike or motorbike all over the splendid island with its lush vegetation, its relative prosperity, and its variety of people and animals. As I visited the central town of Kandy, there was the Perahera, a procession with decorated elephants, with priests and whipcrackers and flagbearers, with dancers and drummers and a great revelry of horns and drums from musicians, and with torches and jubilation, slowly moving among crowds of Ceylonese. There were few outsiders, as the British had a colonial disdain for something so native—and for natives in general. It felt as if I was the only non-Ceylonese, and the people loved it and pushed me forward onto the steps into the Temple, right up to their High Priest, who smiled benignly and made me sit at his feet. He clearly enjoyed having me there, and I was willing to fold my hands as they did, sit as they did, and pray as they did as well as I could. That was in the holiest part of the temple, where a tooth of the Buddha is kept in a golden cage, and it is therefore called the Temple of the Tooth. It was an impressive experience, and the marvel is that it can be repeated even now by anyone willing to go along with the people and behave like the people, for every morning at 6 o'clock there is such a celebration in that Temple of the Tooth. While many tourists nowadays come to hear the horns and drums and to see the religious going in, very few of them dare to enter with the Ceylonese, and bring a small offering of flowers or money, put their hands flat together underneath the chin and walk into the holy of holies up to the golden cage with the Tooth and the smiling High Priest, who will surely bless the stranger and beckon him or her to sit down and relax at his feet.

We got into some rascalry too. When I told a pilot that there was something peculiar about my ears and sense of balance that made it impossible for me to get seasick or airsick, he put it to the worst possible but highly enjoyable test by looping the loop rightside up and backside down, rolling the plane till the wings seemed to crack, and diving through our camp down to its telephone lines and along the beaches through the ocean spray with the wings below the palm tree tops. What a small world it seemed to be when a year or so later I heard that this British Lieutenant A. H. Bender had crashed and died in the Netherlands near the red-roofed farm where I was born.

A nasty commanding officer became for us a challenge to outwit, because he decided to keep us in Ceylon instead of allowing us to go on to Java.[2] Doing some unspeakable thing to his drink at a party and to the gas tank of his staff car did not solve the problem. Something clever had to be devised to exploit his weak spot and force him to let us go. He claimed to be solidly married back home and to be a devout member of some church or other, all of which gave him the right to restrict our liberties and love lives; but we had a suspicion that here lay some hypocrisy and thus a possible chance to create a situation in which he would not wish to see us any longer anywhere near him. I drew the lucky lot to prove the point, and this was to be done in one of the bedrooms of the Galle Face Hotel. I was to enter in military correctness reporting to him some frightful emergency that would justify the intrusion, but secretly in high hopes of finding him in an awkward position. It worked. He was in bed with a secretary. We were on a ship within days.

How lovely a slow boat through tropical waters! The hazy, sultry atmosphere seemed to make the voyage a gentle float on endless seas or along balmy isles, in and out of lazy harbors such as Sabang, north of Sumatra, and into Singapore, which was not yet very active either. I hitchhiked to Johore on the mainland of Malaya and walked across the causeway to see the famous Palace of the Sultan of Johore, with friendly people waving me away in a westerly direction where the palace was to be found. The countryside was beautiful, all neat with bushes trimmed, and flowers and greenery everywhere, but so quiet, walking for what seemed to be hours without meeting a soul. Finally, a distinguished looking gentleman appeared who seemed to be out for a stroll. I threw him a smart salute and asked politely where might be the Palace of the Sultan of Johore, whereupon the answer came with a smile: "These are the gardens of the Palace of the Sultan of Johore, and I am the Sultan Himself." Actually, that was not entirely certain, for they had their problems in these turbulent days trying to sort out who would be the Sultan, especially in the eyes of the British, as these were rather allergic to anyone who had cooperated with the Japanese. In any case, he was curious about Europe and I about occupation and independence, so we got along fine.[3] There was a luncheon to which he invited one of his wives in my colonial honor, a beautiful Javanese princess with whom I could unfortunately communicate only with smiles, she not knowing English, and I not speaking the court language of Djokjakarta.

Batavia, the present Jakarta, was a mess. It had parks, canals, gardens, stately Dutch homes, and an expansive white palace for the Governor General. But then there was confusion, with shots in the night and corpses bloating in the canals, and no one seeming to know who was killing whom. What was going on here? This could have nothing to do with the Japanese anymore because they were long gone from this city.

My first patrol was with Lieutenant Raymond Westerling, an amazingly strong man who could go up and down long climbing ropes using only his hands, but he and others could handle Indonesian prisoners with cruelty. What was I doing here? It finally became clear to me that in addition to rising communism and terrorism, what we had here was a struggle for independence. And we Dutch were to suppress that? We, who had just finished four years of fighting for freedom? So I impulsively wrote an Official Petition to Her Majesty the Queen to be relieved from Her Service. Astonishingly, the commanding officer to whom I was supposed to submit the petition tore it up without explanation. This was illegal, a violation of well-defined protocol. For the time being, however, there seemed nothing to do but to obey orders which had me quickly whisked away to faraway New Guinea.

The assignment was a challenge: Build a parachutists' training camp on the red-soiled, forested peninsula that overlooks the Pacific and

Humboldt Bay east of Hollandia, now Jayapura. It was under the command of Lieutenant N. J. de Koning, a man of ethics and a Commando of great fame. We found plenty of equipment that had been left by the Americans in the docks of Hollandia and Biak, an island farther west off the north coast of New Guinea. We built our camp, and soon we were in training, organized in fighting units, called Troops, of about forty men each. I was fortunate to have in my Troop Sergeants D. de Rijke and D. Garritsen, who had experience in the Royal Army of the Dutch East Indies, and who had survived years of Japanese imprisonment and slave labor on the Burma railroad. Their reports regarding Japanese cruelty were astonishing, such as having to stand at attention for 27 hours. They survived because they were tough and resilient. When they worked on a bridge, possibly over the River Kwai, the Khwae Noi River, they made a Japanese sing out for their heaving together: "Tojoiseenklojo, eentweedriehup," "Tojo [the Japanese general and Minister of War] is a bastard, one, two, three, away!" They would find food in the jungle, such as sambal, hot peppers that provided an essential supply of vitamins. If they got rice they would not chaff it, for similar reasons, but eat it as "red rice." Near Hollandia they made a fish trap in an estuary, and they taught us how to grow bean sprouts, again for vitamins and food value. By rule of caste, I was the young lieutenant in command, but they advised. A bond grew, and our Troop became the best of all. Do not listen to what the other commanders claim.

We had varied duties, because so much had to be improvised quickly. The only task I did not care for was a stint as Military Police, but a quick cure for that was to give a speeding ticket to the Base Commander, who then promptly drafted another volunteer. We were being trained in commando techniques, which included boat landings; the jumping was done later in Java. New Guinea had sandy beaches for practicing boat landings with rubber rafts and larger landing craft (LCM). One of these ships we baptized in 1946, with a gun salute and standing at attention, *Harry Wijnmalen*, after a young agent of whom it was said that he had shot himself while standing at attention before his Japanese captors. In 1942 Harry had gone by submarine from Ceylon to Sumatra for reconnaissance and guerilla warfare with the Japanese. When he was caught, he was subjected to extreme torture. When it became unbearable he reminded his interrogators of the Japanese tradition that an officer may kill himself when his mission has failed. Reportedly, they allowed him to do that.

Near Hollandia in 1946, we were at a location that was interesting and often mysterious. We built a barrier in a stream to make a clearwater swimming pool; sometimes there were snakes in it, especially at night, but we paid little attention to them. It gave me sensuous pleasure to swim out alone into the ocean, diving down deep, seeing colors in the coral canyons and waving at big fish that would come right up to me. They may have been sharks, but I did not know about sharks. During a practice boat landing, however, two of our men fell down at that beach;

we saw them go down, but we did not see them come up again, and we searched but never found a trace of them. The land was called in Malay *Tanah Merah* (Land of Red), after its red soil. We lived near and above the ocean, and we had the farthest of views and the gentlest of breezes as a backdrop for our hectic schedules of bugled roll call at six, hard physical exercise and speed marching, storming steep slopes while yelling and shooting, learning tactical techniques and surprise maneuvers, and practicing it all with live ammunition—which tends to make you keep the head down and to hug that red soil for dear life. Sunday afternoons were free, unless there was some punishment with a road-running to supervise, but that was needed only rarely, for the spirit was high and the discipline was crackerjack.

One Sunday afternoon I was napping in my officer's hut, but awakened by a man standing over me and another closing the shutters. They were from a rival Troop: it clearly was a kidnapping, with jubilation to follow on their side if they should succeed, and an embarrassment on mine. Two strong, big bullies had been selected for the task and one was already forcing me down, choking. But now the secret agent's training came to try a rescue—and mark you, it might work with a burglar or rapist just as well—first to make them believe they had won, that I had given up and was giving in to them, to put them at ease and off guard, to give them a feeling of success and relief, for they also were tense and scared. But in that relaxation—careful not to show any alertness at all—to look for an opening. . . and then to go all out, screaming, kicking for the balls, fingers into the eyes, and to race away, yelling all the time. The embarrassment that day was within the other Troop.

Friday night was celebrated with a drunken brawl in an officer's club somewhere along a beach, mostly stag, as there were few women in Hollandia. Towards morning we might have to crawl away from that club on a narrow walkway above the coral, which is sharp and poisonous, and sometimes I'd have to drive them back over slippery graveled mountain roads, back to the camp in time for the 6-AM roll call. I did not appear to be able to get drunk. But these driver responsibilities in the morning were frightening: having to rehearse to yourself every move of limb and foot, alerting yourself that you are stupidly stoned, but with the responsibility for the lives of these clods who are all you have and your closest of friends, while the only thing you really want to do is to curl up with them and sleep off that drunken burden of imbecility.

We used Japanese prisoners of war for construction tasks, they worked hard and were easy to handle because of their slavish discipline. If there ever were a misdemeanor, all one needed to do was to tell their highest superior. He would punish them physically, in front of his troops standing at attention during roll call. Once a ship came to bring more Japanese; they were suspected of hiding a transmitter among them. The order came to find it and any valuables they had stolen from civilians during their war and occupation years. We took pride in a last victory by finding a radio, transmitter parts, and a lot of loot. The price to pay was to look carefully into hundreds of Japanese rears. Whose victory was that?

There were stories about Japanese still in the jungle who did not want to give up and go home. We did some patrolling for training and found nothing but skulls, as the Papuas apparently had dealt with them before. The jungle was again a special challenge of mud and leeches, infections between the toes called "footrot," and long patrols at temperatures that must have been well over 100 degrees Fahrenheit with humidities near 100%, through steaming, thickly grown rain forests. Leading a patrol in single file brought a training in making quick and firm decisions, as the best path in rough terrain had to be recognized at a glance; there was no time for hesitation or a second trial, since a wrong choice would not be appreciated by sweating men following in these footsteps. We learned to walk without sound, without talking; if necessary one could make a recognition signal in the dark by scratching with a fingernail on a boot. Night patrols had a special fascination, and there was a pride in finding the way in deep darkness through cloudy moonless nights.

My dream was to dam Sentani Lake. American veterans will remember it as the grand lake west of Hollandia, among rolling green hills surrounded by higher mountains, as seen from General MacArthur's Headquarters on one of those hills. An area of about 400 square kilometers drains into the 87-km^2 lake which lies about 70 meters above sea level. A canal 7 kilometers long would have to be built towards the east to carry a stream of some 15 cubic meters per second. For comparison, the Oebroeg generators, operating at a 75-meter level difference and a flow of 10 cubic meters per second was supplying the power for most of West Java in the 1940s. A Sentani power plant therefore seemed a great opportunity. I did not know about technical publishing then, so the study disappeared into the files of bureaucrats.

The Papuas were friendly. We were tactful about stealing their chickens for supplementing our simple patrol rations of rice and dried fish, and we could usually find an interpreter in order to know them a little better. During the night of full moon the Papuas in their widely spread villages would stay up all night and beat their drums incessantly because stopping or falling asleep would be looked upon with disfavor by their Gods. Anything bad that happened during the next month or beyond would be the fault of such a lapse. I was reminded of what we had learned of the early tribes in Western Europe, who also feared their Gods and did their utmost to please them such that punishments would not occur. They would be rewarded by going to a Valhalla Heaven after death, not having to die but living on forever. Was this, finally, the origin of Calvinism and all religions? Had they sprouted, many thousands of years ago, out of fear for the unknown?

Fear is deeply rooted in our background. In earlier animal stages, life and survival must have been frightening. When people began to reason, to explain the world, there were still many terrors that remained incomprehensible—thunder and lightning, hunger and sickness, and there always was the fear of death. Where one could not explain, superstitions arose. The unknown danger was given a name—"God"—and a tradition in order to placate it and to minimize its bad effects or its presumed anger. The tribes of Western Europe, Saxons and Frisians, Gauls and Etruscans, Greeks and Romans, expressed their fear of lightning and thunder with stories and rituals concerning gods. The concepts were handed down through the ages, details were added, and religions grew around the core of fear for the unknown. We can still recognize the names of the Sun god for Sunday, the Thunder god for Thursday, etc. Lightning, thunder, and winds are now understood, but there is still some praying when they occur. A Dutch farmer in the 1930s stood with me watching a thunderstorm and recited:

Het mensdom staat	Humans stand
en staart verwonderd,	and stare in wondering,
wanneer de Heer	when the Lord
der Heren dondert.	of Lords is thundering.

Nowadays, the unknown has shifted from what has been scientifically explained to parts that are still not understood. The fear of death still is a mainstay of religions, which provide hope that life somehow goes on.

Twenty-seven years later, in 1973, I was invited to return to these very same places and people, but now to lecture at their newly founded Universitas Cenderawasih, University of the Bird of Paradise. We had both come a long way, from gods and chickens to planets and universities. What a delight to see that the old camp had become a blooming suburb of Jayapura, while our outlines of streets and choices of locations were recognizable still!

The Dutch kept control of New Guinea some 15 years after Indonesian Independence, but then they had to leave. The area is still unstable with Papuas struggling for independence, which is resisted by Indonesia. Immigrants from Java are settling in, having left their own crowded island at the government's instigation. Foreigners are not usually allowed into New Guinea, the present Irian Jaya, other than to work for Indonesian oil and copper exploration. Our two sons Neil and George were forbidden in 1973 to hike to my favorite old villages deeper inland, away from Jayapura.

In 1947 it was time to take our first unit of about 200 fully trained soldiers to Bali and Java and to report to the authorities in Batavia what had been done in faraway New Guinea. The original instructions had been to build using the American stores, spending up to a million guilders, with the only accounting being a verbal report. A chairborne Colonel was to inspect our airborne expenditures, and he was generally satisfied but for one item. "Why this large number of rubber contraceptives?," he asked, looking sadly at me, still looking young for my age. He seemed dubious about the explanation that we used these things in the jungle for keeping our watches dry.

We were volunteers, willing to work hard, and there were good relationships between troops and commanders. I was referred to as "Sparky" because I drove a jeep with that name. It brings one pride to train crack troops and to march with them. On the way into town after an exhausting night patrol, I'd try to make them sing. If even that was too hard, I had learned from an old sergeant a more basic stimulant: "Men, the women are looking. Make them think 'from that man I must have a child.'" Up would come the chests, feet resumed precise cadence, and in we'd march in fine formation!

These were some of the brighter notes, in addition to the beauty of Java. The contrasts were great in this lush and lovely paradise, with its banana and coconut and rubber and palm trees in fantastic panoramas on the slopes of volcanic peaks, with water running everywhere along terraces of gardens and rice fields. But then there was fighting all over the archipelago.

How could the Dutch resist a movement that was obviously for Independence, so soon after a painful struggle for freedom of their own? What gave us the right to kill Indonesians? There still was a colonial conviction that these tribes were not ready for self-government and should be protected and controlled as immature children. The Dutch held that the best form of government for the islands was a federation of states, following the example of the United States of America, because of the variety in regional conditions and ethnic types of people. This point was easy to see especially for the Papuas, with their black skin and curly hair, who obviously are different from the Javanese with a lighter tan, bronze-colored skin and smooth running hair. On the other hand, there were commercial interests of the large firms for oil, shipping and minerals; these hoped to keep the country divided and under control through economic ties and political federation with the Netherlands; in their schooling the managers had learned a motto from the Roman Empire: "Divide et Impera," "Divide and Rule." The Independence movement had fine idealists among its leaders, but Sukarno, its popular hero, was a man of erratic behavior. The communists also had a strong and rising influence, so that the situation became similar to what the United States was to experience later in Viet Nam. Ironically, it was the federated and anti-communist United States that put pressure on the Dutch to stop the fighting and to free Indonesia. The continuing debate was over federalism, with its relative autonomy of ethnically diverse peoples, versus a stronger united Indonesia that would be centrally controlled by the government in Java.

In the 1940s, the Dutch seemed simultaneously right about seeing a federation solution, and wrong about fighting Independence to begin with. The issues are easier to see in retrospect now than it was within the turmoil then.[4] But it slowly dawned on us in the field, as it would to the American soldiers in Viet Nam that there was something wrong about the situation. At one time action would be needed with vigor, and next it would be called off. In August 1947 a "Police Action" was declared with rapid advance by the Dutch, but soon there were peace negotiations again. Pressure from the Soviet Union, the American press, and several Third World countries in the United Nations, caused the Dutch government to stop the war and to accept a United Nations Commission to mediate.

When issues are not clearly defined, war itself becomes the enemy. When troops are not used in full action and warfare, their mistakes stand out more clearly. During one dark night, our men fired at each other in confusion, which was astonishing, since we had been trained to avoid such incidents. One day we had rounded up a group of about twenty Javanese to be interrogated by our expert of local Chinese origin. They were found guilty of destroying a high-arching bridge. The sentence was to be shot on the spot. The Captain ordered me to have a machine gun brought forward. I was well trained—had a reputation for disciplined operation. But, somehow, there was in me a warning to scrutinize my actions. I had learned before, in touchy situations, that it had to be listened to and never ignored. . . . I refused the order. This was never done: We were Commandos. Rebellion, in front of troops! The Captain could have used his pistol. He did not. . . . It was a frustrating moment. . . . I have a vague but haunting memory that the twenty were run off what was left of their tall bridge to fall to their deaths.

But maybe this is just a nightmare, for I did not see it happen and our paratroopers were not cruel or careless with lives. It was a first-rate unit, proud of its red berets, but not at its best performance as yet; a year later it was to fight heroically in full action. But now I was not doing well either. I was locked up in house arrest for two weeks for another rebellion, having protested the selection of what seemed to be a less capable commander. Finally, I caught jaundice, malaria, and amoebic dysentery, all at the same time.

And tapeworm too. It comes from insufficiently cooked pork. The little head digs into the intestinal wall, and then the tape can grow to several meters in length at the expense of the host, and therefore dangerous in the end. Once diagnosed, the patient would get some foul medicine and nothing else to eat for days, and the tape would come out. But one has to make sure that the head comes out too, or else it was all in vain, so I was not allowed to go to the privy, but equipped with a chamberpot right by the hospital bed. At the supreme moment of delivery, with uncanny timing, a General came on inspection. I was trained in the rules of respect: in bed, one lies straight; and sitting, one gets up. So the worm and I stood at attention and the General and the other patients had a jolly good time.

For me it was time again to request permission to leave. This time, I was allowed to go home. The sickly lieutenant bought a ticket on KLM and found himself back in the Netherlands, where everything looked small.

9. Comets and Liberty

It took me most of 1948 to get back into balance. Malaria takes months or even years to disappear. How strange to feel lonely! Ank had married our resistance commander, Ben Doppen. A friend tried to help, but I fell in love with his wife. Calvinism saw to it that nothing happened, but he did not appreciate that, so that a love and a friendship were hurt at the same time. Another tender relationship ended because she was devoted to her religion and I could not turn back. My weaning from Calvinism was incomplete, the search for the Truth still on, and the time had come to make it professional. But how? I was an experienced observer, organizer, and junior commander. Could any of this be useful? The training had exercised the determination to reach a set goal, and the variety of experiences had produced a liberated liberal, good for a scientist. But much more was obviously needed. What would be the technical training to learn about origins and the existence of God? Astronomy seemed the right discipline, but what would be the daily work? My schooling had been poor because of the restless war times, and the army period had been entirely without books. Would it be possible to sit behind a desk and concentrate? Did I have the brains? Intelligence Quotient was not tested in my days at school, and definitely not in the army.

In Utrecht there was a Psychological Institute that tested people and gave them advice regarding their future.[1] First we made an agreement that no questions would be asked about what I wanted to be. The Institute would find the aptitudes and capabilities. That was done in a week of tests, assignments, and interviews of amazing variety and cleverness. The Institute delivered a detailed judgment: Becoming a medical doctor was ruled out, as I would not have the patience for patients not seriously ill. But the appraisal did include the possibility of studying abstract physical sciences! I then told them the dream of becoming an astronomer, and they supplied practical advice on how to reach that goal.

What a liberation from frustration! I entered into the haven of Oort's Paradise. Professor Jan Hendrik Oort of Leiden University carries the Dutch tradition in astronomy from Huygens and J. C. Kapteyn (1860–1920) into the modern era. He pioneered the understanding of the structure of the Milky Way, our Galaxy, and of the motions of its stars and gaseous clouds. Born in 1900, he still is a central figure, leading new science, keeping fit by rowing a skiff on Holland's windy canals. In 1948, I saw the frail-looking scholar ruling Leiden Observatory with a firm hand hidden in a glove of gentle voice and manners. Students were not encouraged to go into astronomy because the dedicated ones would come anyway and the best would survive through difficult tasks and tests. Undergraduate students were not allowed near the telescopes but advised to complete introductory courses first, and they'd better be good!

The astronomers opened a new world for me. It was like a family: a few professors, a staff of assistants, and students, for a total of about forty. Astronomers were poor in those days, holes-in-the-soles poor; a change would come after Sputnik in the 1960s. Now the Leiden astronomers have been moved into a steel and glass structure, but in 1948 they still were housed in their own observatory building, which contained telescopes, workshop, lecture room, cubicles for graduate students, offices for astronomers, and the home of Director Oort. The most memorable part was the library. Stacks of books going back to the beginnings of astronomy and updated with modern material, all crowded together into files with narrow passageways. Between two of these rows, at a small table by a window overlooking the garden and canal, I spent happy hours, weeks, and months absorbing calculus, basic physics, thermodynamics, celestial mechanics, dynamical astronomy, and Oort's new theory of comets.

Professor Oort did not lecture *ex cathedra*, but stood with his back toward the class working at the blackboard, speaking in such a low voice that we could hardly hear him, so we would move forward, close to the blackboard; this was no problem for a class of only about eight students. In this manner he seemed to develop his new theory that was published in 1950 in the *Bulletins of the Astronomical Institutes of the Netherlands*.[2] It was a rare experience for students to see new research being worked out before us and with our participation.

Comets come from far away in the solar system; they are first seen when they come toward the Sun, close enough for the Sun to heat up the material and drive it off as a streaming tail. As the comet gets closer to the Sun it gets hotter and the tail gets longer. And it does get hot. The planet Mercury, for instance, has surface temperatures exceeding 320 degrees Celsius. The rapidly escaping gases come up from the depths of the comet's core through vents and cracks in powerful jets, taking along molecules, ions, and particles; we see these as ion tails and dust tails trailing away from the comet like smoke from a chimney.

The questions Oort addressed were: "Where do the comets come from? What is their origin?" Other astronomers had already speculated that

An evening celebration. I am the face immediately to the right of the waving fist.

comets come from the outer reaches of the solar system, but Oort could prove that firmly from their orbits. He showed us that the trajectories of the newest, seen for the first time, indicated that they originate from a distance of some 20,000 AU. (One AU, an "astronomical unit," is the average distance from Earth to Sun, 150 million kilometers, or 93 million miles.) For comparison, Neptune orbits the Sun at only about 30 AU. Oort suggested that there is a spherical cloud of comets, 20,000 AU away, all around the solar system; it is now called the Oort Cloud. Occasionally some of the comets are disturbed in their orbits by the gravitational pull of a star or a massive gas-and-dust cloud passing by. The perturbed comets may thereby be ejected from the solar system, or they may come toward the Sun and be seen from Earth in their evaporative glory of gas and dust tails. Oort's model would withstand scrutiny for decades. We were most impressed with the manner in which it was presented and worked out in detail.

The motto of Leiden University is *Praesidium Libertatis*, "presiding over liberty," and that was celebrated in daily practice. Physics Professor C. J. Gorter started out by saying that one did not need to attend his lectures. He would follow a certain textbook and, in addition, one could borrow notes from upperclassmen. That suited me fine. I came to his lectures only occasionally to check that the method of studying from books and notes would work well, which it did. It was interesting, however, to see great scientists such as Gorter and H. A. Kramers presenting a formal lecture, with only rarely an interrupting question from the audience.

Lectures were posted to start on the hour but by long tradition, as everything followed tradition in Leiden, there was a "professoraal kwartier," "professorial quarter-hour," which meant that the lecturer would come in at precisely 15 minutes after the hour—a gimmick that assured accurate timing. Examinations were oral and could be held any time. The student would simply call on the professor to tell him that he felt ready to take his examination for a certain class, say Oort's comets. The professor might reply, "Well, that is nice. Why don't you come for coffee tomorrow at the house." So at 11 in the morning, coffee time in the Netherlands, the nervous student would sit with the professor's family, drink coffee, and eat *koek*, Dutch ginger cake, and talk about the weather or whatever was the topic of the day. After a while, the maid, the children and the Mrs. would leave, and the discussion became more technical. If the student had mastered the subject, that was clear within some 10 minutes, and the professor would ask for the grade card, enter his evaluation, and that was that. If it did not go quite so well, the interrogation might last a little longer. Worse, the professor would ask the student to come back in a day or two, for a certain part in particular. Worse yet, one would be asked to come back in three weeks. If, however, the suggestion was made to come back in a year, that meant never. After about eight prescribed courses had been completed in math, physics, and astronomy—humanities had been covered in high school—a final verbal examination was held with a committee, but this was more of a formality. The result would be announced immediately and it was important to listen with care because certain words were used as *with pleasure* or, even better, *cum laude*. That was the end of undergraduate studies, and produced the proud Dutch title of *Candidaat*, "Candidate," equivalent to the American Bachelor of Science.

For those going on to graduate work there would be another six or so lecture topics, with oral examination, and a research assignment resulting in the title of *Doctorandus*. This was not quite equivalent to the Ph.D. The Dutch had a further advanced program with a large task to complete, a major research project with a dissertation as big as a book, in order to become a *Doctor*. It was not always done, because it was a sizeable undertaking, larger than that for the American Ph.D.; for jobs such as high school teacher it was sufficient to be a Doctorandus. If one did succeed in becoming a Doctor, there would be a final examination, held as a formal ceremony with friends and family in attendance. The nervous candidate sat in front, facing a long, green, velvet-covered table. To his left and right a close friend, never to say a word, but useful in case of fainting. All three in tails. At the assigned time of 14:15 the *Pedel*—Latin titles were used—would command the audience to rise for the entry of the Professors. About seven of them including the Rector Magnificus, a rank equivalent to President of the University, would come striding in, formally gowned, they would place their caps on the green table, and seat themselves. The inquisition could be surprisingly technical and difficult, but at precisely 15:00, the Pedel would interrupt with

Inspection by the upperclassmen. How could we *do* that, in 1948?

"Hora est," "Time is up." The Committee would file out, to return with the verdict after a seemingly long time for their deliberation.

For me as an undergraduate student, the evenings before an oral examination were memorable. To leaf through the notes on thermodynamics and the theory of gases gave a sense of integration, of putting it together, with a glow of knowing it all. I learned to want to learn, or even to pretend that I wanted to learn. Sometimes I may have been fooling myself, but it worked, because it is the interested mind that absorbs new knowledge faster. Feet on the table, notes in the lap, a shaded light on the paper, with the hearth aglow and a cold wind blowing rain against the windows; it brought a feeling of belonging, an association with the people who in this old university town had thought about thermodynamics, argued it with colleagues, and had thereby completed it into an elegant discipline. Studying superconductivity and the properties of matter at low temperatures was the specialty of Leiden's physicists.[3]

What about this motto, *Praesidium Libertatis*? Prince Willem de Zwijger, William the Silent, the "Father of the Fatherland," is considered the founder of the university. The city merited the privilege of having a university because of its persistent resistance during the long Spanish siege of 1574, during which the city had run out of food and people had eaten rats and shoes.[4] Those shoes must have been rather hard to digest. There are no such stories from the hunger winter of 1944–45, but that history is still recent. In any case, the water-expert Dutchmen had flooded the area, and the Spaniards eventually gave up and left. When the siege around the muddy waters lifted, on October 3, 1574, the city fathers were careful to provide for these starved stomachs only simple foods, such as white bread and herring. On the anniversary of that liberation, white bread and raw herring are still freely distributed early in the morning. It is a special day for students as well as the people of Leiden. On October 3, 1952, I sent a cable to my old student club in Leiden to tell them that our first son was born on that illustrious day; for special effects I claimed that he was named after our club "Momus," the dual God of Wisdom and Jolliness. The reply challenged the name, the date, and my ability to produce offspring.

Leiden had no dormitories or fraternity buildings, so that students had to make do by renting rooms, whatever they could afford, all over the old city. The university buildings were ancient and spread over the town, not together on a campus as one sees at American universities. Four of us students resided in a home, each to a room, with the owners living in the attic. John Happee was a medical student, later to become a distinguished physician in Los Altos, California. Pierre Mathijsen studied law and became a leader in the European Economic Community, and a learned professor: Prof. Mr. Dr. P. S. R. F. Mathijsen (the Mr. is a law degree). Hans Struve studied geology, but was already quite accomplished, being able to say "I love you" in twenty-four languages. We were members of a club of about twelve and there were some twenty such clubs per year, all of them under the auspices of the Student Corps which was a snobby affair for higher or wealthier classes, but for a while it was fun. There was a lot of singing, the noblest of which went something like this:

Sara Leander!	Sarah Leander!
'Tis niet om't een of ander,	It's not for one reason or other,
maar de ene bil en de andre bil	but the one ham and the other ham,
die lijken op malkander.	they resemble each other.

And for the next one we would stand as for a national anthem:

Io Vivat! Io Vivat!	A toast to Life!
Nostrorum sanitas!	This is for our health!
Dum nihil est in poculo,	When nothing is in the cup,
jam repleator denuo.	we will refill it anew.

The hazing was rough and not without accidents. Hair came off the head completely; as a veteran I did not have to do that, but I wanted to go through this phase and not have to talk about past soldiering—about which I had not so much a feeling of pride, but more of frustration. Anyway, it was fun, with a great deal of crawling on the floor, shouting and some beating, and training in drinking Jenever, strong Dutch gin, but never any beer, as that was for the vulgar plebeian types, the students of Delft and Utrecht. It was considered part of a classical education to observe one of the ladies of the night, who had the temerity to strip on top of the bar for a howling mass of unmanly manhood. After three weeks of hazing and a humiliating initiation, one was supposed to do nothing the first year and, ideally, not much more in the following years either, other than regularly spending all of every night debating everything and mostly nothing, and drinking that awful Jenever stuff. What surprised me most was that these young and illuminated people were so gullible and conservative. But then, I still will not drink any beer.

UNITED STATES AND SPACE

Astronomers and Their Observatories

10. Palomar Mountain

While still an undergraduate student in Leiden, I was asked to go to the United States in the summer of 1950 and bring back a thousand American and Canadian students on the *SS Volendam*, a ship chartered from the Holland–America Line by het Nederlandse Bureau voor Buitenlandse Studentenbetrekkingen (NBBS, the Netherlands' Bureau for Foreign Student Relations). In Europe the students could choose from an array of excursions or travel on their own, and the *Volendam* would take them back again towards the end of the summer. The sailing across the Atlantic would take ten days on such a veteran of vessels, and during that time the students were to be entertained with social activities but also with lectures to discuss their European plans and preparations. A somewhat older student was needed to supervise all this, especially because the love life of the boys and girls had gotten a bit out of hand during previous sailings. The pride of the Holland–America Line was at stake! That part of the assignment seemed difficult, but there were not too many hiding places and these could get occasional cleaning with a high-pressure water jet.

The trip gave me an opportunity to follow the example of Erik Hazelhoff Roelfzema, who wrote a fine tale during his student days in the 1930s, *Rendezvous in San Francisco.*[1] Loafing on the North Sea beach near Leiden he had met a girl from San Francisco who teasingly made him a wager to come and look her up within 40 days, starting out with not more than 10 dollars in his pocket. He worked on board a ship and hoboed and hitchhiked to California with an umbrella and practically no baggage. My timing and finances were similar, taking inflation into account, with 60 days away from Leiden, and a total of 60 dollars spent.

It was not difficult to find work from Rotterdam to New York on the *Volendam*, as meals had to be served and dishes washed. The ship was loaded with emigrants, whom the Dutch government had given a free passage, one way only, to ease the problem of overpopulation. Most of the families would do well in the New World: One could see their determination to succeed and overcome the hardships of stormy weather on a crowded ship. Most of the student help, however, got seasick as did the Head Steward who left me with a stern instruction: "Remember the pride of the Holland–America Line. The dishes shall be perfectly clean, no matter what." The emigrants would not skip a free meal even if they didn't feel too well, so that the dining hall became quite a mess to clean up, practically alone, and then I still had to do those dishes. The kitchen, fortunately, had an open port through which a lot of dishes went flying into the ocean; on the long way down they must have become clean even by the standards of the Holland–America Line.

For hitchhiking in the land of opportunity and advertising I used a big sign "Holland to Washington, D.C.," with the destination updated each day. A change of clothes, a sleeping bag, and not much else were stowed neatly inside a rucksack; on the outside there was a little Dutch flag. Hitchhiking was an effective means of travel and a challenging sport. The rules of the game were:

1. The place to stand must be where the traffic is slow and free of obstacles so that the driver can see the hitchhiker far in advance, has time to decide favorably, and space to pull off the road.
2. The hitchhiker and his equipment must look clean; he should look the driver straight in the eyes and give a safe and cheerful impression; this should be a personal and appealing communication.
3. Before hopping into the car one should have a brief chat. Does the situation seem reliable? Will this ride end at a good place for hitchhiking again, without getting stuck along a fast or lonely road?

In the days of two-lane highways this went so well that it was possible to choose the fastest cars and rides for as long as 900 and 1200 miles at a time, with me doing some of the driving.

A traveling salesman had driven himself to exhaustion from Florida to Houston and he asked, "Dutchie, can you drive?" He watched me a while and fell asleep. Drive I did, that narrow road from Houston to Midland—scared, actually, because he had urged me to hurry, and the oncoming headlights seemed so bright. But I had him home in time for breakfast, cheered him off to an important meeting, and caught a ride to Pasadena, California. This driver made a deal: I would provide all the beer he could drink, and he would pay for everything else. A lot of beer was consumed and evaporated right out again because it was hot. Other rides were not so safe, with drivers who were actually drunk, but then the rucksack stayed in my lap to buffer the crash, which never came.

Sleeping was done in my bag along the road. Offers of lodging were not accepted, as I had not yet learned that to give can be as gratifying as to receive. Pride and independence were mixed in with awkwardness. Four college girls picked me up in Tennessee and invited me to come with them all the way to California. Their trip was a graduation present,

Palomar Mountain, 1950. The dome of the 200 inch is in the background.

but they were getting a little bored with driving and sightseeing only. I did not dare. Two maybe, but four!?

Americans then were still used to hitchhiking, as a lot of military men on leave had used it to get around during the years of World War II. It was forbidden in some places, however, and one of those was near Salt Lake City. As I was walking along the highway on an early morning, a policeman pulled up, rolled down the window of his car and ordered me, in an unfriendly manner, to show identification. "Why?" He repeated the command, louder. "Why?" Then he flung open the door, ordered me to get in, pulling his gun, so that this was not the time to say again "Why?" Instead, I gave him a furious tirade with the thickest of accents, saying that an Amsterdam policeman would not behave this way (!?) He apologized, showed me around his beautiful city, we became friends, and he finally delivered me to a good spot to continue the hitchhiking.

Some strong friendships came, and love a few times too, but it all went too fast and was too furtive to become hitched in Houston or Nebraska. The girls. Oh, those American girls looked so beautiful! So a male student took me to a sorority house early in the morning and called out his girlfriend and her friends. What did I see but the most bizarre collection of curlers and rollers, and their owners dressed in the shortest of shorts! I was also always in shorts, which was common for hikers in Europe, but just not done by men in the US. A barber in Washington, D.C. threw me out, with the instruction to get dressed if I wanted a haircut in his shop.

The shorts actually helped, for they served as a trademark of an obviously foreign visitor out for a sporty venture.

People and country went rolling by as in a fascinating film. The Smokies made a gentle impression, while the grandeur came in Texas and the Rocky Mountains, awesome in their vastness.

I learned much about this great country and its people during the long rides, talking with drivers and other passengers. Since we would probably never meet again, single drivers could be surprisingly open and frank, discussing the most private matters and sometimes even asking advice. My goal was to get an impression in the short spell of 40 days of what Americans think and makes them tick, and to visit the Palomar Observatory in southern California.

The observatory had been put into operation only two years before and was closely guarded against inopportune visitors by its Cerberus of a Superintendent, Byron Hill. Hill had built the place and he would not even let astronomers in on a visit unless instructed to do so by the Office in Pasadena. The astronomer George Van Biesbroeck tried once to get in, knocking on the glass cage for visitors in the 200-inch dome, but he was refused entry. I was naively unaware. The last ride in the evening had been from an Indian woman of the Pala tribe; at the base of the mountain she had pointed out the long, lonely road, and there I was walking, up and up for hours. Deep in the night came the only car, and it stopped for the unusual sight and sign: "Holland to California." After some introduction came the question of where I would sleep, and the driver said not to worry, which seemed fine as he was a farmer or something and I would be comfortable with my bag in his barn. He was not just a farmer, however, but Hap Mendenhall, one of the Mendenhall Brothers, ranchers on the top of the mountain who had helped the observatory in many ways. So he drove through the back gate straight to the house of Byron and Tina Hill, rang the bell and said to a sleepy and surprised Superintendent: "I have a guest for you."

This was a most fortunate introduction to the observatory and its astronomers, the Fame of the World! Byron was pleased to have a rider for his horses, and so we became friends, and he a good host. We rode beyond and above the observatory for a glorious overview of the area. One of these horses was killed a few years later when a plane crashed into the mountain close to the Hills' residence.

The greatest of wide-angle cameras was being used for making a sky survey, photographing star fields in blue and red light. It was a Schmidt Telescope, the Big Schmidt, named after the designer of this combination of a mirror with a correcting lens, Bernhard Schmidt. The principle of the photography is the same as for a hand-held camera, except that instead of film the Schmidt used large, thin plates of glass, with the emulsion on one side. These were developed in total darkness, one by one, gently rocked in trays, first for developing, followed by hypo-fixing; finally they were washed, and dried in a dust-free enclosure. The plates, 14 inches (35 centimeters) square, would be inspected the following day and taken anew if any imperfection showed up. The work was being paid for by the National Geographic Society and it therefore was called the National Geographic–Palomar Sky Survey.

In a much bigger building was the 200-inch reflector, so named because its giant mirror had a diameter of 200 inches, 5 meters, then the largest in the world.[2] That "200-inch" seemed the greatest structure I had ever seen, more impressive than even the skyscrapers of New York. Byron took me all over that giant on stairways and elevators to the very apex of the dome and on top of it, to look around almost as free of obstacles as from a parachute, such that one feels the urge to spread out the arms wide in jubilation of freedom and power. To have equipment as great as this for observing the universe!

The astronomers were taking photographs for detailed studies of star-cluster populations, fantastic pictures of thousands of stars clumped together in a configuration called a globular cluster. The senior astronomer was Walter Baade who clapped his hands with glee that this chap had come hitchhiking from Leiden all the way to see the 200-inch. "You surely must be interested in Astronomy! Come and be one of our students here." But no, the arrogant Dutchman said that Leiden Observatory was doing just fine in astronomy, thank you. "Well, you will change your mind some day, and then you just write to Baade who was an immigrant himself and who will help you get started." We left it at that.

I was taught by the locals never to call the place Mount Palomar, as that does not exist on any map or registry. Palomar Mountain is not high as observatories go, only 5600 feet (1706 meters), but it lies close to the Pacific coast, and a cold ocean current off the shore causes a stable layering of the air above it and thereby a steady atmosphere. Similarly there are a cool current and stable air along the coast of Chile, where astronomers have now also built major observatories on the coastal mountains west of the Andes. The reason telescopes are built on mountains is primarily to obtain the most stable air, the least amount of turbulence, so that the star images are fine pinpoints, staying steady and not spread out into diffuse disks. Turbulence and currents of warmer or cooler air make the stars shimmer (just as the hot air rising from a radiator or hot road makes things shimmer), and their images on the photographic plates are spread out. Because the steadiness of the atmosphere has such a strong effect on the quality of images, it has been given a peculiar astronomical name, "seeing." Seeing is often discussed by astronomers. At dinner they will talk about how the seeing was last night, and over the years they have studied it to determine how it can be improved by going to higher altitudes or by building the domes around their telescopes with greater care and ventilation. The continuing challenge is to work in air of even temperatures; it may be windy, as long as the temperature does not fluctuate. A telescope is protected from the weather by a dome that usually is hemispheric in shape. There is a slit in that dome, with large shutter doors that are opened for the observing. The telescope is pointable in the sky, so that dome and slit have to be rotated in the proper direction. Astronomers must ventilate their domes, open them a few hours before the start of the night in order to let warm air out and to equalize the temperatures inside and outside the dome. This explains the tradition to have the dome shutters towards the east in the daytime so that they are heated up the least in the late afternoon by the Sun in the west.

The ground cover around the observatory is also important. One can easily understand that rocks, bricks, and roads are bad, as they collect and store heat in the daytime which they continue to radiate during the night, causing parcels of warm air to come shimmering by the telescope. The ground cover at Palomar had grasses, small shrubbery and larger trees. Quite a few trees had just been planted, as the astronomers knew what would give the best seeing. The very best is, of course, where there are no air effects at all: in space. But only few astronomers were thinking about telescopes in space in 1950; my own budding interest was of an adventurous nature, and space flight was one of the things I wanted to ask about in the United States. The only astronomer who reportedly was involved in such futuristic explorations was Professor Fritz Zwicky, a senior scientist and imaginative maverick who worked part-time at the Jet Propulsion Laboratory in Pasadena. The greenhorn foreigner didn't catch the amusement with which he was directed to go see Zwicky. So he innocently trundled off to Pasadena, found the house, rang the bell, and waited. Finally a tiny opening came in the door and a female voice asked what I wanted. "I would like to see Professor Zwicky about space flight." The little opening closed for what seemed a long time and then opened again for a booming voice, "Waddaye want?" "I would like to see Professor Zwicky about space flight." The door swung open, and a big man appeared in it saying, "But don't you know? It's secret!"

Ten years later I would have another great encounter with him. I was assigned to the Big Schmidt for observing asteroids, making the "Palomar–Leiden Survey of Faint Minor Planets." It required using that beautiful instrument during the middle of the nights, while Zwicky had been assigned to the early evenings, and again in the mornings. It is a schedule no astronomer likes, for once he is up there he would just as soon work all night, especially if it is beautifully clear. So here I came as a beginning post-doc astronomer with some trepidation to this Holy Mecca. I stopped at the "Monastery," the place where one stays for food and lodging. Zwicky must have seen me walking up; before I could get to the steps, the door swung open and here the big man stood again, booming, "Who are you?" He knew, of course, precisely who I was and what I was going to do there, but this was part of the act. So I politely introduced myself and told him that I was going to share the night with Professor Zwicky. At which he turned around and walked away, saying that was all right, I could come at 2 o'clock in the morning or later and get the tail end of the night. Something had to be done, quick. The ex-soldier's voice came to the rescue, thundering, "Wait a minute, Fritz," and you never saw a giant turn around so quick. I would come at exactly

Yerkes Observatory in the winter. AIP Niels Bohr Library

10 o'clock and leave at exactly 2 o'clock. Or else! He was furious, strutted to the phone to call the Director in Pasadena, who, after some of Zwicky's yelling, wanted me on the line in order to check the prearranged schedule and whether or not I was willing to let go a few minutes here and there, which I was not. So then Zwicky got back on the phone, louder and louder, finally barking, "I won't talk to you anymore, young man," the young man being Ira S. Bowen, a few years his junior, but still the Director, and me thinking this was going to be a godawful time on this mountain, and Zwicky slamming down that telephone. . . . Then he was turning around with a naughty grin on his face, saying, "Now let's have dinner." We were the best of friends the rest of his life, and oh, yes, the times of switching were exactly 10:00 and 2:00, with courtesy. He would have loathed me if I had yielded.

Back at the doorway in Pasadena in 1950, there was the question of non-secret civilian spaceflight, and Zwicky did take the trouble to tell me about Willy Ley in New York. Ley was of the old school of German rocketry, of Hermann Oberth whose book I had read, and of Arthur C. Clarke and Wernher von Braun and others who knew where the future lay. Willy Ley was not going to turn away anyone interested in space even if it was only an undergraduate student from a small country. I had to stay for dinner at his house near New York, and there was a touching of kindred souls, and he kindled the spark.

The sailing with the 500 girls and 500 boys was a frolic. Talking and singing and dancing and flirting and cheering and lecturing all day and all night for ten short days and nights on an old ship that couldn't be too slow. And there, in the dock at Rotterdam, eyes that hadn't had much sleep and had seen so many gorgeous gals close up on that crowded ship, now laid sight on the most beautiful girl of all, a distinguished-looking cool kind of a person, very much a challenge to get to know, by the lovely name of Liedeke, "Little Song," a Dutchie of course, but one who would not insist on religion.

The United States were thus quickly forgotten and it was back to studies and exploration of body and soul, of comets and stars and of physics and math, until one day in the windy autumn we were walking together, Liedeke and I, tightly arm in arm, as Europeans will do, by the university's bulletin board and saw an announcement of student assistantship in the United States. In front of that board my American pilgrimage was remembered and told again: about the Palomar 200-inch being the greatest of telescopes, and how Baade said that he would help me into a good school whenever I decided to come to the United States, as he knew I would. And so we worked out a great plan in front of that bulletin board, and it all seemed logical and the right thing to do, which was to get married and start a new life in the land to which Willy Ley and Walter Baade had emigrated, and to make that for us, too, the country of our choice.

When I tripped into the office of Professor Pieter Oosterhoff and told him that we were going to the United States, he made me calm down for an unforgettable lecture of wisdom for all people about to change countries. "To adapt to the different circumstances will be a serious challenge, but it is absolutely essential for success," said Professor Oosterhoff. Liedeke and I therefore spoke nothing but English from then on. Once in the US, we tried to forget what lay behind us and jumped happily into the new society. After the requisite trial period for immigrants, we became citizens in order to share the burdens as well as the privileges of the United States.

The great attraction of the United States was not astronomy per se, because Leiden was doing well, using new developments in radio astronomy to find the structure of our galaxy, but the American universities seemed less structured than those in Europe, with greater independence for the young, and opportunity to branch out into any direction at will. At my level of Bachelor of Science, the European and American students from good universities had acquired about the same knowledge, the former having learned most in high school and the latter in college. Still, the US was a hard country where students worked long hours to study and to make ends meet. For us it was a particular challenge because Dutch law at that time made it difficult to bring money in with us, and we had decided anyway to do it on our own and not with parents' inheritance. The first school I attended was the California Institute of Technology in Pasadena, where I did not do well, from insufficient math training perhaps, or from lack of appreciation of the competitive atmosphere. We therefore moved to the University of Chicago, which was more like Leiden University. It had the Yerkes Observatory about 80 miles northwest of the Chicago campus, out in the Wisconsin countryside, near beautiful Lake Geneva.

Yerkes was one of the great American observatories founded by George Ellery Hale (1868–1938). Van Biesbroeck told me the story. Hale was born in Chicago to wealthy parents who had provided him with a telescope of 12-inch aperture, a sizeable instrument for an amateur astronomer. George Ellery had greater dreams, however, and he was able to bring them into reality. After graduating from MIT, he went to the University of Chicago as an associate professor of astrophysics. It was about 1892 when he heard of a set of 40-inch telescope blanks that could be purchased for $20,000. They had been ordered from a French firm by a group of prosperous businessmen in southern California who must have been interested in astronomy but who had another reason for wanting a 40-inch: the Lick observatory in Northern California had a 36-inch, the largest in the world at that time, and out of rivalry they wanted to build themselves a larger telescope. Before the project could be completed, however, their boom went bust and the blanks were for sale. George, together with his father and President William Harper of the University of Chicago, approached Charles Tyson Yerkes, a streetcar magnate, who was found willing to pay for the large telescope that would be named after him. When Mr. Yerkes made a first grant, the Hales and President Harper saw to it that this became widely known, with the Chicago newspapers telling the story of the Yerkes Telescope. Ah yes, but the 40-inch refractor needed an office building with smaller domes for other telescopes including Hale's 12-inch. So George Ellery would not let Mr. Yerkes forget about their astronomical venture.[3] The grand Yerkes Observatory was built away from the smoke and the lights of Chicago, such that by the turn of the century a first-class astronomical institution was underway, with George Ellery Hale as its first director. He was an outstanding astronomer who initiated the study of magnetic fields in astronomy for which he had designed the first equipment to observe the effects of magnetism on the surface of the Sun. He also was an imaginative and restless organizer driving himself and others, with tact and good manners, but driving all the same.

His next dream emerged because Wisconsin does not have the best climate for astronomical observing in the United States. The southwest has the driest conditions and therefore the largest number of clear nights, while in California there are accessible mountaintops along the coast close to the cool ocean current where the best seeing was to be had. Hale went to Mount Wilson north of Pasadena to set up an instrument for the study of the Sun, with funding supplied by himself and wealthy friends.

Hale must have known how at Cincinnati, Ohio, one of the first observatories in the United States had been financed through many contributions of small bills and checks from thousands of people. It had been such a success that President John Quincy Adams had come to open the observatory in 1843 with a resounding speech. The Cincinnati process, however, was too slow for Hale, not his style. He had an aptitude for obtaining the confidence of donors who could provide large grants, operating in elegant surroundings, such as can still be seen in the Athenaeum Faculty Club of the California Institute of Technology in Pasadena. Caltech was another institution for which he helped in the initiation and fund raising, as he did for the Huntington Library in Pasadena; he also organized national and international organizations to promote and publish scientific research.

Hale brought from Yerkes Observatory the design and components for a 60-inch reflector, and it was erected in short order on Mount Wilson. But he wanted to reach fainter stars to allow studies deeper into the universe. The principle of large telescopes is to collect more light by having a larger aperture, just as more rain is collected by a larger bucket. But when a lens larger than 40 inches is held horizontal, when the telescope points straight up, it sags under its own weight. The figures of the lenses that are required to bring the 40-inch beam into a pinpoint focus are thereby distorted. A mirror can be supported from behind by a rigid framework, so it has less of a problem with gravity. Hale's 60-inch telescope was a successful experiment with a reflecting mirror instead of a transparent lens.

One day in Pasadena, Hale was socializing with Mr. John D. Hooker, with whom he had already discussed on various occasions the astronomical necessities of a larger reflector. Perhaps Mr. Hooker had gotten a little tired of the scientific elaborations, for he asked, "Tell me now. Why do you want a 100-inch telescope?" And Hale caught on to the spirit of the moment by answering simply "Because I like it," upon which Mr. Hooker agreed to pay for the Hooker Hundred-Inch.[4]

Now the challenge was again to the technical side, and the big question was if such a large reflector could be successful. Although a 70-inch built by the Earl of Rosse in England had been in operation already as early as 1845, it had a long focal length and could be moved practically around one axis only. Could a 100-inch mirror be supported properly so as to keep its perfect optical figure, while still being able to point anywhere in the sky?

For the first night, for "First Light," several friends came up the mountain with Hale while the contractor rushed all day to make everything ready for that evening. When the gentlemen looked at the stars, they saw nothing but ugly blobs, and a gloom settled over the observatory. Hale's ambition had surpassed his ability—no such large telescope could be built.

He could not sleep. He went back into that dome later in the night, for he could not believe in failure as it all had been too carefully prepared. The images were perfect! Thus it was found that one cannot have the Sun shining in through an open dome on such huge blocks of glass, as the contractor had done, or they will become distorted and may not settle back into their proper figure until many hours later.

George Ellery Hale was encouraged to aim for a still larger telescope: a 200-inch reflector. But this would require some ten million dollars as well as carefully worked out designs and engineering. He was no longer in the best of health, having driven himself relentlessly for many years in utter dedication to observational astronomy. His intense worry for the problem at hand would cause insomnia and severe headaches, for which long rest periods were prescribed; even these he would turn into foreign travel, with visits to colleagues and laboratories or interesting archeological sites in Egypt and elsewhere.

People liked Hale; they loved him as a fine, sensitive person. I interviewed an optician and a machinist who worked for him, and they both vouched for the fact that he was one of the kindest people to be with and work for. Astronomers who had known him confirmed his reputation for tact and gentleness, in addition to his drive and dedication. All of which makes him one of the finest scholars and greatest organizers in science. Hale obtained the support for the 200-inch from the Rockefeller Foundation, and by the time of his death he must have known that it would be built to allow others such as Walter Baade and following generations to penetrate deeply into the universe. It was named "The Hale Telescope" at the dedication in 1948, its institution being called "The Hale Observatories." When our second son was born in 1956, at the time I got my Ph.D. degree from Yerkes, Hale's first observatory, the boy was named George Ellery Gehrels.

11. An Observer's Molding and Training

Being a student at the Yerkes Observatory was like living in Plato's Academy. Here resided the greatest astronomers and we could not get away from them. We practically lived with them in the isolation of this cathedrallike building in the countryside.[1] To astronomers the names were impressive:

- William P. Bidelman, the spectroscopist
- Adriaan Blaauw, my professor and advisor at Leiden and Yerkes both, who had moved to the US at about the same time we did
- Joseph W. Chamberlain, expert in planetary atmospheres
- Subrahmanyan Chandrasekhar; I will say more of him below
- Daniel L. Harris, able photometrist of the solar system, who died young
- W. Albert Hiltner, one of the best among observers
- Gerard P. Kuiper, the planetary astronomer; more of him later
- Aden B. Meinel, the initiator of new ideas and institutions
- W. W. Morgan, who discovered the structure of spiral arms in our galaxy and got so excited and worked so intensely that he had to be hospitalized for a while
- Nancy G. Roman, who left later to monitor astronomical research at the National Aeronautics and Space Administration
- Bengt Strömgren, who had been strongly encouraged to go into astronomy by his father, a distinguished astronomer in Denmark, and who later returned to Denmark
- George Van Biesbroeck, an observer of double stars, asteroids, and comets; I will say more about him later
- Robert Weitbrecht, a genius with optics and instrumentation; he will come up again later

and there always were others visiting from elsewhere.

Chandrasekhar had a strong influence on my later life. I now dare to call him Chandra, but then it was formally "Mr. Chandrasekhar" (at the University of Chicago the only title used is "Mister"). Chandra watched students closely. He was in charge of the colloquia, the special lectures held once a week by a visitor or a faculty member (never by students). This required some organizational effort because it was not easy for visitors to pass by out-of-the-way Yerkes. Their lectures had to fit their schedules, not ours; at least one of the colloquia occurred on a Sunday. Before the speaker started, Chandra would look around the room to see that every one of the students was present. If not, he would come by the next day for a friendly chat about the weather and things, but there also would be a question on the topic of the visitor's lecture. Story has it that for one of these students the same question came up years later in the final examination for the Ph.D. degree. That final exam was the goal toward which we were striving: it was the only examination other than the final testing for each course—no qualifying, no preliminary exam. It was oral, lasted at least half a day, and was held as a discussion by all members of the faculty with the new maybe-Ph.D. about anything the questioners could think of, sometimes their own research problem of the moment. The idea was to test knowledge, but also the approach and the ability of this new researcher to solve astronomical problems. Not everyone passed on the first try.

Chandra came from a great tradition in India where his uncle, Professor Sir Chandrasekhara Venkata Raman, earned a Nobel prize in physics in 1930 for his studies of the scattering of light within gases and transparent materials and his discovery of what is now called the Raman effect. I began to understand what Chandra had to overcome when, many years later, I was asked in India to lecture "Chandrasekhar style," that is, to be rigorous in homework and with the starting times for lectures. With Chandra the lectures started precisely on time, and he made it clear that latecomers were not welcome, so we were not late.

These were just small exercises of the essentials to success. A scientist has to be precise and hardworking even if he or she is endowed with as high an intelligence as Chandra's. Since Chandra would come in early in the morning, we would too; since he would come back after dinner, so would we. To succeed in science one needs a drive, a personal eagerness, and a reasonable amount of competitive spirit, together with a curiosity for the beauty of nature. Competition can be healthy and stimulating as long as it is friendly and does not exclude cooperation—a sensitive balance is needed.

There is a division between theoretical and observational astronomers.[2] The best is to be both, theoretically posing a problem in detail, making one's own observations at the telescope, and working them out toward a conclusive interpretation. Chandra is a theoretical astronomer and mathematician. He switches his topic of study every nine years or so; the work on a certain discipline usually results in a book, a complete treatise that will be leading the field for years.

I was to become an observational astronomer, and Chandra's lectures were mostly on subjects that I would not be concerned with in later life, but even so they were interesting. One particular exam question was on the Reciprocity Principle, which says that an optical system (lenses, mirrors, and so forth) is, in a sense, reversible: if light from one point forms an image at another, then putting a source at the second point will give an image at the first. The exam asked us to name examples from daily life, not mentioned during the course, where there is reciprocity. For instance, when one treats people nicely, the kindness will generally be reciprocated. Another example I remember from a hiking trip, where people were camping near the base of a curved, hollow rock face of a mountain. We could hear their voices clearly from a large distance because the mountain was shaped somewhat like a telescope dish and the people were near its focal point. When we met the people later, they said that they had also been able to hear our voices from a surprisingly great distance: The mountain had reciprocated by acting as a huge receiver.

The staff and students at the Yerkes Observatory were dedicated to astronomy and worked hard to live up to its demands. My student assistantship was for as much as 30 hours per week and none of it was forgiven for holidays or examinations; if a student fell behind, he made up for the hours in his own time in addition to the required 30 hours per week and the full load of courses prescribed by the University of Chicago. But there was the excitement of finding out new things, of understanding what had not been understood before. By being so close to people who spent most of their time on research we were educated to investigate rather than teach. Results of research were published in scientific journals, and that included the dissertation; contrary to the European tradition and that at most American universities, it was considered a waste of time to make a separate thesis book. The beautiful garden on the hilly terrain of the Yerkes Observatory has a tree I sat under most of a long day, trying to start my first research paper. I felt like the Buddha sitting under his Bodha, the Tree of Wisdom, but all I could come up with was a technical sentence, of interest to only a few: "Light-curves were obtained of 20 Massalia at phase angles ranging from near 0° to 20°."

Writing a paper was considered the Supreme Act. The University of Chicago publishes a *Manual of Style* which shows the format: a title and an abstract, next an introduction (sort of a huddle with the reader), and then the observations and interpretations described in the finest detail. Acknowledgments to the colleagues who had helped and to the agencies that had given financial support, references to previous papers on the topic, and tables and figures were to conclude the paper. This sounds dry and routine, but the emotional effort and the amount of work needed to get a good scientific paper out is sometimes compared to the pain and joy of childbirth. The paper would be submitted to a more or less prestigious periodical, the *Astrophysical Journal* being superior to all others, as its editorial office was at Yerkes. Then came the refereeing stage, in which one or two colleagues would criticize the paper, making comments for its improvement, and advising the Editor to publish it, or not. After modification, usually, the paper was accepted for publication. What a relief! Most papers are read by only those few who are working on the same topic, but each article contains a stepping stone in the edifice of knowledge. The reputation of a scientist is often measured by the number of papers published, taking into account the prestige of the journal, whether one was the leading author or a coauthor, and the real importance of the work.

At the Yerkes Observatory, one had access for astronomical observations not only to the 10-inch Bruce, 12-inch Hale, 24- or 40-inch Yerkes telescopes at the observatory in Williams Bay, but also to the 82-inch reflector and a few smaller telescopes at the McDonald Observatory on Mount Locke in West Texas. The story of the beginning of that observatory was told to me by Mr. Van Biesbroeck.

William Johnson McDonald was the banker of the little town of Paris, Texas. When he died in 1926 he left his estate, nearly a million dollars, to the University of Texas for building an astronomical observatory.[3] McDonald's next of kin brought the will to court, trying to prove that he had not been in a will-making capacity. Testimony was provided by a barber who quoted Mr. McDonald as saying that some day an observatory would be built to look into Heaven, and, as all normal people know, only God Himself can do that. A settlement was eventually made with the family, but the University was still left with most of the capital. This was a challenging problem, for although the president of the University had been trained in astronomy, there were no astronomers on the faculty. So the president contacted the University of Chicago. The Chicago astronomers had already felt the need in the 1920s for a telescope larger than the Yerkes 40-inch and in a better climate, and they saw their opportunity. Finding a good site was the essential next step. In 1933 Van Biesbroeck and a few others went in a Model-A Ford to West Texas. One has to imagine the arduous trip in those days of mere horse trails there; Van Biesbroeck showed us, years later, the rocky terrain and boulders that had to be removed for their car to pass. Then there was a memorable day on which they decided that Mount Locke in the Davis Mountains was the Chosen Place.

A detailed cooperative arrangement was worked out between the universities of Texas and Chicago for building an observatory with an 82-inch telescope and staffing it; the agreement was in effect until 1962, when the University of Texas did have astronomers who could take over the operations and enlarge the splendid facility. When I was a student in the 1950s, however, there was not much money available for travel, so that sometimes weeks or a whole month were assigned to a student at the 82-inch, then the third-largest telescope in the world! That situation would change after the funding that followed Sputnik came into astronomy. With more astronomers requesting more time at the large telescopes, the assignments would be made for only a few nights at a time after careful scrutiny of the observer's proposal—and in stiff competition with other proposers, since the requests would sometimes total twice or thrice the available time.

In the 1950s there was also no money to have an assistant at the telescope on weekends, and on such glorious occasions I was sometimes alone on the mountain with this powerful instrument and not another soul within miles. How exhilarating it was to sing all night when it was still and clear with the observations coming in as planned! The singing was an exuberance that, once started, might go on all night, flowing from one tune to the next. I felt a relief from the frustrations of waiting for a clear night, of fighting sleep and weariness, or of having had to struggle with new electronic equipment. It took me a few years to grow into this new environment and to appreciate an equipment problem as a challenge that makes it worthwhile for the astronomer to be there. It is nice work, most of the time. Toward 3 AM the tired brain could have a vague dream of charming girls rushing up the mountain, tearing their gowns on the thickets, to be with this Great Astronomer. It has not happened yet, but who knows?

A most intriguing professor at the Yerkes Observatory was Gerard Kuiper (1905–73). He was also born in the Netherlands, in the municipality of Haringcarspel, now called Harenkarspel, about ten miles north of Alkmaar, way north of Haarlem and Amsterdam. It was a remote area largely consisting of high-water moats and ditches for digging peat; it is now pumped drier and not so isolated anymore. Gerrit Pieter Kuiper was the son of a tailor, the oldest of four children. Young Gerrit Pieter was an outstanding student in grade school, and his self-confidence must have gained a good start within his loving family.[4] In such a community, what example can a bright young man follow? The decision was apparently made to become a teacher. There was no high school nearby; he went to Haarlem to a special school for training schoolteachers. There was a marked class structure in the Netherlands; some of it still exists. His school was looked down upon by the upper classes of students who would attend the high schools called "Lyceum" or "Gymnasium," that led to university admission. Kuiper, however, knew by then that he could do as well as anyone, and he passed the hard special examination that allowed him into the university anyway. Such outsiders were rather snubbed at Leiden University, even in my days twenty years later, and openly called *knor* ("nerd") if they could not afford to belong to the Student Corps. Kuiper had to rely on himself; his fellow student Bart Bok later described him as already quite self-assured at that time. The university was a rather autocratic world controlled by an upper class of senior professors; on only those few was the title of "Professor" bestowed, and these were recognized with veneration in town as well as at the university. To Gerrit this must have seemed a world of distinction, science, and reason. He found a guide in Descartes,[5] who had a strong influence in the Netherlands still in my days: *Cogito ergo sum*, "I think, therefore I am." Reason was crucial, while subjective judgment was mistrusted and the affairs and feelings of human beings were considered rather inferior to knowledge and facts.

As for his astronomical career, Kuiper went in 1929 on an expedition to observe an eclipse of the Sun in the Dutch East Indies. Later he accepted a position there, at the observatory near Lembang, Java, but he did not actually go there because an opportunity arose in the United States; if he had gone there, it turned out, he might have spent the war years in Japanese camps. Instead, he adapted to the US, changed his names to Gerard Peter, embarked on a brilliant research career in astronomy and became a senior professor at the Yerkes Observatory.

Kuiper offered me a student assistantship in 1952 to make brightness measurements of asteroids on photographic plates taken with a 10-inch, wide-angle camera at McDonald. It was to establish a new set of brightness or "magnitude" values for all known asteroids; the old system was chaotic and far from the truth. However, two previous assistants had already tried it, and they were no longer there. Kuiper simply told me that 2337 photographic plates were upstairs in a vault, that I was to obtain the magnitudes from them somehow, and "Good Luck." That was a challenge! Just what I had come to the United States for: to be left alone with a major assignment. Next followed years of measuring asteroid images—bright ones appeared big on the plate and faint ones were hardly noticeable. The observed brightness was, however, greatly affected by the distance of the asteroid from Sun and Earth, decreasing with the squares of the distances, and a multitude of other phenomena came into play as well. One of these is the phase, the angle between the Sun and the Earth as seen from the asteroid; the full Moon for instance is ten times brighter than a half Moon. The transparency of the sky also affects the measured brightness, and so does the seeing, especially on photographic plates, since poor seeing spreads an image out and thereby makes it appear brighter than it really is. I had to make measurements with a photometer, to develop methods to calibrate the asteroid images with stars of known brightness on the plates, and to compute distances and phase angles from the orbits, all of this for thousands of images. For other aspects of this massive program, such as identifying the asteroids on the plates and measuring their positions, there was a team consisting of Van Biesbroeck, Kees van Houten who had come from Leiden to help, Ingrid Groeneveld (later Mrs. van Houten) from Heidelberg, and Mrs. Helen Thorsen from nearby Williams Bay.

George Van Biesbroeck had an interesting history. He was born in 1880 in Ghent, Belgium. His father had insisted that he first obtain a practical education as a civil engineer, to be able to support himself before pursuing his real interests in astronomy. In the evenings George volunteered at an observatory in Brussels, and he became a tireless observer of double stars, their motions about each other, and of comets and asteroids to derive their orbits.

After taking the job of determining asteroid brightnesses, I had to select a topic for my thesis research. Solar system objects such as the

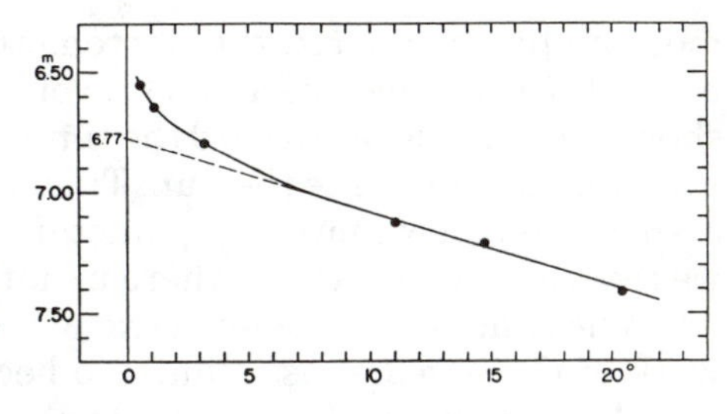

The opposition effect for asteroid 20 Massalia. The graph plots the astronomical magnitude of Massalia as a function of phase angle.

minor planets, the rubbly rocks in interplanetary space, were not highly esteemed among the astrophysicists who studied distant stars and galaxies. So, for a dissertation topic, Mr. Strömgren suggested I work with him on the modeling of stellar interiors, following routines of computing step by step the increasing pressure, and therefore temperatures too, deeper and deeper down into these spheres of hydrogen gas. Mr. Blaauw suggested work on his new discovery of motions in stellar clusters, how the stars move away from a central location where they were formed together; by precisely measuring recent photographic plates and comparing them with plates taken 50 years earlier, one could tell the speeds of the stars and figure back when their birth had occurred, millions of years ago. All fascinating, and an honor to be considered favorably by such great astronomers, but I had this bread-and-butter job on asteroids, and I was in trouble. There was a mystery in the results. They would not hang together. For asteroids observed more than once, at different phases, the derived magnitudes would not agree. Here was another assistant about to be tossed out.

It was a challenging situation. Previous literature on asteroids said that the brightness depends linearly on the phase angle. But did it? The only way I could see out of the predicament, to make the magnitudes jibe and get my Ph.D. in a reasonable time, was to show that the magnitude–phase relation was not linear and that there is an exceptional surge in brightness close to opposition, that is, close to the phase of exactly full moon. Proof was needed. It had to be sought discreetly, for my credibility was low. Bob Weitbrecht was the resource; this lovely man was deaf and we had to communicate with writing and lip reading, but he had given an excellent course in electronics and now he was the genius who provided a light-measuring device, a photometer, for the Yerkes 24-inch reflector.[6] Nancy Roman advised on photometric procedures, how to compare and calibrate with known stars.

A suspenseful winter followed. Asteroids rarely come close to that "full moon" geometry in which the phase angle is zero. There would be only three suitable candidates. The first one was clouded out. The second one was clouded out. All that was left was gentle Massalia, asteroid number 20 in the international catalog. That night was clear. And here she came, to within 0.6 degree of exactly full phase. I observed her on five more nights, compared the photometry in minute detail, and made a plot of the brightness as a function of phase angle. It was not straight! It showed a sharp surge toward the time when the asteroid is seen from the Earth exactly opposite the Sun. Thus was the "opposition effect" discovered. It was a new concept in planetary astronomy, but now it is an effect we recognize nearly everywhere in nature, an enhanced reflection back when no shadows can be seen (we will note it again in Chapter 15). The work of three years' assistantship became coherent, and the International Astronomical Union could publish my magnitudes for 1622 asteroids. That system stood for 27 years, until 1985, when it was replaced by a different definition proposed by other astronomers. In the meantime, at the Yerkes Observatory, the determination of asteroid magnitudes and their phase relation became the subjects of my final Ph.D. examination.

Then there followed a rather pressured period of five years' postdoctoral fellowship, with hard work and fortunate findings, but little understanding or interpretation. I was offered a position as Research Associate at Indiana University in Bloomington to work with Professor Frank Edmondson and Mrs. Dee Owings and others who had a loving interest in minor planets. They allowed students to work in their asteroid program, making observations with a 10-inch telescope and performing reductions of the data with desk calculators and a computer. At Indiana, Marshall Wrubel had established the first Computer Center in Astronomy, with an IBM 650, and Bell Labs Routine as the language. It was a pleasure to have the run of that place and to experiment freely during nights and weekends. I learned, however, that I should leave computer programming to the experts—the languages were still complicated and one needed to do a lot of programming to become good at it. My work was mostly a follow-up on the dissertation, namely gaining more information on asteroids.

The largest of these rocks in space have cross sections as big as the states of Indiana or Arizona, and even the smallest we can see with our telescopes still have the sizes of mountains on Earth. But they are far away, so they appear as pinpoints in our telescopes, looking like stars, which is why they are called after the Greek, aster, for star, starlike. They do, however, show variations in their brightness, that is the amount of sunlight they reflect, because they have irregular shapes and as they rotate in space they present to the Sun and to us on Earth cross sections that vary with time. The result can be visualized by rotating a book and looking at it edge-on: now one observes a maximum because the long side of the book is seen as a large area, next will be a short end and therefore a minimum in the lightcurve, then another long end, but perhaps the spine of the book is not of exactly the same color or reflectivity, so this second maximum in the lightcurve may not be exactly as bright as the first. Next we see the other short end of the book and finally return to the original view of the first long side. Thus there is a periodicity in the variation. With a photoelectric photometer one can measure

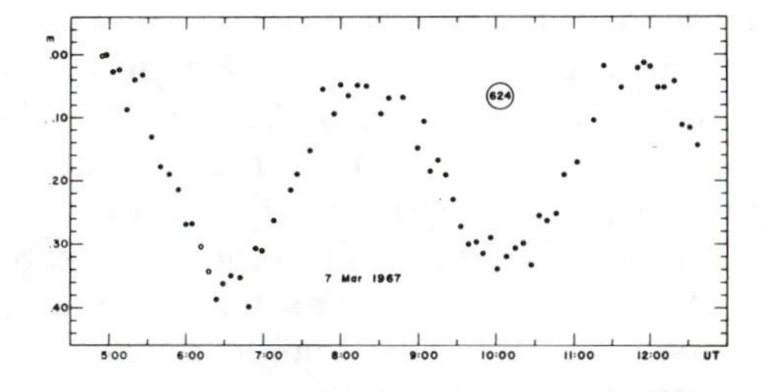

A lightcurve for asteroid 624 Hektor. Astronomical magnitude is plotted as a function of time, given as Universal Time, or Greenwich Mean Time.

the intensity of the changing light electronically. If we see the intensity cycle repeating in 8 hours, we conclude that 8 hours is the period of rotation of the asteroid about its axis in space, as the Earth rotates in 24 hours about its polar axis. From the shape of the lightcurve we can model at least some information about the asteroid's shape.

One research project was to check if a distant group of asteroids, called Trojans, might be lost satellites of Jupiter. In his theory of the origins of the planets, Kuiper had suggested that Jupiter had lost much of its original mass and thereby its gravitational strength, so that it could not hang on to all of its original satellites. One way to check this idea is with the rotation periods of the Trojan asteroids. A planetary satellite generally has a slow rotation, with roughly the same rate as it orbits the planet, because it has an irregular shape and its spin is then gravitationally locked to its revolution about the planet; the Moon, for instance, rotates about a body axis once in the same 27 days as its monthly cycle about the Earth. If the Moon could be set free, occur alone in space, it would still be rotating with a period of 27 days; we could observe this rotation with a lightcurve because the Moon's surface does not have a uniform reflectivity. If the Trojans have rotation periods of many days' length, unlike the usual asteroid periods of about 8 hours, this would be a confirmation of Jupiter's loss of mass and satellites, and it would yield a spectacular discovery. In 1957 the Trojans were observable in the Southern Hemisphere, and I could get time with a 74-inch telescope in South Africa. But we did not have much money. A letter was rushed to my old friend and teacher, KLM Captain Abspoel, and he helped indeed by sending me and my photometer on a special low fare to Johannesburg. The observations of the faint Trojan asteroid Hektor were made with the Radcliffe 74-inch reflector, which has now been moved to a good site at the Sutherland Observatory, but then it was still located near an ironworks plant by Pretoria. Every 20 minutes they would pour out the slag and I would have to shutter the photometer. However, I did manage to find Hektor's period of rotation: 7 hours. Back to Indiana.

A long-range research idea emerged: I became interested in the dust that floats between the planets. It would be a fine lead into understanding our origins from the interstellar clouds of dust and gas that I will describe a little in Chapter 16. The dust that floats between the stars would surely also occur between our planets and perhaps even on the

Moon and in planetary atmospheres. How to recognize it? How to determine the characteristics of small grains? They scatter and obscure starlight, as one can see in dark patches covering parts of the Milky Way or in the reddening of starlight that passes through dust clouds. Dust grains near the Earth reflect sunlight shining on them, giving rise to the "Zodiacal Light," a faint glow that can be seen with the naked eye in the evening after sunset and in the morning before sunrise. One can learn about the sizes and composition of these grains from their brightness, from the colors of the scattered light, and especially also from their ability to polarize, that is to cause a preferential orientation in the planes of vibration of the waves of the light. The most sensitive technique for analyzing these small grains would to be that of polarization measurements. A new discipline to learn and develop!

French astronomers had already made successful polarization studies of planets. Bernard Lyot, for example, performed a classic program of observations between 1922 and 1928. He was one of those dedicated astronomers who would construct an instrument with his own hands and money when he needed it. He died young and unexpectedly, and it then appeared that he had spent much of his income on astronomy. Friends and colleagues then held a special collection to help support his widow. Lyot was the first to succeed in observing the solar corona without an eclipse, using an ingenious instrument he designed, the coronograph. In his polarimetry he learned most from comparing the measurements at the telescope with those on various samples in the laboratory. At the Yerkes Observatory, Mr. Hiltner had taught a class in observational techniques including polarimetry of stars, a field he had pioneered and was famous in. However, polarimetry had an aura of being difficult. It was said that the chance of observing zero polarization, without spurious effects, was zero. And now I was daring to pursue subtle phenomena of light scattering by molecules and small particles!

The sizes of many of the grains in space are close to the wavelength of visible light. Chandrasekhar had taught us the properties of light scattered by molecules and small particles. Even objects in our daily environment do polarize the light: the reflection from a tarred road in the afternoon and the sunlit blue sky are strongly polarized, for example; one can see that nicely with Polaroid glasses, by simply tilting one's head. The molecules and dust grains I wanted to observe act like small antenna dipoles, like small vibrators in the lightwaves from the Sun. Since they are sensitive to visible light, they scatter and polarize differently in lights of different colors, different wavelengths. If I could make such observations diligently over a range of wavelengths, the reward would be learning new characteristics of the grains: perhaps their sizes, shapes, and compositions. My new photometer–polarimeter would therefore have an array of seven filters: for ultraviolet light, and violet, blue, green, orange, red, and infrared. I appropriated an old photometer from a program of Kuiper and Daniel Harris that had been completed. It eventually became the Minipol photopolarimeter, and it is still in oper-

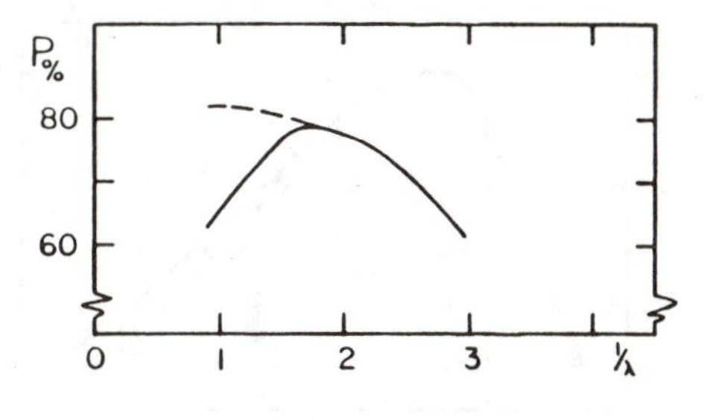

Polarization of the clear blue sky as a function of the frequency of the light. The percentage of polarization is plotted for a phase near 90°, for the sun 4° above the horizon (solid line), and 3° below the horizon (broken line).

ation; we may later put it in a museum, because it has provided millions of observations. Tom Teska, a young genius in the laboratory, who was then a student at Indiana University, was willing to refurbish that photometer into a wavelength-dependence photopolarimeter, and Bob Weitbrecht made some exceptional electronics for it.

The first observations with that polarimeter occurred at the McDonald Observatory in Texas on April 17, 1959, an unforgettable night! I had worked frantically during the previous night and day on a solution for the calibration, to ensure that what was measured was true, and just before the start of that first night on the 82-inch I had completed the final assembly of a depolarizer designed by Lyot. There was no tiredness that night, however, because everything observed showed new effects, polarizations so strongly dependent on color as had never been seen before: the Moon, the interstellar grains, Venus and the molecules of the sunlit blue sky. Juan Carrasco, the night assistant, was writing on the strip chart recorder where the data were being displayed, and I kept both of us awake, exclaiming: "Look Johnny, look, it's polarized like crazy!"

The original goal, to compare interstellar with interplanetary grains, was shelved for later years. The faint Zodiacal Light presented too difficult a problem that was not solved until some twenty years later with the Pioneer spacecraft described in Chapter 12. After the first nights' excitement quieted down, the discoveries became a matter of scholarly pursuit. There followed some of my happiest days as a scientist: observing planets and stars all night, and then in the daytime using the same telescope on the sunlit blue sky above the observatory. The work became even more exciting when I found that the observations could be fit to scattering theory derived in the nineteenth century by Lord Rayleigh in England, which had been given a new treatment by Chandrasekhar just a few years before my observations. On a visit to Chandra I could show him the new results, at which he asked critical questions, and I had most of the answers! Except for one peculiar effect: a decline of the polarization at longer wavelengths, deep into the red, which was not expected and which I could not explain. Upon which Chandra, the proper theoretician, presumed that there might be something wrong with the new photopolarimeter.

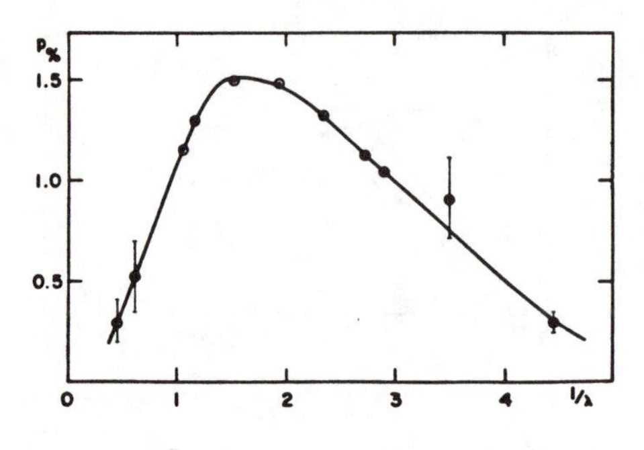

Percent polarization of the light from the star zeta Ophiuchi as a function of frequency, as measured by the reciprocal of the wavelength λ. The line shows a typical fit to van de Hulst's theory for interstellar particles of impure icy or icy silicaceous composition and diameter of about 0.3 micron.

It took months to find the right answer. The sky was less polarized at deep red wavelengths because there was another source of illumination in addition to the Sun: The ground all around the Observatory is much brighter in the infrared than at shorter wavelengths because plants produce chlorophyll, which strongly reflects the infrared radiation of the Sun. Chlorophyll is always present in all plants even after they die, so that in a picture taken in the infrared, gardens, trees, and cacti would all look bright compared to water, roads, and buildings. Perhaps this infrared radiation was scattering into my measurements. There was one scientist who had made observations in orange light, on Cactus Peak in California, and his data surprisingly did not show the decline of the polarization. It was Professor Zdenek Sekera of the University of California at Los Angeles, and I knew him to be an excellent scientist, so there probably would not be a fault in his data. Could there indeed be something wrong with my polarimeter? It was an unforgettable moment when I realized that all would be well and consistent if there were no chlorophyll, no cacti, on "Cactus Peak." I had to know the answer right away. One of my friends at Indiana University, John Irwin, is a great hiker and he surely would know Cactus Peak. So I called him with an inquiry about Cactus Peak, without first explaining to him the reason for the call, and he laughed and said that, yes indeed he knew Cactus Peak well, but it was a misnomer because there were no cactus nor any other vegetation on the top of this mountain. A great moment in the life of a scientist! The final proof came by letting the Sun set during careful observations with the 82-inch; after the Sun was no longer providing the other source of illumination (the bright light reflected by the grounds around the Observatory), the polarizations in infrared light came back to normal. A striking change occurred within 30 minutes, from 65% to 85% polarization. What a delight to report that to Chandra!

The exciting results had to be reported to other colleagues as well. At the meeting of the American Astronomical Society in Toronto in 1959 I was to make a verbal presentation, "give a paper," on the new discoveries, including those on interstellar grains. Starlight can become polarized, show a preferential direction of its vibrations, due to scattering of the light by the small particles in interstellar space. Magnetic fields

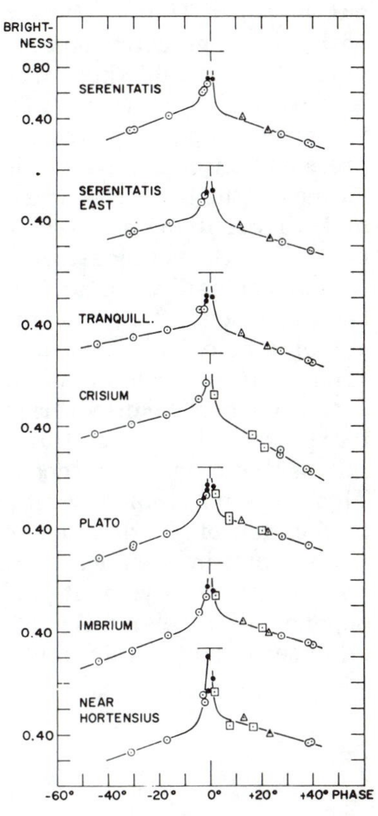

Intensity as a function of phase angle for regions of the lunar surface. Each curve comes from a region about 23 arcsec in diameter.

within the galaxy cause some alignment of nonspherical grains and thereby a slight preferential alignment of the light vibrations as well; polarimetry therefore had been a tool since 1949 to study the magnetic fields. My discovery was that the amount of polarization depended on the color of the light. This was exciting because it meant that the characteristics of the grains played a role, and one could now use polarization to study the grains as well as the magnetic fields. The theory of light scattering by such elongated particles had been worked out by one of my professors in Leiden, Henk van de Hulst. Now for the first time his calculations were fitted in detail to observations. The fit to the color profile was nearly perfect, so that I could derive diameters of the particles between 0.2 and 0.5 microns, one-hundred-thousandth of an inch. It was a thrill to be able to do this for objects so small and remote but so important for the formation of stars and planets.

Instead of celebration over the new discovery, however, there was a jealousy over who was the discoverer. My AAS presentation was to be on one of the afternoons, but while I walked back to the meeting with a small group of astronomers after lunch, Mr. Kuiper in his capacity as Director of the Yerkes and McDonald observatories, forbade me to present the paper. That was a most unusual command! He was busy talking with the other astronomers, and an explanation was therefore not made. I knew that there were severe internal difficulties among the Yerkes faculty, but I simply felt that these were none of my business. As for my polarimetry, it was done openly, there was nothing to hide. The presentation of the paper thus had a fine suspense, what with Mr. Kuiper in the front row, the objecting astronomer in the audience too, and other AAS members realizing that this could be an interesting moment: Was the young astronomer willing to hurt his career, go against the edict of his director, and probably never observe at McDonald again? Long ago, our history teacher Mr. Bax had told us how Martin Luther, standing before the Diet at Worms to defend his faith, had declared, "Here I stand. I cannot do otherwise." I thought of that, a little immodestly maybe. The discoveries were exciting, and the paper was therefore presented with zest. During the coffee break, Mr. Kuiper came by to tell me that it had been a good paper, tactfully presented. That was nice! There was no problem with getting telescope time in the future, but publication of the detailed results in the *Astrophysical Journal* was made nearly impossible. Oh well, there still was the *Astronomical Journal*, and over the years it published a series of our papers on the color dependence of polarization of planets, stars, and nebulae as well as another series on asteroids and related objects.

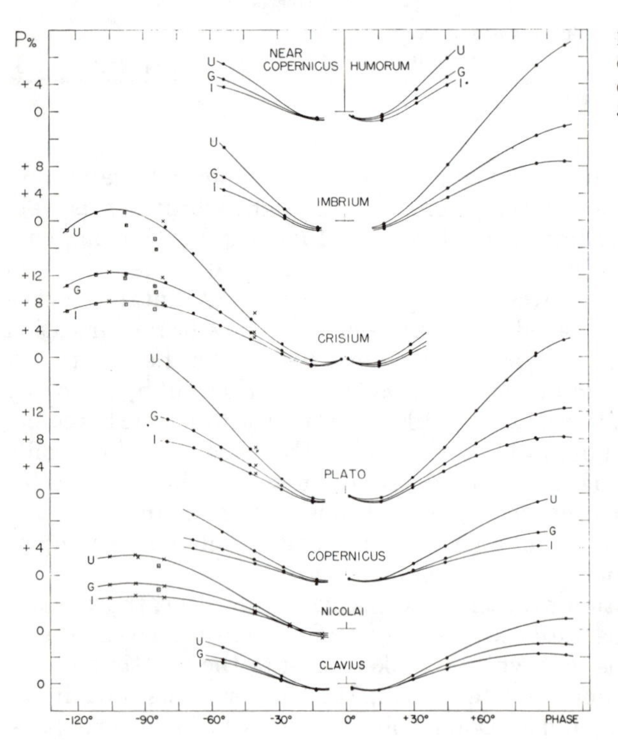

Polarization as a function of phase angle for regions on the lunar surface about 5 arcsec across.

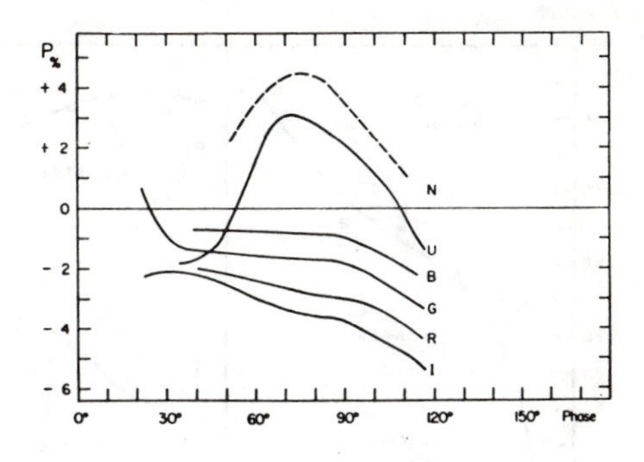

Polarization of light from Venus at various wavelengths: I infrared, R red, G green, B blue, U ultraviolet, N far ultraviolet (taken with a filter containing a nickel sulfate solution that eliminates stray red light).

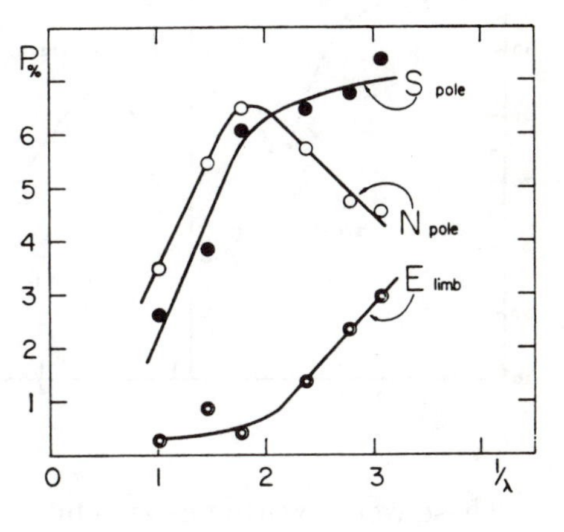

Polarizations of light from 5-arcsec regions on the disk of Jupiter.

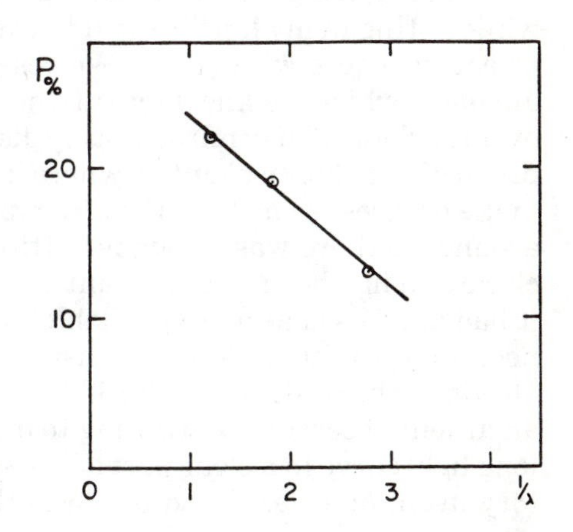

Unusual polarization as a function of frequency for starlight reflected from a nebula, that is, starlight scattered sideways.

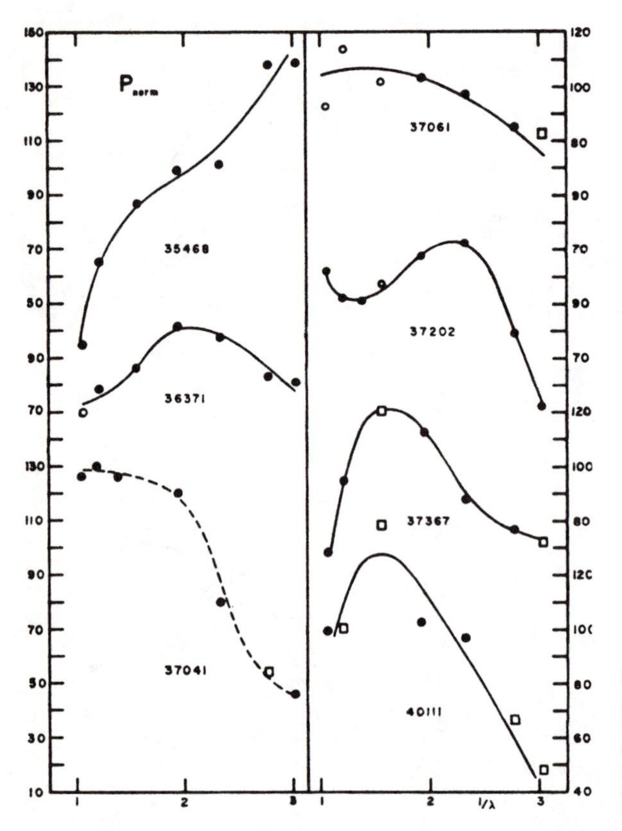

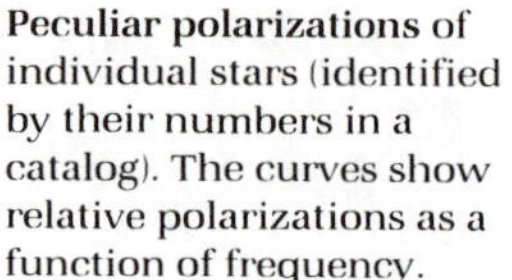

Peculiar polarizations of individual stars (identified by their numbers in a catalog). The curves show relative polarizations as a function of frequency.

These were exciting days, a bit wild perhaps. In retrospect I can see the army training still playing a role. Driving the 1360-mile trip from Indiana to the McDonald Observatory without stops would take 23 hours, police having no radar yet. Even the family enjoyed those non-stop runs, during which we'd sing, play games, and let the boys steer while sitting in my lap. We could not stop anyway, as we would not stay or eat in places with signs "We reserve the right to refuse service to anyone," which we knew meant "no niggers," and those signs were all over the South. They miraculously disappeared about 1960, and we were proud of our chosen country, where integration and human rights could make progress so fast! At the observatory I was not the only old soldier around, so there was a tougher attitude than now. In the army, a machine gun had been as important a property as a telescope was now, but it had been used as an expendable tool, dragged through the soil when necessary; the attitude was transferred to the telescope. If the wind was blowing to bring dust into the 36-inch dome, that was considered proof of an ardent observer not wanting to quit. If one had precisely observed a long lightcurve in a clear part of the sky while it drizzled at the observatory itself, one was asked only whether the telescope had gotten wet, which it had not, as the curtains in the slit of the dome had been used with care.

A new observatory was to be built in Chile, and its proper location had to be found. Mr. Kuiper asked me if I would be interested in that job, my military experience in the field would come useful. Ah, but that was exactly what I wanted to get away from. Science and discovery were now my goals. So there was no hesitation to say no, and I knew of a better person anyway. In 1957, while observing in South Africa, I met Jurgen Stock, who had been a German Feldwebel (junior officer) on the Russian front; he was highly respected by British and Afrikaander observers, and he knew astronomy well, particularly the procedures to be used in testing sites for new observatories. The result of my recommendation was that Stock was given an appointment at the McDonald Observatory and sent to pioneer an observatory in the Chilean coastal mountains west of the Andes. This involved carrying a telescope for testing the seeing by mule to various peaks, investigating the possibilities of making roads and supplying water and electricity, and, of course, checking on a suitable lack of cloudy weather in the meteorological records. Cerro Tololo was chosen, thanks to Stock's competent work, and the observatory there is now called the Cerro Tololo Interamerican Observatory, CTIO for short. Other astronomers from Europe and California have located major observatories on other mountains in Chile, La Silla ("the Saddle"), and Las Campanas ("the Bells"), that Jurgen found for them. I once went to CTIO to observe and was awed by the privilege and responsibility: "Who am I, with my own little study project, to be allowed to use such facilities, for which so many people have worked so hard? Better make use of every available minute of telescope time!"

The stories that are told around astronomical observatories! The superintendent of McDonald loved to tell of Guido Munch, a bright theorist and effective observer. Stars would trail along the slit of a spectrograph, typically for half an hour, during which there was nothing for the astronomer to do. Guido would rest in a chair and have the telescope, which was automatically moving to compensate for the Earth's rotation, nudge him alert after 25 minutes. One morning, however, the superintendent saw the dome still open, and he ran as fast as he could, for surely another astronomer had fallen off the platform. Guido was sound asleep in his chair! He had put it on the wrong side and the telescope was running away from him.

There were tales about my singing, which was said to have sounded like a mountain lion in heat. And there is a curious story of an astronomer who had come to make an urgent observation with an eyepiece whose focal length was much too short to fit the McDonald telescopes. The 82-inch focus mechanism jammed against the end stop, and its motor burned out. He next requisitioned the 36-inch from a student and the same thing happened there. The McDonald Observatory was thereby stilled a few days for the rewinding of motors. Another focus mechanism, at the 60-inch in Chile, apparently did not have an end stop, and when the mechanism went past the end, the mirror fell out—fortunately not all the way down to destruction on the concrete floor, for it got caught by the frame of the telescope. That story taught me a lesson to never stand underneath a moving telescope or dome shutter.

I had an interesting dilemma once because of observing too fast and running a platform into a borrowed photometer, thereby denting it. I had the dent fixed perfectly, so that the damage could not be recognized. But then I had to face the question: tell the owner, or not? Fortunately I did, but unfortunately it brought me the reputation of being a reckless observer. I denied that label at the time but now recognize some of its truth.

Much less frustrating was what happened on the Palomar road years later. Studying plates back in Tucson I discovered a comet! But was it real? It was 2 PM on what became a memorable day. A telephone to the Palomar Director brought his astonished reply: "If it is important enough for you [translation: if you are crazy enough] to come all the way from Tucson, we will give you an hour on the Big Schmidt Telescope this evening." So I caught a plane at 3, rented a fast car at the San Diego Airport, skidded it against the mountain on the Palomar road, hitchhiked without delay to the Schmidt, and called for a wrecker truck after inspecting the plates next morning. There was no comet.

Gradually I learned that scientific pursuit needs more than dash and determination, that telescope mirrors need to be kept clean, that careful interpretation is as important as serendipitous discovery, and that curiosity and respect for the beauty of science are the most essential of all. Colleagues and observatory staff were gentle enough to show me the way.

Night Assistant Juan Carrasco was a lively and lovable person who is well known among astronomers; he later moved to the Palomar Observatory, where he is still employed. The job requires working with sometimes tired, stubborn or excited prima donnas. Back at the McDonald Observatory in the 1950s, Juan was new at the job and I was one of the first to break him in. We had great times together, as he was interested in everything and you could not wear him down. After a long night he would complain that he had not seen Venus in a while, so what could we do but look at Venus and other objects in the early morning, instead of going to bed? His custom was to repeat astronomers' requests, which was nice for an old soldier to hear. Come to think of it, Juan probably had been in the military too. So in the deep of the night one would hear: "Platform up." "Platform up." "Turn off the light." "Turn off the light." "Turn off the Moon." "Turn off the Moon." At sunset and sunrise we would stand outside on the catwalk that goes high around the dome for checking the sky. We looked for the green flash that reportedly can be seen at the instant of sunset or sunrise, but I have seen it rarely and never convincingly, in hundreds of tries. Evenings and mornings we would look at the shadow of the Earth, a dark blue band along the horizon (Chapter 15), and we'd dash out from our work inside the dome to observe the Zodiacal Light as a cone of pale glow standing up from the horizon. Around midnight we did see the counterglow opposite the Sun, a faint circular patch the width of four fingers spread at arm's length. Because it is so faint, it can be seen only when the eyes are fully dark adapted—which takes 40 minutes—and with averted vision, looking a little away rather than directly at it, because the peripheral vision is more sensitive to faint light. The cause of the counterglow is the same as that of the Zodiacal Light, the reflection of sunlight by the grains orbiting between the planets; its extra brightness is due to the opposition effect.

Juan would tell me the names of mountains nearby and one of them he called "The Bambino." Later I mentioned that to Van Biesbroeck, who was annoyed since it is supposed to be called "The Van B Knoll." Van B would take us on hikes all over the area around the McDonald Observatory. He was an observer of the finest detail of plants, rocks, valleys, and mountains as well as planets and stars. He told us how our eyes get tired of seeing colors and contrasts, how the pigments responsible for vision bleach out, and that the same thing causes after-images. The monotony can be broken by turning the head sideways, with the line through the eyes perpendicular to the horizon for a while. Right away, the colors return fresh and in contrast.

One evening when Van B was showing me the operations at the "prime focus" near the top of the telescope, I could not help remarking that it was a long way down and that a fall would be fatal, at which he said with a gleam in his eyes: "What a way to go for an astronomer!" He did not quit astronomy, working full time and setting an example for us younger observers until he died in 1974 at the age of 94. For the night of his 84th birthday he had requested the use of the new 84-inch telescope on Kitt Peak, which was granted of course. Six years later the 90-inch came on line, and he was there for that birthday. Van Biesbroeck enjoyed going to extremes, working so low in the sky that the narrow platform behind the telescope had to rise to its greatest height, with the astronomer balancing on it, in absolute darkness! It seemed too dangerous to have him alone on that platform high above the concrete floor. I could not possibly suggest accompanying him, as he would not tolerate such a bother, but we could make him accept that an engineer or an assistant would come with him, because the new telescope needed watching.

All went well during his birthnight, but he got himself the next night assigned too, and that went fine until he sent the engineer to bed at 3 AM and Van B fell off the platform at 3:15. He was picked up for dead, or nearly so. The next day I went to the emergency ward of the hospital, but the Head Nurse would not let me in because he was dying. Well, if he was dying I could not hurt him, and I had to say goodbye or something. So, as she looked one way I sneaked in the other. Stood by his bed in shock and despair. He looked dead indeed, with his little white beard and white hair on a white pillow and everything so sad and pale. But somehow he seemed to notice me. What to say? He was not a person to ask, "How are you?" or even discuss anything personal about him. That was trivial and not tolerated. Astronomy and other people were his concern. So I blurted out: "Mr. Van B, this morning I observed Comet Tago–Sato–Kosaka. . . ." At this, his eyes opened and he pointed a finger up, saying clearly, "Did you notice that its nucleus has split in three?" He was back at work soon. We now have a "Van Biesbroeck Award," a small cash prize given on his birthday, January 21, for which the carefully selected candidates have to be as young at heart as he was and known for their concern for other people as well as for good science.

One Van Biesbroeck Awardee was Marc Aaronson. He was a devoted family man and a respected friend for his students and associates; and he had an inspiring passion for finding out the age and evolution of the universe. So he was intensely involved in several research projects with different types of techniques and telescopes, one of the most energetic astronomers at work. In the early evening of 30 April 1987 the weather on Kitt Peak was fluctuating, and then it rather suddenly cleared; we rushed to ready our equipment for what might become a valuable night, and kept a close look at the sky by hurrying out on catwalks or watching from high on our domes. In the freakest of accidents, a protrusion of Marc's rotating dome slammed a steel door shut just as he opened it to look at the sky, killing him instantly. The first and only astronomer to die while working with a telescope.[7] Until then our science and domes had looked attractive and protective, but this dreadful accident made us all a little slower and older.

12. The Fascination of Space

After World War II, rocket scientist Wernher von Braun and his Dora–Peenemünde pals and some German V-2 rockets had been brought over to the United States; these were primarily used for military purposes. Although rockets had been used for some science experiments, only a few astrophysicists saw an opportunity for research at high altitudes or even space. So, at first, astronomers considered the Soviet launch of Sputnik on 4 October 1957 mostly a military stunt. These attitudes changed slowly, and chiefly with the emergence of planetary science as a separate discipline. One of the exceptional pioneers was James Van Allen of the University of Iowa, who studied cosmic rays and particles—electrons and atomic nuclei—trapped by the Earth's magnetic field. With other scientists he initiated the International Geophysical Year for a wide-ranging study of the whole Earth from July 1957 through December 1958, and he participated in it with instruments built by him and his students at Iowa. His experiment on the first US satellite, Explorer I, launched in January 1958, discovered the regions of trapped high-energy radiations around the Earth that are now called the Van Allen Belts.

One of the first explorations planned by NASA in the early 1960s was called the Voyager mission, to Mars. I submitted a proposal for a photometer–polarimeter to measure the brightness and polarization of sunlight reflected by the surface of Mars and by the dust in its atmosphere. The observations were to be made during various parts of the mission when the planet would be seen from the spacecraft at a variety of phase angles, of angles between incoming and reflected light. Our experiment was an extension of what we were doing with ground-based telescopes, from which we were learning the characteristics of planets and atmospheres. I proudly thought it was an extensive proposal; it was about fifty pages, long by my standards, but much too short by theirs, and

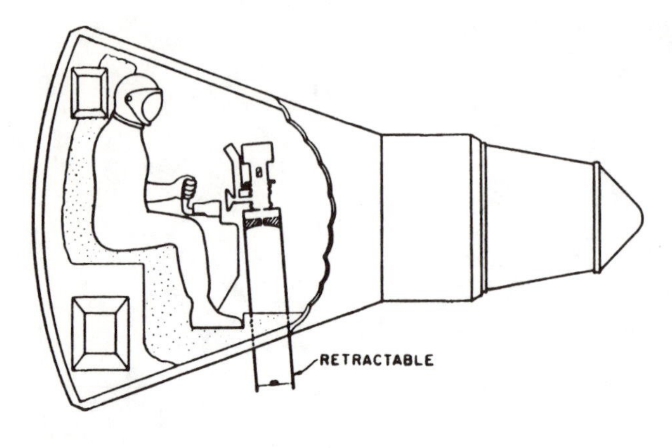

The Mercury capsule with our proposed polarimeter in place of the original periscope.

probably a bit ahead of its time, so NASA turned it down. That Voyager mission never took place, probably for the same reason, that it was too ambitious for its time. Later on, of course, there would be two highly successful Voyager spacecraft that are still sending useful data from the outer reaches of the solar system, one of them flying by Neptune in August 1989.

In 1963, we proposed another polarimeter, to be used by the astronauts on Mercury spacecraft; the proposal[1] was put together with the help of research assistant Tom Teska, and with the technical support of the McDonnell Aircraft Company in St. Louis, Missouri, the manufacturer of the Mercury capsule. The US was lagging behind the Russians in the race into space, and we therefore suggested to NASA officials to make the US program more sophisticated by flying a scientist with the instrument. I volunteered myself of course. By then, however, the astronauts appeared to be in control of the program, maintaining that highly trained jet pilots were required to operate the spacecraft (even though a monkey also flew), and that the pilots were too busy for science. It became clear to me in any case that one would have to invest years, if not a lifetime, away from science if one wanted to stay with the program for an eventual trip to the Moon. Gerard Kuiper talked about volunteering for a one-way trip to the Moon, provided he could make observations of its surface and be heard for at least one hour. We were adjusting to a new era. What ways were open to us? Where might astronomers and planetary scientists fit in? Gradually, we became more politically adept, and our dreams became more realistic and were put into action.

A wise science monitor at NASA Headquarters in Washington, D.C., Roger Moore, who later joined the Rand Corporation, explained patiently that we would have to learn how to do experiments remotely, with equipment on distant spacecraft operated by radio command, instead of the close control we were used to within our telescope domes. He suggested that we consider using high-altitude balloons or rockets as our

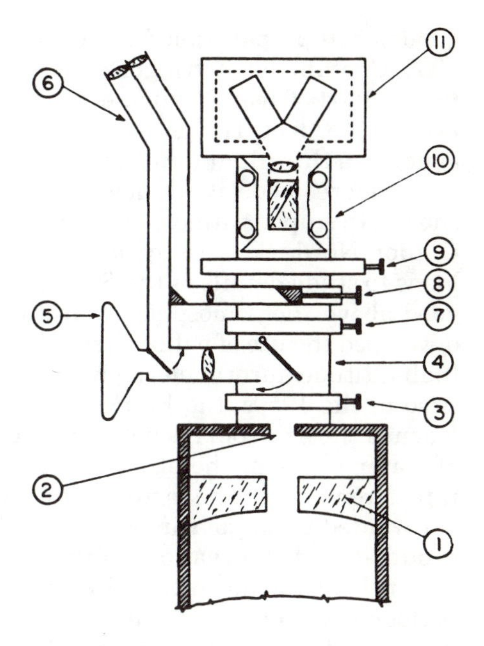

Components of the Mercury photopolarimeter, a device similar to the original version of Minipol: 1 eight-inch mirror; 2 pressure-port window; 3 Lyot depolarizer; 4 viewing mirror; 5 photographic camera; 6 eyepieces (schematic); 7 focal plane apertures; 8 viewing mirror; 9 color filters; 10 rotating assembly for polarizers and the like; 11 refrigerated photomultiplier tubes.

first spacecraft, and NASA would provide financial support. Mike Belton of the Kitt Peak National Observatory took me to White Sands Proving Ground in New Mexico to see their experiment on an Aerobee rocket. We stood outside, not wisely and definitely against regulations, to watch the thunderclap launch, followed immediately by another loud bang right above the launching tower and, soon thereafter, bits and pieces of rocket and equipment came whirling out of the sky bringing mortal danger—and misery, because the Kitt Peak crew had worked devotedly on that instrument for years. Back to ballooning.

Ballooning has a great tradition. In 1783, man—in the person of the Montgolfier brothers—left the Earth for the first time, with three balloon flights in quick succession. The old dream of Leonardo da Vinci and others was finally being fulfilled. Man could fly! The balloon launches were made before the eyes of the world, from the squares of Paris, and thousands of people came out to be thrilled. Liberated from Mother Earth, what could man not do? A few years later, he liberated himself from aristocratic tyranny, in the French revolution. A similar psychology must have worked through the 1960s. Before the eyes of the world, man was freeing himself from the Earth, going into space, to the Moon! All over the world, millions watched in fascination as the launches were covered live on television. What could man not do? Like the revolutions of the late eighteenth century, the rebellions of the 1960s may have had a deeper background too.

The United States had an active balloon program long before Sputnik. Starting in the 1930s, a few military men flew inside narrow, closed gondolas in preparation for the spaceflights they expected to come. In 1959, Captain Joe Kittinger made a parachute jump from a balloon at nearly 100,000 feet, 30 kilometers, falling free until he opened the chute near 15,000 feet. He had guts! The United States Navy had a balloon gondola, a thin aluminum sphere that could maintain a pressure of one atmosphere inside; it was flown a few times with two people in it to study the Earth's upper atmosphere. In the 1950s Martin Schwarzschild and Gordon Newkirk were flying remotely controlled telescopes with balloons to study details of the Sun; the seeing is always guaranteed to be good above the atmospheric turbulence. Van Allen and his students developed the use of balloon-launched rockets, "rockoons," for studying high-altitude auroral electrons, magnetic fields, and cosmic-ray fluxes. Also in the 1950s, the French astronomer Audouin Dollfus, who had learned the art from his balloonist father, measured the concentration of water vapor in the atmosphere of Venus, from a gondola, without interference from the terrestrial water vapor.

I wanted to do polarimetry from above the atmosphere because my ground-based observations of the polarizations of Venus and interstellar grains had shown tantalizing results. I therefore proposed that a balloon should carry me, a pilot, and a telescope in the Navy gondola above most of the atmosphere and, particularly above most of the ozone layer that absorbs ultraviolet radiation. I planned a single flight, pointing my telescope by hand at Venus and a few bright stars.

The first step was to convince the Navy that I could point a telescope from a gondola even when it would be rotating and swinging underneath its balloon. For this purpose we flew from Minneapolis to Oshkosh at an altitude of 7000 feet in an open basket underneath a helium-filled balloon. Four men in a tub: a Navy pilot, his understudy, a Navy doctor, and myself. We had a great time pointing the telescope, which was easy, and playing around a wide hole in the bottom of the gondola; it reminded me of the hole in the balloon gondola for training at Ringway. We were floating with the wind, without a breeze, on a sunny day. It was dead still. Even over towns the sounds did not come through, and though we could see a locomotive blow its whistle, we could hardly hear it. Apparently the thermal layering of the atmosphere prevented most of the sound from reaching us. As soon as we landed near Oshkosh, I took off my helmet, but our pilot, Malcolm Ross, sharply commanded me to put it back on. Within seconds a heavy bolt fell from high up with a hard impact on my helmet; without it I would have been killed instantly.

Next there was to be a test flight in May 1961 to an altitude above 100,000 feet, again in an open gondola, launched from the aircraft carrier *Antietam* on the Gulf of Mexico. Only two men could go. Who would accompany the pilot? To me there was no doubt, but the Navy doctor wanted to go too. To settle the matter, we had to go to a hearing in Washington, D.C.; it was a memorable session, with admirals and other Navy personnel, and poor me, with my foreign accent, and without Navy experience.

Balloon astronomy near Oshkosh, Wisconsin. The picture was taken just before the connection between the balloon and the parachute was severed.

So the pilot and the doctor went.[2] They made a record-altitude flight in manned ballooning, to 103,500 feet, and all went well until the landing. The gondola floated on the water, the pilot was safely lifted off by a helicopter, but when the doctor had been hoisted up part way, he fell out of the sling. His space suit was similar to that used by NASA's astronauts, but it did not then have a drawstring around the neck. Water came pouring in, forcing the top part of his body down, and he drowned. Soon a special collar string was installed in the space suit, and it was said that this saved astronaut Virgil Grissom's life when he floated in the water after his spaceflight in the Mercury capsule on July 21, 1961. The doctor's name was Victor Prather. For the space program he did not die in vain, but he left a young family and many friends behind.

Because of the accident, our proposal was not doing well. We had asked the National Science Foundation for financial support of the ballooning. But the NSF monitors wanted to be careful, not to be held responsible for the death of an astronomer. So they sent it to one referee after the other. They all said that it was too dangerous and that Gehrels should not drown, twelve of them, although rumor had it that a thirteenth had approved. Then, of course, I had to find out who number thirteen was. Occasionally I still meet him, he is my old friend Vincent

Lally, and we cheer the value of my long life. Nevertheless, the final NSF judgment was that it should not be done by a human, but with instruments.

This question of flying instruments or people was an interesting debate, already then, and it was strongly revived at the time of the Challenger disaster of 1986. For making observations in space, most scientists feel that instruments alone can do it better: they are simpler and cheaper to operate, without the complex and heavy systems that are required in the hostile environment of space to support human life and return it safely. Space scientists were therefore never excited about the Shuttle's carrying their equipment, and they have had to compromise some of their missions to fit the Shuttle. The Galileo mission to explore the satellites and atmosphere of Jupiter, for instance, will take six years to get to Jupiter in the 1990s because the Shuttle can accommodate only a weak booster. So Galileo has to be accelerated gravitationally from a flyby of Venus and, twice, the Earth. Compare that to the Pioneers, which were accelerated by powerful Atlas–Centaur rockets directly from Earth such that they reached Jupiter in twenty months traveling at 15 kilometers per hour. It is typical for the resilience of scientists that Galileo was rescheduled to make observations of Venus and the Earth as well as of the Jupiter system. In the 1960s there was enough money for both manned and unmanned spaceflight, so the natural urge of people to want to fly themselves was taken care of as well.

In the case of my ballooning, however, it would take a long time and much effort as well as money to develop complex instrumentation in order to replace a person who was willing to go whatever the risk, and a gondola that was nearly ready to fly. The agonizing choice was either to drop the idea or to see it through with a large engineering project. I decided for the latter, provided competent people could be found to do the engineering and leave me alone to do science. The University of Arizona showed an interest in our plans for a complex balloon gondola and for participation in space programs, so we moved to Tucson in 1961. It was a hard task to develop that Polariscope gondola, with its telescope, special filters, detectors, televisions, remote control, and pointing devices. The gondola would not be in sight during its flight, and we would have to monitor its obedience to our commands via a microwave radio link. The craft would swing and rotate, but the telescope had to remain stabilized to a precision of arcseconds. A dedicated crew was therefore assembled, though we had to replace some engineering faculty who turned out to be distinguished only on paper. Mel Diels designed the aluminum-framed gondola. For the electronics we had Don Brumbaugh, Jack Frecker, Jyrki Hameen-Anttila who came from Finland, with Dale Hall, Louise Hess, and Mike Arthur doing the wiring. David Coffeen, Ed Roland, Tom Teska, and Rene Toubhans solved unheard-of problems for providing frictionless couplings and constructing the ultraviolet polarimeter. These people slaved day and night almost literally; at mealtimes and during our long field trips they discussed engineering, and in a bar at night they still discussed our engineering.

The Polariscope gondola in its dome.
LPL photo by D. Milon

I worked by delegation, keeping other research programs active as well and being too lazy to get involved in engineering problems. Besides, I trusted them. A basic capitalist concept is to let each person do his own thing, whereby one obtains the best and the most from each. It is, in any case, the most pleasant and thereby the most efficient procedure for everyone to do what he or she likes best. Furthermore, in a university environment one has to work by delegation and encouragement within a team, or it will break apart as its members leave for other jobs at higher pay once their reputations are established. They will stay only if they are fully involved with the project and its challenges. As a team leader, or "principal investigator," one searches for a balance between being friends with the team members and demanding efficient performance, running a tight ship—and that does include some firing as well as hiring. I once discussed these principles and our program with Martin Schwarzschild, a senior Princeton astronomer who ran a balloon program called Stratoscope II for high-resolution imaging of planets. He told me that the scientist in charge should be a manager too and be involved in all technical details and decisions, including those of balloons and launch procedures. But our crew did well, and with expert help from Al Shipley and Bob Kubara and the people of the National Center for Atmospheric Research, our Polariscope gondola flew without failure.[3]

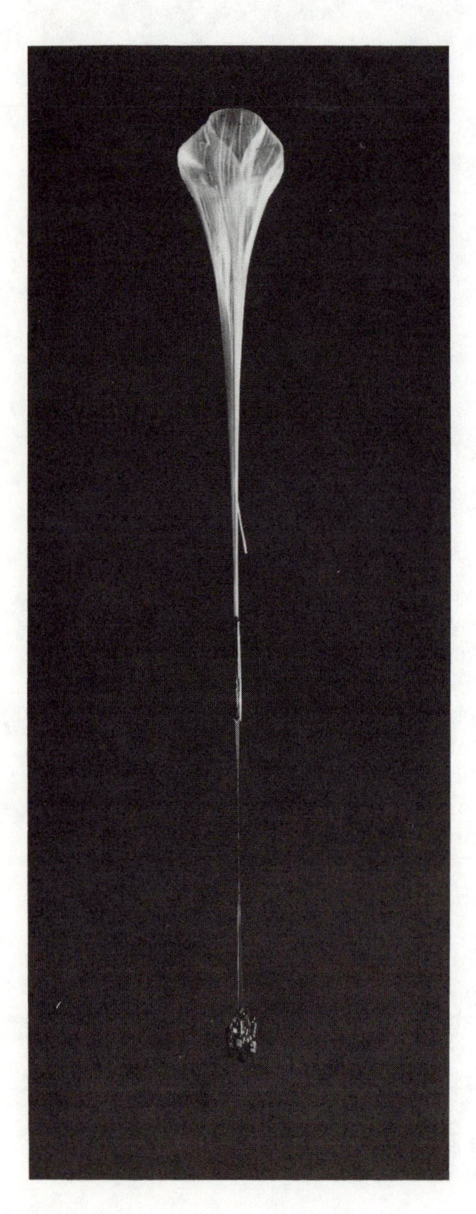

The **Polariscope** gondola soon after take-off.

Ballooning is, however, a hard way to do science. The balloons were made of thin plastic and filled with helium.[4] The thin air in which our equipment was to operate made it difficult to deal with the high voltages used in photoelectric detectors, nearly 1000 Volts; the electrical arcs and glow discharges were spectacular during tests. I insisted on testing the whole gondola with its equipment in a huge chamber for vacuum and low temperatures, before every flight. As in an orchestra rehearsal, each instrument had been tested before, but how would they play and interact together? So we tried the equipment out under conditions similar to those of flight and we found problems, every time, to make the test worthwhile.

I had made another request, namely to install a kick-plate and hammer in the gondola. In those early days of electronics, mechanical relays did much of the switching, and their contact points would get rusty or dirty. If there was a problem during a flight and remote trouble-shooting did not help, a good swift jolt to the gondola would bring the desired result. It still is a grand tradition, because contacts in computers and other electronics can still get dirty or corroded. So I asked that a commandable kicker be flown on our gondola. That was too weird for our electronics experts, but they agreed to put in at least a mechanical end stop for the elevation rotation of the telescope tube. It saved at least one of our flights.

Before the system could finally be launched, we would have to wait for the proper conditions at our launch sites of Page, Arizona, or Palestine, Texas. The winds had to be low at launch and landing; they had to be of known direction during flight, but they never were. The art of predicting the weather up at 120,000 feet, 36.6 kilometers, was too new at that time. No end to anecdotes on that. We had a second ground station many miles downwind to command the gondola overhead during flight, but we never had it in the proper place, as the gondola was always being blown elsewhere. The gondola once landed in a snowstorm in Nebraska, scaring a farmer who heard our electronics verification beeper. So the sheriffs came out with red lights flashing and guns at the ready. They must still be telling stories to their grandchildren about the weirdly beeping contraption they encountered during that blizzard! For one of the flights, in 1967, I kept delaying a launch from Page because the predicted trajectory kept heading toward the Grand Canyon, a landing site that would have guaranteed the termination of the project. We had been in Page a long time already, and the ballooning crew wanted to go home to Palestine, Texas. They were big, tough men; except one, the weatherman who was a tiny, timid fellow. Finally, one morning he reported that the wind direction had changed and the flight would be safe. I trusted him, too—Martin Schwarzschild would have been more clever—so up she went into the evening, away from danger at first, but the winds near 120,000 feet were such that the next morning the gondola floated over the heartland of the Canyon. At noon a preset timer would separate the gondola from the balloon. Planes on the Chicago–Los Angeles route were diverted in anticipation. We could see it all with field glasses and were frantic but helpless. Down she came from 120,000 feet, leaving behind a shattered balloon that had frozen solid and could not flex with the sudden change in loading. Down, and down, in a crazy free fall, with the enormous cargo chute fluttering behind in thin air. By 40,000 feet the parachute started holding. And there the miracle happened. A quirky jet stream sent the gondola toward the South Rim of the Canyon. She landed in the desert, hard but safe, ready for us to fetch and fix and fly again the following flight. For all her interesting destinations, that good gondola never did visit Tennessee, which is fortunate, for I would have been jailed. We could not obtain a license for our radio link, and I had therefore approved pirating on the frequency belonging to a television station in Tennessee. It would have made for an interesting evening.

Four times she flew, that gondola with her 28-inch telescope, successfully making polarization measurements at 2250 and 2750 angstroms wavelength in deep ultraviolet light, from above most of the ozone in the Earth's atmosphere.[5] At that altitude of 36 kilometers, about half of the starlight at 2250 Å came through the ozone; farther below we would have seen nothing. Frankly, not much progress was made in the understanding of Venus and the small dust grains that float between the stars. A local newspaper reporter, Carle Hodge, was asked by his editor to expose Polariscope as a boondoggle, wasting taxpayers' money. The conscientious man quit his job as he could not do what he believed unjust, but his editor may have had a point as far as the scientific returns were concerned. Yet the project was a milestone because this type of observation above the ozone had never been made before and we showed how it could be done. Furthermore, without that Polariscope experience we would probably not have won the competition to get our imaging photopolarimeter on the Pioneer spacecraft, and no imaging might have been done on these missions at all. By that time, we had learned the new techniques and had also flown a prototype of the polarimeter we proposed for Pioneer; graduate student Sam Pellicori was in charge of that. We were thoroughly experienced in making checks and calibrations as well as in using a pointable telescope to make observations over a range of phase and illumination angles.

The outer planets, Jupiter, Saturn, Uranus, and Neptune, would be lined up in the 1980s, so that a spacecraft could conveniently pass by all of them in a single mission. Such a favorable lineup would not happen again until some 200 years later. Scientists and engineers connected with NASA therefore proposed to make a "Grand Tour." (That term was common in England and the United States in the 19th century, when young men of wealthy families would round off their educations with a Grand Tour of the capitals of Europe.) NASA's plan was eventually approved, namely to fly a Voyager spacecraft with television cameras and a variety of other instruments. To reconnoiter the way, however, especially through the radiation belts of Jupiter, the Pioneer missions had already been initiated at the urging of Van Allen and his physicist colleagues. The name was appropriate: they would be pioneering. These were the days of the US space program at its finest: preliminary plans for an "Asteroid and Jupiter mission" were announced in 1968, and the first launch, of a new spacecraft with new types of instruments to distant Jupiter never visited before, occurred already in 1972.

It was not originally planned to have imaging devices on Pioneer; that was to be done later by spacecraft like Voyager, which would be equipped with the best of cameras on a stabilized pointing platform. In our own proposal of a photopolarimeter to study the characteristics of the Jovian atmosphere and the Zodiacal Light, we also described how we could make at least a preliminary reconnaissance of imaging by putting a small aperture in the focal plane of our telescope and letting the spin of the spacecraft accomplish scans across the disk of the planet. By making small incremental steps in the pointing of our telescope we could obtain adjacent scan lines and put them all together in a computer. We called it "spin-scan imaging."

Our Pioneer proposal of 1968 was 800 pages long, and it was evaluated in tough competition. The closest competing proposal was from a theorist expert on planetary atmospheres—I was frankly told they much preferred to have him!—who had combined with a group at one of NASA's spaceflight centers. They had used the maximum allowable weight to design the largest possible infrared and optical telescope. However, it could not be pointed. We set as the first requirement that the telescope be pointable for observing our objects over a range of phase angles as the geometry changes during the flyby; this would be crucial in our studies of atmospheric constituents. That, our promise of some imaging, our insistence on checks and calibrations, and our experience in ballooning, made us win the competition.

At this time also, Harold Johnson and I, with the help of competent engineers at our university and in industry, worked out a large proposal for infrared and polarization measurements on a spacecraft in the series of the Orbiting Astronomical Observatories. Either our proposal was not good enough or the OAO budgets declined, for that plan was not approved, and we had wasted much time and effort.

For the Pioneer proposal, I had visited five aerospace firms to find the best manufacturer for executing in space-qualified hardware the basic designs we had tested during the ballooning. We made a good choice, the Santa Barbara Research Center in California, an affiliate of the Hughes Aircraft Companies. Next we asked NASA's approval to write the proposal jointly, with SBRC as a partner rather than as a subcontractor; it apparently was a new concept, in that there would be no competitive bidding, but the idea was approved, and we got the very best of engineering support for our ambitious instrument. We worked together for three months on that proposal, SBRC putting nearly a dozen experienced engineers on it, and we described exactly what we hoped to observe on the planet Jupiter and its satellites and how we would analyze the data to

Charlie Hall

learn about the atmosphere. During the long cruise from Earth to Jupiter we would be passing by the asteroid belt, but none of the known asteroids was found to come within reach. We would, however, observe the sunlight scattered by the small particles that produce the Zodiacal Light. It seemed to be a complex instrument, this Imaging Photopolarimeter (IPP). The NASA engineers were concerned. Space was at that time considered a hard environment for instruments. Our rotating telescope, and inside the instrument a rotating disk—with a small aperture for imaging, an intermediate one for photopolarimetry, and a large one for Zodiacal Light—all this was considered complex and likely to fail. We were even more concerned because we knew how difficult the balloon experiments had been. That had been good training—in an environment of thin atmosphere that was actually more difficult than the vacuum of space—and we could describe the problems and their solutions in convincing detail.

The IPP became a fabulous device, a complete observatory weighing only 4.2 kilograms. The low weight was achieved by adopting magnesium as the primary material; magnesium is light, but also toxic, so that only an especially equipped shop could fabricate it. The incoming light was reflected by a primary mirror of one-inch aperture and a secondary mirror so small it had to be mounted on a glass plate near the opening of the telescope. Inside of a miniaturized compartment were the optics to analyze two colors of the light, blue and red, and two polarizations. This

Investigations for Pioneer 10

Instrument	Experiment Objectives	Principal Investigator
Helium vector magnetometer	Interplanetary and planetary magnetic fields.	Edward J. Smith, Jet Propulsion Laboratory
Fluxgate magnetometer	Planetary magnetic fields.	Mario Acuna, Goddard Space Flight Center
Plasma analyzer	Interplanetary solar wind; planetary bow shock and magnetopause boundaries.	John H. Wolfe, Ames Research Center
Charged particle detectors	Distribution of galactic and solar particles. Planetary trapped radiation. Identification of nuclear species.	John A. Simpson, University of Chicago
Cosmic ray telescope	Spectra and distribution of galactic and solar particles and their presence in the planetary magnetospheres.	Frank B. McDonald, Goddard Space Flight Center
Geiger tube telescopes	Energetic charged particles and radiation environment in planetary magnetospheres.	James A. Van Allen, University of Iowa
Trapped radiation detector	Energetic corpuscular radiation trapped in planetary magnetospheres.	R. Walker Fillius, University of California at San Diego
Meteoroid detector	Population of interplanetary grains.	William H. Kinard, Langley Research Center
Asteroid–meteoroid astronomy	Characteristics of interplanetary particles.	Robert K. Soberman, General Electric Company
Radio transmitter and Deep Space Network	Celestial mechanics. Mass distribution in the planets. Masses of satellites.	John D. Anderson, Jet Propulsion Laboratory
Ultraviolet photometer	Interplanetary and planetary hydrogen and helium including glows from satellites and rings. Hydrogen-to-helium ratio of the gaseous atmospheres.	Darrell L. Judge, University of Southern California

Investigations for Pioneer 10 (continued)

Instrument	Experiment Objectives	Principal Investigator
Imaging photopolarimeter	Light scattered from interplanetary particles. Photometry and polarimetry of planetary atmospheres, satellites, and rings. Spin-scan imaging. Determination of the spin of the spacecraft.	Tom Gehrels, University of Arizona
Infrared radiometer	Thermal energy flux from planets and rings. Hydrogen-to-helium ratio.	Guido Munch,* California Institute of Technology
Radio transmitter and Deep Space Network	S-band occultation. Characteristics of planetary ionospheres and atmospheres.	Arvydas J. Kliore, Jet Propulsion Laboratory

* When Munch left for a position at one of the Max Planck Institutes in West Germany, Andrew Ingersoll took over.

would need four photoelectric detectors, while there was room for only two tubes; each tube therefore had a double detector structure inside. Behind that motorized, pointable telescope compartment there was a box that contained all of the electronics for signals and commands.

The Pioneer spacecraft was small and simple compared to Voyager. It was stabilized like a bicycle wheel by the spin of the spacecraft, turning at nearly seven revolutions per minute about the axis of the radio dish, which was kept pointed toward the Earth. For power, there were two small nuclear generators on booms that extended from the spacecraft to minimize any radiation interference with the sensitive instruments. There also was a long boom to keep the magnetometer away from magnetic effects caused by moving components within the instruments and spacecraft.

The communications link was impressive; it was called the Deep Space Network. There were three large radio dishes, of 64-meter diameter, distributed in longitude over the Earth: one at Canberra, Australia; one near Madrid, Spain; and one at the Goldstone Station in the Mohave Desert of California. I was sent out to one of them when there was a rumor of a strike to be staged during our encounter with Saturn; it was a touchy assignment, to convey our interests and enthusiasm, but in the end, no strike occurred, and I in turn was deeply impressed with their part in this enterprise. From the DSN the signals were relayed via communications satellites to the Goddard Space Flight Center near Washington, D.C., from there back up to another satellite and down to the Jet

Propulsion Laboratory near Los Angeles, and finally via microwave dishes to the Ames Research Center near San Francisco. For imaging there was a special telephone connection to the computers at the University of Arizona, and the processed pictures were then sent back to the Ames Research Center for display to scientists, reporters, and national television.

How much I learned from the Pioneer manager, Charles Hall! He quickly grasped even the finest detail of our instrument and all the others, as well as of the spacecraft. He could be tough, as he was, for instance, when our team was lagging in the production of good images after the first encounter. But he maintained the nicest atmosphere I have seen in any project. He had gathered a team of about fifty-five people, only fifty-five to handle such a big project, and they did it with expertise and dedication. They all deserve to be mentioned, but this is not the appropriate place; let Richard Fimmel and John Pogue be their representatives here, those tough negotiators of our contracts and finances. Every year we would make a formal proposal to them, inflating our budgets just a little here and there—for a total of about 10%—for we knew the game they would play. Then they would come to Tucson for two days of review and we'd all put on our best shirts and make new viewgraphs and talk about the past and the next year. It was useful and interesting even for ourselves to review our work, and many helpful suggestions would come forth. Friendship ruled the proceedings—until the moment of Contract Negotiation. They would leave the room and stand whispering in the hall, and we'd act scared for our future. They would come back in and describe all our excesses and demand a 20% cut. Great consternation on our part. Then some dickering and we'd offer a 5% cut, and they'd come down to 15%, but that would be their Final Offer. So with great indignation I would walk out. . . . After a few minutes of that, we'd compromise at 12% or so, and we would be friends again. They could report to Charlie Hall that those Arizonans had been given a hard time, and that they had saved the Tax Payer a lot of money!

Charlie showed himself to be a genius as a leader. Every morning he would gather about twenty-five people, investigators and a few of his own crew, in his own office for a "stand-up meeting"—standing up, for then they would not talk too long. The program had begun with a planning conference of some 200 engineers and scientists. At that time it looked as if we imaging people were going to have a serious conflict with the groups studying energetic particles and magnetic fields, particularly with John Simpson of Chicago and Van Allen, both of them my seniors. Each side wanted most, if not all, of the data link during the encounter with Jupiter. Both had excellent justification. Tension was building in the packed conference hall. How would one resolve this? Most other programs would have set up a committee, with panel meetings and reviews by competent referees, to do a study—twenty-five copies or more—at useless expense and loss of time. Not Charlie Hall, who thrived on problems. First he declared a coffee break to relieve the ten-

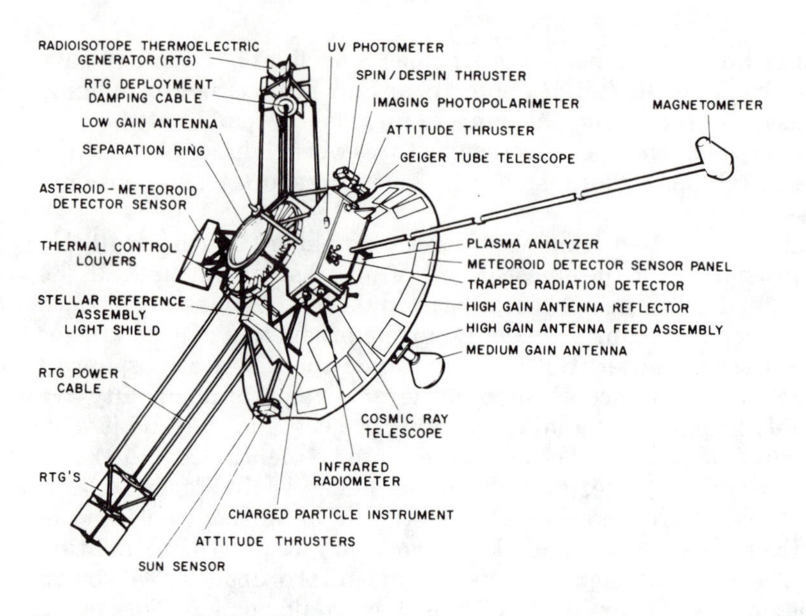

Pioneer spacecraft. This sketch shows the layout for both Pioneer 10 and 11.

sion, suggesting I chat with Van Allen and Simpson. After a while he quietly joined us and said, "How about each of you—imaging and particles-and-fields—getting half of the data link?" What was there left to say? If he had offered 60%, we would have fought for 65. But half each had a touch of fairness, equal rights, and King Solomon's wisdom. So we proceeded, relieved, to the next problem.

Some eighty-three scientists from twenty-one institutions participated; the overall organization was shown in our Table of Investigations (page 123). Within each team there were co-investigators, each with a particular expertise. For the photopolarimeter we had David Coffeen, Lyn Doose, and Charles KenKnight of our own group at Tucson, and Bob Hummer from SBRC. The photopolarimetry would be analyzed by Martin Tomasko and his students. For the imaging there were Charles Blenman, Jim Burke, John Fountain, Bill Swindell, and others of the Optical Sciences Center at the University of Arizona. Several other colleagues cooperated in various stages of the program.

The informality of the Pioneer program was useful as well as attractive; the following example shows that it allowed effective response to an emergency. One afternoon, as Pioneer was flying toward Jupiter, I received an alarmed phone call: a comet was seen, continuously at the same bearing, that is, in the same direction from the spacecraft. It meant that it was on a collision course; I recognized that from my sailing days. Would I come immediately to check that image? I caught a flight to San Francisco and we agonized for hours that night. At the Palomar Schmidt Telescope I had seen "comets" that were actually caused by

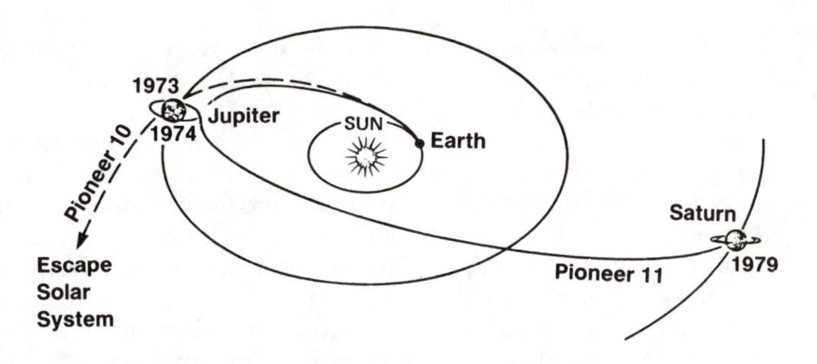

Flight paths of the Pioneer spacecraft.

stray light from Jupiter or a bright star reflected by some bolt or part of the telescope at just the right angle of chance and glance into the tube. Late that night there were just a few of us left, but we were able to command the spacecraft to point in a slightly different direction. The "comet" disappeared. It had indeed been stray light from the Sun reflected into our telescope by some part of the spacecraft, possibly the edge of the radio dish. I was too excited to go to sleep and caught an early flight back to Tucson.

It had been a risky decision to add the Zodiacal Light mode to our instrument, but I could not see this complete astronomical observatory, however small its aperture, sitting idle most of the time during the long cruise between the planets. I also had an urge to compare interplanetary and interstellar particles, which I've mentioned in Chapter 11. We therefore added additional baffles to the telescope and a wide focal-plane opening for observing that faint glow. Jerry Weinberg, an expert on Zodiacal Light, joined our team, working mostly at the University of Florida. The study turned out to be successful, but the added capability brought a risk that could have ruined our instrument and its primary mission, the observation of Jupiter, in December of 1973.

The flyby of Jupiter by Pioneer 10 on December 3, 1973, the first encounter by a sophisticated spacecraft with the giant planet, was one of the most impressive experiences in the lives of its participants. Because of extremely strong magnetic fields, Jupiter was expected to have destructive belts of high-energy radiation, similar to the Earth's Van Allen belts, but much more dangerous. These belts consist of electrons and atomic nuclei ejected from the Sun; this stream of charged particles, called the solar wind, interacts with magnetic fields, such as those around the Earth and Jupiter. Some of the particles are trapped by the planetary field and forced to move along the magnetic field lines between the North Pole and South Pole. Near Jupiter, because of the strong fields, it was expected that the radiation belts would be dangerous for any man or spacecraft that came their way. With data from radio observations made from Earth, NASA had given a model of the radiations to our engineers, for them to design the instruments with sufficient safeguards.

However, as our first spacecraft, Pioneer 10, was already on its way to Jupiter, the plasma physicists and radio astronomers held another study session to include additional data and derive a new model. Misery of miseries, they now predicted a level of radiation about a factor of 10 worse than had been specified for our instruments! We therefore feared the following scenario for the day of the first flyby: our imaging photopolarimeter with its subtle and sensitive optical components would send confused signals at about noon, pacific time, and soon thereafter die forever. The other instruments would give up similarly later in the afternoon, and by about 4 PM the spacecraft transmitter too would garble out its very last signals. The closest approach to the planet would occur near 7 PM, but we would not know about it because by then the Pioneer spacecraft would be silent, dead. That Monday started at NASA's Ames Research Center in an atmosphere of doom and gloom, with perhaps a secret hope or two. . . .

At noontime the IPP sent us frightening messages. Instead of sampling the bright radiation of Jupiter, as it had been commanded to do, its signals said that it was looking for the faint Zodiacal Light with the wide-open aperture. If by chance it were to scan across the disk of Jupiter, our detectors would be destroyed by the bright light coming through the wide opening. Later that day, wild signals also came from the other instruments and from the spacecraft's transmitter. That evening however, miracle of miracles, Pioneer came around the disk of Jupiter with strong and proper signals telling of the wondrous world of the planet and its environs of fields and particles. Thus we learned that engineers deal with instrument specifications differently than astronomers do: We would have designed to the original specifications, taking the uncertainties into account, but the engineers apparently treated the design parameters with great reverence and used such a large safety factor that a failure could never occur. They saved the day. Pioneer 10 could go on to explore the outer reaches of the solar system and we could send a second similar spacecraft, Pioneer 11, past Jupiter and Saturn and eventually out into deep space.

Jupiter is by far the largest planet in the solar system, with a radius eleven times that of the Earth and a mass that is almost three-quarters of that of the entire solar system outside of the Sun. Both Jupiter and Saturn seem to consist mostly of hydrogen and helium; one surmises that only deep down near the center is there a core of heavier elements such as iron and silicon. On Jupiter, the enormous weight of the atmosphere presses down on the layers of gas below, such that the pressures may reach as much as a million atmospheres and the temperatures 20,000 degrees. At these pressures, the atoms are squeezed so close together that electrons are no longer bound to their atomic nuclei; the material behaves like a metallic conductor. It is this "metallic hydro-

F ring of Saturn and the satellite S1 1979. This photo is from the scanning imaging photopolarimeter of Pioneer 11. At lower right one sees sunlight scattered through the A ring, the bright Cassini Division with a narrow ring at its center. The bright spot is the satellite Tethys.

gen" in which electric currents flow, producing the magnetic fields that cause the strong radiation belts.

Near the visible outer surface of Jupiter, the hydrogen and helium behave like normal gases, an atmosphere with clouds and haze caused by some ammonia, and perhaps hydrazine or phosphorus, while at deeper levels there appears to be a layer of water vapor. There are other gases as well, some of which have colors and thereby produce the variety of circulating clouds that we see with glorious plumes and ovals, brighter bands and dimmer zones, red spots and dark polar regions. However, according to Toby Owen who is an expert in the chemistry of planetary atmospheres, it still is not known exactly which of the possible trace gases cause the subtle colors seen on Jupiter. Our detailed measurements of brightness and polarization, starting with Pioneer 10, have provided data for years of study, so that one begins to understand the atmospheres of Jupiter, Saturn, and its satellite Titan. Martin Tomasko and his associates have derived models of the structures and compositions for the zones and bands, polar and equatorial parts, and of spots and dark brown regions. In their models small crystals of methane, ammonia, and other substances interact at various altitudes and latitudes in intricate arrays and interactions. They are continuing their modeling using the Pioneer data and also laboratory observations of the brightness and polarization scattered into a range of phase angles by a variety of droplets and crystals.

By aiming Pioneer 11 at the proper place near Jupiter, mission control was able to use Jupiter's gravity to redirect the trajectory for an encounter with Saturn five years later. Saturn is cooler than Jupiter because it is farther away from the Sun, and its atmosphere has more of a foggy haze at the top, so that we cannot see as much detail as on Jupiter. But Saturn is the Lord of the Rings—flat rings of ice and snowy rock particles orbiting the planet—and master, too, of an array of satellites.

The various instruments on Pioneer 11 collected large amounts of new data for Saturn as well as for Jupiter. A major discovery was that Saturn also has a magnetic field and radiation belts. Detailed models were derived for the shape of the fields and for characteristics of the radiations. Precise numbers were obtained for the dimensions of the major rings of Saturn. By long tradition, the rings are labeled with letters: The brightest outer ring, easily seen in small telescopes, is the A ring; it is separated from the next one inside, the B ring, with a gap that was discovered by the French astronomer Cassini and is therefore called the Cassini Division. Farther in there are the much fainter C and D rings; and there is a tenuous E ring, found by Earth-based astronomers, far outside the well-known ring system.

As Pioneer approached Saturn, we carefully looked for new rings with our IPP at the highest sensitivity whenever possible. It was a team effort to prepare the observing sequences and to staff the consoles around the clock for sending commands and receiving the data. The Pioneer management was so helpful and flexible that we could change our commands within minutes before their transmission to accommodate our findings.[6] Outside of the A ring, our imaging photopolarimeter discovered the narrow F ring. It was exciting to see it come in on the projection screen, line by line from our spin-scan imaging. That ring was immediately confirmed in absorptions observed by the particles-and-fields investigators, who in turn discovered another thin ring a little farther outside, and then we made a detailed inspection of our data and could confirm their G ring. We call the gap between A and F rings the Pioneer Division.

Beyond Saturn there was not much left to do for our IPP other than making additional observations of light scattered by interplanetary and interstellar particles, and providing the "Doose Correction." The latter is a service that is being provided occasionally to the particles-and-fields investigators who must know precisely the rotation and orientation of the spacecraft at the times of their measurements. Charlie Hall and his people usually determine the orientation using equipment that is part of the spacecraft, but Lyn Doose worked out a method to calibrate this important function by checking when a known bright star shines through the focal plane opening of the IPP.

The Pioneer program was exceptional in its friendly atmosphere among the participants. They were eleven Great Years. In September 1979 I came away from the Saturn operations in California in a daze, having been in a wondrous world of snow and ice, of particles and fields, of satellites and jet streams. I did not want to leave this fascinating world, not even to go home. But the pioneering stage was finished.

The third Pioneer spacecraft with its instruments—nearly all flight-tested hardware—never flew; it was hung in the National Air and Space Museum in Washington, D.C. The Ames Research Center would not design and execute another major low-cost efficient mission to the planets. Charlie Hall was allowed to retire. My experience was not used in later missions.

My last consulting for NASA was a frustrating experience. I served on the study committee for the Grand Tour to the outer planets. The committee consisted of chairmen of subcommittees, in my case a team of experts in polarimetry. The primary committee meetings were a lot of squabble, very different from the productive atmosphere of the Pioneer project. In order to improve the priority of their own proposals, some of the other chairmen kept asking difficult questions on the merit of polarimetry, which indeed is somewhat dubious regarding the solid surfaces of planetary satellites, but not at all in the study of atmospheres. Our subcommittee provided an extensive defense and documentation, and it was finally decided to fly a polarimeter on Voyager.

When NASA made its Announcement of Opportunity to participate in the Voyager flight, our subcommittee wrote a proposal for the polarimeter. I insisted that we would again work with the Santa Barbara Research Center as partners, because the IPP instrument on the Pioneers had been so successful, Sam Pellicori was at SBRC, and our experience of years of balloon testing had gone into that IPP design, just as Roger Moore had wanted. An accusation was then made that I had a financial interest in this SBRC involvement. I dismissed it as a joke, but something happened to teach me that there was a rough world out there. Two men from a major aerospace company asked for a confidential appointment. They came to suggest that I would be paid as a consultant to the company in order to pass on to them our subcommittee's polarimetry studies so that they could better compete in case they were to bid on making the spacecraft and its instrumentation. This seemed inappropriate. They assured me that they had just signed up the chairman of another subcommittee. At that, I could only stumble to my office door and open it for them to leave.

Next, our subcommittee was astonished to learn that one of our own team members had submitted a competing proposal from his laboratory, which wanted to get the business of building the polarimeter. Using the knowledge he had gained within our team, they underbid us by a million dollars, and their polarimeter was selected—even though we protested that it was an inferior instrument and even warned that it might fail because some corrosive material was going to be used in the wrong places. It flew on Voyager and its polarimetry mostly failed.

But a chapter entitled "The Fascination of Space" should not end on negative notes. Pioneers 10 and 11 are still gathering data on the charged particles in the outer solar system. I am the proud parent of those two one-inch IPP telescopes in space. The IPP became the precursor for a similar instrument flown by David Coffeen and his colleagues on Pioneer 12 around Venus, and for the design of a more advanced instrument on the Galileo mission to Jupiter. The imaging and other equipment on Voyager performed with great success, and even its photopolarimeter gave spectacular results in occultation observations of planetary rings.

We now know the truth about space, with its hard vacuum, as an environment for instruments: it is great. Many of the spacecraft launched years ago and designed for only a few years' operation are still sending valuable data. The Pioneers and Voyagers are on their way to the stars, still working, and I have seen other NASA missions succeed as well as those early in the space program. We will soon have the Space Telescope, Galileo, Magellan, Ulysses, and other missions. More and more nations are getting involved in space, and international cooperation is increasing. New generations of people and spacecraft will fly. There is no doubt about that. As the Earth becomes smaller because of growing travel and computerized communication, space has come closer to all of us. Space has become not something remote that concerns only a few, but a natural frontier for humans and their tools to explore.

Of Passengers and Pilots

13. From Tasmania to Tenerife

The story of the Pioneer missions *had* to be told, and I traveled around the world to tell it. The audiences reacted with enthusiasm. There was an unforgettable afternoon in Shanghai in a lecture room with a few hundred lively young Chinese. My talks usually last fifty minutes, but they asked for the whole afternoon. Their knowledge of English was only slightly better than my zero knowledge of Chinese, but our host, Mr. Feng Da-Sheng, seemed a good translator, and he certainly stimulated their interest. A part of the presentation was a filmed computer simulation of Pioneer 11 flying underneath the rings of Saturn, which was so spectacular that all other audiences kept dead still in fascination. In Shanghai the people kept pointing and talking and shouting with enthusiasm, and at the end they demanded, "Again!" So I made our Pioneer go backwards and forwards around Saturn again.

These talks were sponsored by NASA and requested by various science academies, high officials, and members of the US Congress. Partly because our project scientist was ill, I wound up giving more than a hundred lectures and radio and television presentations. I also talked about Pioneer informally, for students and teachers and common people, as during one night on a train, the Orient Express, through Yugoslavia. Its conductor was an amateur astronomer; he had assigned me to a compartment by myself, and then he brought in a crowd and found an interpreter, and we had a marvelous time with questions and answers, seemingly throughout the night. People had a fascination for that small spacecraft, organized and built in a hurry to execute a dangerous reconnaissance mission, and doing it splendidly. I found it an educational experience to adapt my presentations to a variety of audiences and circumstances and to give an account evenly distributed over the results of the fourteen Pioneer experiments. Our daughter Jo-Ann often came along on these travels. She would take care of the slides, and she would also grade the success I had with the audience, on a scale of 1 to 10, but usually ranging from failures with a grade of 4, to successes near 9. I usually agreed with her evaluations. The range of grades was surprisingly large: the same lecture went wrong in Surat, and right in nearby Baroda on the same day. The finest compliment was once made by a senior scholar in Sri Lanka who quietly said: "That was not only science, but also poetry."

I once spoke in Phoenix, Arizona for a group of wealthy donors to the University Foundation, an organization that solicits support for various activities at the university. I was presenting our new results and theories with great delight and confidence. In the question-and-answer period, however, I was challenged by one of the donors, Howard Leker, who is quite knowledgeable as an amateur astronomer: Do we speakers in the space sciences not present many of our ideas as if we know them for certain, while in fact our science is still in its beginning stages? Great theories of today may be rejected tomorrow, he added. Howard was right, of course, and I have remembered the lesson.

Formal presentations to distinguished persons, such as His Royal Highness Prince Charles of Wales, would scare me at first. The long tall hallways of the palaces are indeed imposing, but the distinguished persons were always friendly and interested, they had usually been briefed on Saturn before my appearance, and our presentations were handled in a professional manner. At Buckingham Palace, Jo-Ann and I were received by the Prince's Equerry, his personal secretary. (The title refers to a traditional responsibility for the horses.) He reminded us of some protocol: the first time one uses the full "Your Royal Highness" and after that one uses "Sir." We next made a long walk through broad and high-ceilinged halls hung with large paintings, mostly of royal ancestors. At an intersection of corridors we had to wait and check that we would not run into the Queen and her entourage returning from a formal engagement. I saw the palace gardens and remembered how I had walked outside the walls in 1944, wondering how I could jump over to surprise her, then Princess Elizabeth, as I felt in love with her. I might even have succeeded—in jumping over the wall, that is—for in the early 1980s someone did the very same thing and visited her, and he got away without much punishment. Anyway, we proceeded to the lecture room, prepared the film and slides, and waited for the Prince, who was coming by helicopter from another engagement. When he arrived, we heard him dressing down the Equerry for not having had preparatory material regarding Saturn on that helicopter. Poor man, for there was so little known about Saturn before the Pioneer flights that it would have taken him a difficult search to find such material. His Royal Highness seemed puzzled at the presence of Jo-Ann, but he was genuinely interested in Saturn and the Pioneer spacecraft, so that we had a good time together for nearly two hours, running over the schedule arranged by the Equerry—for whom this was not the best day in his life.

I also have the good fortune to meet occasionally with Prince Bernhard of the Netherlands, our Commander during times of Resistance, who shows clearly the human side of royalty in his concern for people and their problems and in a willingness to help. His son-in-law, Prince Claus, described to me a method of evaluation that can apply to us all: At a certain point in life we come into adult consciousness, we find ourselves at a starting position—be it as a prince or whatever circumstances we are born into. We are to be judged only by what we make of our lives from that situation and that point onwards.

One of my first official lectures occurred before the Royal Swedish Academy of Sciences in an ancient hall with a high ceiling and large pictures of distinguished past members hung on the walls. I remember Svante Arrhenius (1859–1927), the famous chemist, looking down at me shaking in my shoes, as if he was wondering what I would make of it. About fifty members had come, in tails, and were seated at a U-shaped table covered in green cloth; I was trying to hide behind a tiny lectern placed at the wide opening of the U. About ten minutes into the lecture I had a sinking feeling that it was not going over, I was not getting through to them; they were sitting so still and erect in their formal clothes! But that instant I decided to forget about them and just have a good time for myself with the beautiful new data of the very first flyby of Jupiter. Afterwards came the question period and they still did not move—just with a nod of the head, I think, a member would indicate to the President that he wished to ask a question, who then gave him the word. But there was a record length of questioning, and it kept the dinner waiting, so the President told me later. In fact, it was so good that the next time, for Saturn in 1979, they especially invited King Carl XVI Gustaf of Sweden. Dinner was formal too, but nice; they were pleased that I knew their custom of the President's making a toast first and then my making a toast in reply, as I was seated to the right of the President. I hope I did not, and certainly did not intend to, say anything stupid in these toasts for such kind and interested people. I had brought a framed picture of Jupiter with its striking Great Red Spot for the King to hang in his palace; someone commented that the giant red spot gives Jupiter an analogy to the one-eyed God Wodan in Scandinavian mythology.

Astronomers are fortunate by being able to experience foreign cultures. My travels for Pioneer gave me a particularly wide experience, but all astronomers have some chance for travel. Once every three years, for example, there is a meeting of the International Astronomical Union, each time in a different country. There are additional opportunities for travel when one has to use a foreign observatory—to observe objects in the southern hemisphere, for instance, or at special times and places in the sky. So far, I have worked at a total of eighteen observatories with about forty-eight different telescopes, our own at Arizona included.

On these trips, I have noticed that success and failure are often separated by the smallest of chances. In the rest of this chapter I will describe two examples of such vagaries, one failure and one success that happened to me, and one disaster that has become a part of my life.

I failed to discover the rings of Uranus. Astronomical textbooks mention that thin rings around the planet Uranus were found in 1977, but for me that discovery represents a failure in sagacity and serendipity by planetary scientists, particularly me. On the 10th of March of that year, Uranus would be seen close to a bright star, and from places near the Indian Ocean, Uranus might even be seen to pass in front of that star; such an event is called an occultation. Uranus is normally seen only as a small disk even in large telescopes because of its great distance; it looks somewhat greenish, but without detail even though the planet's diameter is 49,000 kilometers, almost four times that of the Earth. Through the study of spectra taken with powerful telescopes it was already known that Uranus is a gaseous planet, probably mostly hydrogen and helium; the greenish color is probably due to other gases or particles in the atmosphere.

During an occultation, the relative orbital motion of the Earth and Uranus carries the disk of the planet across the star, and even if the planet is not resolved in the telescope, detailed information about its atmosphere can be obtained from the progress of the occultation. One uses a photometer on the telescope and records the combined intensity of the region around the star and planet. Just before the beginning of the occultation (ingress) one sees both, and the intensity is high. At the start of the occultation there is a dimming, because the light from the star is obstructed and only the light from the planet is received. The duration of the occultation yields the size of the planet. If there were rings around the planet, they might also obscure for a short time the light from the star; the photometer will then record a blip, a dip in the brightness record. If the occulting body has no atmosphere, the ingress and egress are sudden. If, on the other hand, the planet has an atmosphere, one can learn the extent and density of that atmosphere from the detailed profiles of ingress and egress.

The person in charge of our expeditions to observe such occultations is William Hubbard, a theorist who concerns himself with planetary interiors and atmospheres. We have a good cooperation, for I am willing to go to faraway places to acquire the data he needs for his theories, and so we worked together on plans for the 1977 occultation. I had arranged to have four specially designed photometers built by Jack Frecker and others in time for the event. Another part of the preparations is to find observatories willing to have us come as guests at their telescopes. For three of our photometers we selected observatories in India: Naini Tal in the foothills of the Himalayas, Hyderabad in central India, and Kavalur in the south; the fourth expedition went to Perth in western Australia. Through a considerable amount of correspondence we had obtained all the permissions, cooperations, and arrangements, but then, about two weeks before the event, the positions of the star and Uranus were remeasured to give a more precise location and time for the occultation. The results caused us to scramble for locations farther south on the Earth, even though the viewing conditions there would be inferior. If we had ever thought of the possibility of rings, some of our expeditions would have stayed with the Indian observatories, where the weather was expected to be good and the planet would have been high enough in the sky. We would have observed rings, even if the disk of Uranus itself had been missed. This occultation had been proposed, reviewed, discussed, and prepared by at least 100 planetary scientists, particularly since NASA was to provide one group with an expensive high-flying research plane. Not one of us ever mentioned the possibility that Uranus might have rings! Until this time only Saturn was known to have clearly visible rings. My own shortsightedness is the most surprising, as I was one of the principal investigators involved in the preparations of ring observations for Saturn by Pioneer 11 in 1979. More busy doing things than carefully thinking them through, off we went to hastily selected sites on Mauritius and in South Africa, all with poor conditions either from weather or having the planet low in the sky.

I was proud to have selected the Sutherland Observatory in South Africa where the weather during that time of the year gave a high probability of success. Indeed, while we prepared during four days before the occultation, the sky was brilliantly clear. Everything was beautifully arranged; Connie Feast, the Director's spouse, was my assistant. But an hour before the event it was as if a weather god had taken a cloud in his hand and put it over the observatory, removing it again about an hour after the event. This happened to the astonished dismay of four expeditions that had come from England, France, Arizona, and Capetown to observe Uranus on the four Sutherland telescopes.

Within an hour or so after the event I got a remarkable telephone call in that dome from someone at the Sutherland Observatory who had been called by someone in Johannesburg, who had been called by someone in Massachusetts, who had been called by someone who had received a radio message from the research plane flying somewhere high above the Indian Ocean. On the airplane they had observed blips in their recordings, and I was told at that miserable time in the Sutherland dome to look near Uranus for an *asteroid*. Even if the sky had been crystal clear, I would have refused to look for merely another asteroid among the thousands in the skies! Later we learned that the message had originated on the plane as a report of a "satellite belt"; if that had been clear and if our skies had been clear, we would have jumped into action. Astronomers at Perth had made observations from which later supporting evidence for the Uranian rings could be reconstructed. The only other Earth-based observers who had a clear sky were in India and they had gone to bed as soon as they recognized that they were not going to observe an occultation by Uranus itself. Had they been forewarned of the possibility of rings, they would have continued the observations that had already shown some blips and from which, again later, corroborating ring evidence would be obtained. Even for the people flying in the plane it was days before they realized they had made the Great Discovery of Uranus' Rings. In fact, their finding, of narrow rings, was completely novel, as until that time only Saturn's broad rings had been known.[1]

In 1983 there was another occultation of a star, this time by Neptune. Rings were in vogue now, and a principal part of the observation plan was to look for them. Our Arizona photometers were placed on Taiwan, on Guam, in central Australia, and on Tasmania. I went trudging off with Jo-Ann to Tasmania, protesting to Hubbard, our team leader, that we had only about 30% chance of acquiring his data because of the miserable weather in those latitudes—they aren't called the Roaring Forties for nothing—but he gently insisted that Tasmania was where we should go, the point farthest south on Earth for this event. While we prepared near Hobart during four days before the occultation, we could not even open the dome because of the clouds and the storms that came roaring by incessantly. The evening of the event itself brought a thunderous rainstorm, putting the astronomers—including Director Mike Waterworth and our assistants Auda and Jules Geysen and Father John Hayres of the nearby Richmond Presbytery—all in despair, but not idling them. Everything was readied inside the tightly closed dome, the telescope pointing in the precise direction where Neptune should occur. An hour before the event it was as if that same weather god had returned magnanimously, this time to remove with his hand all the clouds away from our observatory for the duration of the occultation. It became still and clear. Ironically, we were the only ones to have perfectly good skies, other than of course our friends in the high-altitude airplane again, Jim Elliot and his associates.

A paper was written by Hubbard and his twenty-one collaborators, Jo-Ann being one of them because she had been the actual observer who operated the photometer and recorder in Tasmania. The paper presented new and detailed knowledge of the atmosphere of Neptune. Two unexpected depressions were also found in the recordings, just before ingress and just after egress, but my theorist friend and team leader dismissed these as possibly due to something wrong with the equipment. I objected with indignation: all had been well as long as the data appeared as expected—"good photometer," "trustworthy observers"—but now that our data were unexpected, we were unreliable. This hit the core of all our observing in life! What is the Truth? All we can do is observe, analyze, and report. After that we must trust what we have learned, to the extent of its estimated precision, and we must build on that knowledge to proceed to the next observation. Especially when the result is unexpected! To waver then would be scientifically unsound, and we would never learn what is new. I won the argument, and the paper reports the possibility of an absorbing layer, a thin cloud of soot in the upper atmosphere. Perhaps I had finally encountered the interstellar grains I had wanted to observe twenty-five years earlier, as mentioned in Chapter 11. But no rings. Or were there rings? A year later, after other observers of another occultation reported some subtle blips, we had a closer look at our data and found a confirmation of at least one partial ring of Neptune, after all.

The following story is of more than a personally haunting interest. It is one of the strangest cases of a thin-edged separation of success and failure; it involves a chain of peculiar circumstances, such that if any one of them had not happened, the worst air disaster in history would not have happened either. The seven links of the chain were: a terrorist bomb at an airport; the overcrowdedness of the alternate airport; an unusual need to return to Amsterdam that night; the erratically varying visibility; a missed taxiway; radio traffic that was not rigorous; and a rare transmission interference. In addition, I wish to tell my view of the horror because of a religiously influenced report of the accident that made a deep impression on my atheist inclinations of that time. Finally, being a pilot had been a boyhood dream, and I still do ride sometimes with the crews in their cockpits. The disaster took place on March 27, 1977, within two weeks after I had come back from South Africa from the Uranus occultation, and its nightmarish image has become engraved in my head. It still reappears occasionally when I am driving a car and come to a stop sign for a busy road and having to decide to go or wait. The image is of my being the Captain of a Boeing 747 of the Koninklijke Luchtvaart Maatschappij, Royal Dutch Airlines, just before its take-off from Tenerife in the Canary Islands.[2]

Captain Jaap Veldhuyzen van Zanten was fifty years old, chief flight instructor of Boeing-747 crews for KLM. All crew members in the two planes involved in the accident were senior, experienced personnel; the Dutch copilot, for instance, had been a Captain on DC-8 aircraft. Their flight was a special charter that had left Amsterdam in midmorning to bring vacationers to Las Palmas Airport (Gando) in the Canary Islands; these passengers were to occupy lodging vacated by others and van Zanten therefore was to return with the others to Amsterdam that same day. If all went well, it could be done in a total of nine hours flying time, within the allowable working hours of the crew, and no spare crew members came along.

Things did not go well. At Las Palmas, a bomb exploded in the ticketing area at about noontime. Five people were wounded. The organization responsible, demanding independence of the islands from Spain, announced that there was another bomb, and the airport was therefore closed. Like all other flights to Las Palmas, KLM was rerouted to Tenerife Airport (Los Rodeos) on a nearby island. The small, one-runway airport became crowded. One of the other planes was a Pan Am 747 from New York bringing vacationers to Las Palmas to board a Mediterranean cruise ship.

van Zanten decided to refuel at Tenerife. The weather was getting worse, and if he could not land at Las Palmas, he would be able to return to Amsterdam. If he could land at Las Palmas, he would not have to spend time refueling and could perhaps still return to Amsterdam within work- and rest-time regulations, on a reasonable schedule for the returning vacationers. He let his passengers go into the terminal; the representative of the charter company had business in Tenerife, but all others came back on board. The emergency lifted before the completion of the refueling. Other planes began to take off, but, even though it had arrived later, Pan Am could not leave because KLM was in its way. Two crew members paced off the space to see if it was enough for them to pass by. It was not. They were held up for at least another hour and a half by the Dutch refueling. There was anger and frustration.

Finally, the KLM plane could taxi away into the late foggy afternoon, on the runway, as the taxiway was not free. They had asked permission to get onto the runway "for take-off. . . ." The tower first told them to get off the runway at Exit 3, and van Zanten and his copilot tried to identify it in the fog. The exits were not marked. Soon, however, the tower changed the instruction and told them to proceed on the runway until the end, and "make a back track." The change may have appeared to them as if they would have the runway free for take-off; they verified the last instruction with "Roger. Make a back track" and double checked by asking if the tower wanted them to get off the runway, at which the tower told them again to stay on and "make back track."

On the KLM voice recorder, when it was recovered, one could hear a busy schedule of checking, but no mention of Pan Am (which seems to confirm my impression that KLM believed the runway to be theirs, free for take-off); they did discuss whether conditions were good enough for taking off, with the copilot saying that if they did not do so, they'd be holding up the other planes as well. The visibility varied. Los Rodeos Airport is notorious for fogs rolling off the nearby Pico de Teide (Peak of the Devil). The tower did not see the aircraft nor could the crews see each other's planes.

Pan Am was also taxiing on the runway, but it was told by the tower to get off at Exit 3. They were also busy with their checklist, and they were following KLM's progress. They had problems with identification of the exits. The tower repeated: it had to be Exit 3, one-two-three. Exit 3 required a sharp turn, about 145 degrees, partway in the backwards direction. Another exit, with a more logical 52-degree turn was coming up, as one could see on the airport sketchmap that crews are provided with. Pan Am kept going toward Exit 4.

Then there was confusion on the radio, and some interference, so that not everyone understood the same thing. Usually one obtains Air Traffic Control instructions, which are for after take-off, separately from those for the take-off itself. But KLM had been busy with the checklist and the two instructions were therefore now combined: KLM reported ". . . ready for take-off and we are waiting for our ATC clearance." The tower replied: "KLM 8705, you are cleared to the Papa beacon, climb to. . .," with which the tower seemed to clear them for take-off. KLM started to take off. The copilot read the instructions back to the tower for confirmation while they increased throttle and released the brakes, and he ended with "we are now at take-off," as the tower understood it. Others who later listened to the flight recorder heard it as "we are now. . . uh. . . taking off," which is what must have been meant, because they just then started rolling. Anyhow, the tower replied: "O.K." There was a pause of nearly two seconds, and then "Stand by for take-off, I will call you." Only the O.K. was heard by KLM, because Pan Am had also understood the O.K. as the clearance for take-off, causing the captain to push his microphone button to say, "No, uh. . .," and his copilot said, "And we are still taxiing down the runway." The simultaneous transmissions of tower and Pan Am caused radio interference, and only a shrill noise or whistling sound was later heard on the KLM recorder.

The KLM pilots believed they had been cleared. They thought that Pan Am had gotten off the runway earlier. If anyone in the KLM cockpit had understood what was said immediately after the "O.K.," by either tower or Pan Am, there would have been enough time to abort the take-off.

Pan Am's crew was not worried anymore, because they had reported that they were still on the runway. The tower was not concerned either, for KLM had been told to wait.

The tower did have a further communication with Pan Am about clearing the runway. This was somewhat overheard by KLM's flight engineer, but he may have misunderstood, so he said to the pilots without alarm: "Is hij er niet af dan?" ("Is he not off it then?") They did not hear him well, so the Captain asked: "Wat zeg je?" ("What do you say?") at which the engineer repeated: "Is hij er niet af, die Pan American?" ("Is he not off it, that Pan American?"). The Spanish report has the Captain replying "Yes;" the Dutch report has both pilots replying with: "Jawel" ("Yes, he is."). They were already rolling fast.

Finally, through the patchy fog, the two crews saw each other's planes! KLM pulled up so hard that its tail made a gash in the runway. The collision was between the undercarriage and fuselage of KLM's 747 and the topside of Pan Am's, which was torn open. There were perhaps sixty seconds for seventy people to scramble out of the Pan Am plane, nine of whom died later. All others perished in a huge fireball. Five hundred and eighty-three people died as a result of this disaster.

Terror at Tenerife by Norman Williams as told to George Otis was published by Bible Voice.[3] It is a religiously slanted tale: God delivered Williams and others because they prayed, while of the Dutch, he says on page 19:

> van Zanten frantically shouted, "*Gott dam*!" But the problem which was in the making wasn't God's doing.

And from page 100:

> "GOTT DAM!" The final words of the KLM captain. "God damnit!" He and every wonderful person in his plane perished.

And, later on page 100:

> If you had a choice of a flight, would you choose a plane of God-blessers or the other?

The book shows more sanctimony than understanding. The Christians I used to know would have treated the terror of Tenerife with compassion and an urge to find clarifying facts and extenuating circumstances. Instead, Williams and Otis apparently made no attempt to understand or even to quote the Captain properly. On the KLM voice recorder are heard only two more words after that "Jawel." It came from the Captain or copilot: "oh godverdomme." The "Gott dam!" in the Otis book is neither Dutch nor English nor German. Both men grew up during Nazi occupation, and neither would have used German in any case. The "godverdomme" exclamation is used in Holland as a reaction to pain or shock. Its guttural sound with which it is said in Dutch brings relief, like "ouch," but more forceful. It translates literally into "god damn me." But that is not the point. The Captain had been raised in a strict Calvinist church, and the copilot probably too as he was born in a small village in northern Holland. They would not deliberately swear themselves into disgrace. It is merely an exclamation, likely to be used in such a case of utter shock and horror. That is all there is to that final word, as any Dutchman could have confirmed for Otis and Williams. They should have checked with one of the inspectors who listened to the tapes, such as Captain F. A. van Reijsen, who told me, "It sounded so shocked. It sounded like a prayer."

The accident has been studied, and improvements have been made in prescribed communications between planes and towers. Airports are adding lights that indicate runway exits and taxiways of airplanes. Perhaps one should also consider simple traffic lights at ends of runways, to display a "cleared for take-off" message. Such a light could have avoided a near miss in Tucson in 1961. I was leaving Tucson Airport in an American Airlines DC-7 and noticed an identical plane leaving the station too. Ours taxied, revved up the engines, started rolling on the runway—and then braked hard, off the runway. From the other side the other American Airlines DC-7 came rising in uninterrupted take-off!

Praying for passengers is effective according to our friend Mrs. Wijesinghe of Colombo, Sri Lanka. Her son Mahendra, an astronomer on his way to a meeting in Arizona, stood in line at the counter of Air India in Bombay on January 1, 1978. A woman ahead of him burst out in tears for she was not on the manifest, the plane was full, and her daughter would be waiting for her at London Airport. She turned around and asked Mahendra if she could take his place. If it had been me, I would have gently pointed out to her that I also had relatives waiting for me (as Mahendra's relatives would be waiting for him in New York), and that I already was running late for a scientific conference in Arizona. But Mahendra is a Buddhist, so he gave her his place. After take-off the plane turned too steeply and fell into the sea. There were no survivors.

Mahendra's mother knew that all would be well with her son because she had prayed for him. Did no one pray for the British woman and for the others on board?

We are defenseless as passengers in these flying machines. By going into them we accept their miracles, and their hazards too. Equipment may fail and the people who control and maintain the planes may make mistakes, or misunderstand each other's language. Like other human predicaments, flying has its contrasts of success and failure. We are but passengers in life's journey. There are lights and darkness, engineering and human emotions, preparations and chance events, and exuberance and sudden death.

14. Russian Dissidents and Atomic Weapons

I first went to Russia in 1958 for one of the triennial assemblies of the International Astronomical Union. Under Khrushchev, the Soviet Union was then opening up to foreign visitors. This was the great opportunity to make a first exploration of Minsk, Moscow, and Leningrad. I tried to have an open mind and was interested in communism and attracted by its principle of equally sharing the goods of the Earth. The only fair thing to do was to find out as much as possible about Russia, trying not to be biased against it, just as I had travelled to America in 1950. I entered in anticipation of meeting a new culture and seeing the beautiful country described in classical literature. I entered by train at Brest-Litovsk, east of Warsaw, where we were processed by Intourist, the official travel bureau of the Soviet Union. We were to go on the train directly to Moscow. My plan was, however, to see the principal cities on the way, Minsk and Smolensk. "Impossible." "Why?" No answer. "I will get out at Minsk. The train will not be locked, will it?" I tried to make it clear, but politely, that I meant it, and the Intourist person replied that, of course, the train would not be locked. My railroad car was locked all the way, but it was unlocked that evening so that I could step out at Minsk, where I was whisked away in a black limousine, which had been driven onto the platform and parked precisely where I would descend from the train.

No communication with the driver. I tried French and German as well as English. At the hotel, the driver now was the Manager as well. Next morning, a Sunday, seemed an occasion to look at some churches. I walked alone. The churches were surrounded by green wooden fences. One of them had an opening. It was a Greek Orthodox Church and a wedding was taking place, with a few people attending. When I came out, there were my limousine and driver again, waiting at that opening in the fence. This time there also was a nice girl interpreter. We went on a prescribed tour and she showed pictures in an album made for the purpose of illustrating what the place looked like before the Glorious Revolution, after the Glorious Revolution, after the Huns' destruction, and how it would soon be rebuilt. The town already was largely reconstructed and it looked nice. Sometimes when I asked a question, she would mysteriously say: "Later."

Soon it seemed time for another experiment, and I asked to be driven out of town. "Impossible." Well then, I would walk. Just as I was ready to step out, it became possible. It was all done in the best of manners, and friendly too, not to be the "Ugly American": rich, noisy, demanding, knowing it all, and seeming to travel only to see how good it is at home. Travel is to learn and never to condemn; this is not only a moral position, but also a practical one because the travelers usually have insufficient data for making a judgment. Anyway, the moment we were out of town, the roads vanished into muddy cart tracks so that the girl thought it better to leave the car for a jolly walk in the woods. When we were out of earshot of the driver who stayed with the car, she recalled my questions and answered them. I thanked her for the information, but asked her why she had not answered in the presence of the driver. "He is a card-carrying Party member." But there was nothing political or offensive about my questions, and besides, "He does not speak English." "That is what you think!"

She had an English textbook with her in which all practice sentences were political. I remembered that ours would run something like "Papa smokes his pipe," or "Johnny rides his bicycle." Her sentences were "John lives in Chicago. His parents are poor. John can therefore not go to college. That is only for the rich." I told her that I had worked my way through graduate school at the University of Chicago. Her reply was that I should not take it seriously, for they did not either.

On the train from Minsk to Moscow I talked politics with East German communists. It became clear to me that when people of different creed or background try to communicate, they need a dictionary to define their terms. It was a lesson for a lifetime: to use a dictionary and to agree first on the definition of a concept. Words such as communism and democracy meant something different to them than they did to me. Our discussion got so violent that they were about to throw me out of the train. Years later the same almost happened on another Russian train traveling through Bulgaria. I had given a colloquium in Bucharest and had to hurry to an international meeting the following day in Varna, but there was no other transportation than the train my innocent travel agency had recommended, which turned out to be a Russian troop train! Definitely forbidden for outsiders. I had no choice but to hop on when the conductor was not looking. I was discovered soon after the train started rolling and thrown out at a small border station—where I would be stuck even worse, so I had to repeat the sneaking-on trick. Then there was a lot of shouting by an impressively heavy female head conductor with a big mouth and a loud voice; she actually flung open a door with the train at full speed to impress me, because I refused to understand the word "dollare," and there was no dictionary.[1]

When I finally got to Moscow in 1958, the astronomers were honored guests and treated royally with a reception in the Kremlin. One evening during a jolly dinner at the conference, several of us were wondering what the conditions were in the parts of the university that we were not allowed to enter. Were they showing us the nice places only? I drew the lot to make an investigation. The safest way seemed to pretend that a fish bone had stuck in my throat such that a doctor rushed in from the forbidden quarters, and that is where I was carried back to. Things were in terrible shape in that little hospital, with torn furniture and gruesome-looking equipment. Of the ten or so doctors and interns who came by, not one spoke any English, French, or German, let alone Dutch. This was news to us, because after Sputnik it was claimed that the Russian education was so far ahead of ours. Then again, only a few of us were learning Russian.

I have returned to the Soviet Union a few times, including trips to make the Pioneer presentations. The people always were such kind hosts, doing their best to take care of us and showing us proudly their monuments and interesting places. My lectures were well received, and the audiences appreciated my starting with two sentences in Russian to say thanks for the invitation, "Balshoye spasiba za priglashenie," and that it is a great honor to talk to you about Jupiter, "ya sjitajoe dlya seebya tjestjoe raskovareevat svamee ob Yupiter." Winter visits were especially impressive. Russia is a world of somber lights and wintry weathers marked by great differences in temperature from day to day. I have had the privilege of walking alone in daytime and at night through old Leningrad among buildings and back alleys that looked as if they dated from before the Revolution. The scenery was like that in *Dr. Zhivago* when Yuri returns to his father-in-law's home and goes into back alleys to find wood for the furnace.

We kept wondering whether the country would have been in better shape without the Revolution. The movement toward social reform in Russia was strong already in the 19th century, but, so many years after the Revolution and World War II, with plenty of time to rebuild, it was surprising how backward life was for the common people. They were well fed and clothed, in the large cities, but everything seemed so drab and burdened. "Nyet" and "not possible" were the predominant responses to any request, without any reason given. It seemed like the ever-present control and the lack of personal incentives had produced a lethargy and an acceptance of deficiencies. One can hope that human nature—desiring color and freedom—cannot be suppressed and that we therefore may have a long-range confidence about improvements in the Soviet Union, but at the time it seemed as if an effective job had been done to keep the system and people in their places.

What did become especially clear during those travels was the hypocrisy of the regime, the lack of integrity. It reminded me of the Nazis.

Claim one thing, and do another! We saw for ourselves that this is not the communism of equal distribution, nor is it a classless society. Class distinctions were strong and became more so during the twenty-six years of my observations. The top caste includes those involved with maintaining Party and Government; they have privileges, such as country homes and limousines. Scientists are just below, with relatively high incomes, and those that are trusted can travel abroad. But many other scientists have tried with careful applications to come to our scientific meetings; they would write to us that this time they would certainly come, but they did not come. Later when we could speak privately, they complained that they did not even know who turned their request down. During the question period of a public lecture, I was asked how Jo-Ann and I could have come while the relations between our governments were so bad. I answered that we had asked no one for permission. Disbelief in the large audience. I therefore told them that on the day before departure I had remarked to the Director of our Laboratory, when we met in a corridor, that I would be away for awhile: "Where are you off to this time?" "Soviet Union and India." "How nice! Have a good trip!" That a Director did not even know of a foreign trip made quite an impression. They could not even travel within their own country without permission.

The second time we were invited to the Soviet Union together, their Embassy in Washington may not have liked that idea, or perhaps there was just sloppy office work. In any case, Jo-Ann's visa had three major mistakes that looked deliberate, and it was sent three days after the date we were supposed to leave the United States; expecting that, I had given them a departure date a week earlier than we planned. Knowing that Aeroflot or any other major airline would not take her to Moscow with papers so wrong and incomplete, we switched to a small airline to enter the USSR in Kiev. Jo-Ann was unconcerned, lying in the nose of what was a converted bomber while the flight engineer pointed out some rocket installations around Kiev; she got a grand tour, as our landing was delayed for nearly an hour because a special government plane had to take off. After we landed, however, the police said they could admit me but not her, and they ordered our crew to take her back. Ah, but the Soviet Union has a class structure; nearly anyone can be overruled. We whispered to a nice customs girl to run upstairs and see if there were not a small welcoming committee, as there usually is, with an Academician or KGB member among them. Imperially they descended down those stairs, and Jo-Ann was promptly welcomed into the Soviet Union. Getting out again, later, in Tashkent, was another story. Some KGB clod wanted to see all our notes and papers, every sheet of them; I was yelling at him, and he was threatening to hold us there. Since the next plane out to India would leave a week later, he had the upper hand and got to see almost everything.

The upper classes' privileges can be ridiculous. As special guests we were taken to the Bolshoi Theater, and our heavy winter coats had to be checked. After the performance we did not stand in line with all the others, but our host said that he was a member of the Academy of Sciences and we wanted our coats first; nasty remarks came from the line, embarrassing to us, but not to our Academician. It seems so unfair to the people that the best shops accept only foreign money. Using the words "Amerikansky" or "Intourist," or being a guest of the Academy, opens all facilities and special privileges. Our bus driver in Alma Ata was stopped for a traffic violation, but he was quick to tell the policeman that he carried Amerikansky, and he could go on.

Such small observations have contributed to a feeling of great pity for these people, who have to live under a regime that is thoroughly ingrained with deceit and that has spies to report any individualisms to the authorities, just like in Nazi times. To live there would be infuriating. Not to be able to speak out!

In 1978 we were honored to receive Cronid Lubarsky as a house guest soon after he was expelled from the Soviet Union. Until 1972, he was an active planetary scientist in Moscow. He was suspected of being a dissident, however, and when investigated by the police, they found *samizdat*, underground writings, in his flat. These are writings, poetry sometimes, copied by hand or on a typewriter, about various topics, usually only mildly critical of the government and its institutions, but they are a free literature, in which human freedom and concerns for the future of the world are discussed. Cronid had several copies of the same issue, so he was obviously distributing them, but he would not tell where he had obtained them or to whom he was bringing them. He was to spend the next five years in prison and hard labor camps. Cronid did not complain much about camp and prison, but described them rather as interesting human experiences. Anatoly Shcharansky did the same when he was released in 1986, as did Yuri Orlov on his arrival in the US. These people seem to have been tempered into noble metal. Their mettle was probably noble to begin with.

What Cronid did believe to be unfair was that he was not permitted to return to work as a planetary scientist after serving his sentence. He was not allowed to live in Moscow within reach of the observatory and was forced instead to stay in the small town of Tarusa some 150 kilometers from Moscow. Nor was he trusted with any responsible job, such as that of a teacher. But it is illegal for a Soviet citizen to be unemployed; the only thing he would be able to do was manual labor, possibly for the rest of his life. So he made a protest through legal channels. He was then advised that he had a choice of going back to camp for ten years, or being expelled from the country. In the event, he was expelled together with his wife and daughter, and surprisingly quickly. It turned out that they had become an embarrassment within the KGB.

Cronid's spouse Galina Salova-Lubarskaja—we call her Galya—was born to a KGB family, her father rising rapidly within the ranks. In their caste, the children go to privileged schools. Soviet schools are described by Vladimir Sakharov and Umberto Tosi in *High Treason*.[2]

Every child is taught that obedience to authority is a virtue, and that personal loyalty to anyone or anything other than the state is a vice. From the beginning, the individual means nothing; betraying one's schoolmates is encouraged. My first-grade class in Moscow was typical of all first grades in the Soviet Union. There were about thirty children. The class was organized along military lines and it was impressed upon us that we were part of the system. We were required to wear gray cadet uniforms with brass buttons. We marched in and out of the classroom and otherwise were expected to behave in a disciplined fashion. We were told that we now were "Oktyabryata" (Octoberites), or children of the Socialist Revolution of October 1917. As "Oktyabryata," we were expected to live up to high standards in order to be worthy of eventually becoming Young Pioneers like our older brothers and sisters.

Like all other Soviet grammar school kids, we had a class leader, one student selected by the instructor for exemplary behavior. In the United States such a child would be called a teacher's pet and be teased for it. Our class leaders, however, were backed by all of the teachers and parents and, although the other children might grouse behind his back, most were too respectful of authority to confront him. The class was divided into three "links" consisting of ten children apiece. Each link had a leader, who was elected by his group on a rotating basis. The link leaders were under the command of the class leader.

Every morning the link leader would go before the class and denounce failings he or she had observed by link members. "Sasha didn't finish his homework. Nina has dirty fingernails. Boris picked his nose. Sakharov, again you are wearing non-uniform trousers." Everyone had a chance at being a snitch. It put each of us into a bind. The children didn't like snitching on the whole, although some took a perverse pleasure in it.

When Galya was about 14, her father wanted her to enlist in a high school for children of KGB personnel. She had, however, come to the conclusion that she did not like the emphasis on Party and politics, and that this was not the environment in which she wanted to grow up. She wanted to be free to develop and express herself. That she concluded this alone, at such a young age, is one of the finest examples of freedom of the human spirit. It was, however, the start of her troubles with her father, because having such a daughter was not conducive to his career.

It is not surprising then that Galya found Cronid, and vice versa, for they were equally free spirits. Galya had to pay for it though. During the years that Cronid was away, she was put under pressure to divorce him, but she was just as tough and determined as he. She was totally disowned by her family.

After Cronid, Galya, and their daughter were expelled from the Soviet Union, we arranged for him to obtain a position as a planetary scientist in the US, but it became clear that their dedication in life had become the cause of human rights. They moved to Munich and Cronid became the editor and publisher of *USSR News Briefs*,[3] a bimonthly that documents cases of persecution all over the Soviet Union. Galya and the daughter work for Radio Free Europe, broadcasting programs to the Soviet Union and other oppressed countries. Cronid also issues a yearly list of political prisoners in the USSR, something like a "Who's Who in the Gulags." How does he obtain up-to-date information, and, in many cases, send support to these people? These questions must not be asked or answered for the sake of the safety of sources. The *USSR News Briefs* point toward cases where writing letters or mailing books and packages may make a difference, and to refuseniks in the Soviet Union who would like to be visited; such visits are rewarding and quite feasible.[4]

Andrei Sakharov is the outstanding proponent of human freedom and initiative. He has written extensively on public issues.[5] His effectiveness is illustrated by the following example. On October 9, 1981, he wrote an open letter to his colleagues to explain that his daughter-in-law Liza Alekseyeva was not allowed to leave the USSR to join her husband in the United States. Sakharov described how his family was made to suffer because of his activities for human rights, and how he had been removed to exile in Gorky without any legal procedure. He mentioned that he had made appeals in vain to the authorities and also to Academicians A. Alexandrov, E. Velikhov, Ya. Zeldovich, Yu. Khariton, P. Kapitza, and B. Kadomtsev. He announced that he and his wife would go on a hunger strike on November 22 to demand that Liza be given permission to emigrate from the USSR. After seventeen days of hunger strike, the wave of international reaction had resulted in permission for Liza's emigration. Sakharov's victory was obtained by careful selection of an issue, courageous determination, and the ability to obtain international support that produced an outburst of letters and telegrams to Soviet authorities and to leaders of other countries who relayed their concerns to the government of the USSR.

If our less conscientious Soviet colleagues continue to ignore issues of human rights, pressure should be brought on them by scientists in the West. Leo Goldberg and I addressed this topic in a Letter to the Editor of *Sky and Telescope*.[6] Our example was V. A. Ambartsumian, a leading Soviet astronomer who had reached his senior position not only because he was a smart scientist but also because he went along with the regime, much more than seemed necessary to us. Our letter ended with:

> It is a pity that a man as politically powerful and scientifically prestigious in the Soviet Union as Ambartsumian has to our knowledge never lifted a finger to aid his unfortunate colleagues. On the contrary, in numerous statements and interviews which have been published, for example, in the newspaper *Le Monde* (April 3, and April 7–8, 1979), and in the *European Herald-Tribune* (article by Leopold Unger, March 30, 1979), he has assailed Soviet dissidents in general and Orlov, a former member of the Armenian Academy of Sciences, in particular.

He has also castigated foreign scientists who have tried to assist their colleagues who are persecuted in the Soviet Union and accused them of pursuing rather mean objectives (interview given to Tass on the occasion of the 30th anniversary of the Universal Declaration of the Rights of Man, December 7, 1978). He was, of course, one of the forty members of the USSR Academy of Sciences who signed a letter condemning Sakharov nearly ten years ago, and presided over a recent meeting of the Armenian Academy of Sciences during which we have reason to believe Orlov was expelled from membership. Furthermore, Ambartsumian did not intervene to obtain a lightening of the twelve-year prison sentence Orlov received for the "crime" of seeking to monitor Soviet compliance with the Helsinki agreements.

There should be a clear understanding by astronomers in the Soviet Union and elsewhere (we do not exclude the United States) that violation of the rights of their compatriots will do serious damage to their own image abroad. We believe that such an understanding may ease the plight of those scientists who are suffering the consequences of their devotion to the cause of human rights.

During my visit to the Soviet Union after this letter had been published it happened a few times that some unknown person would come up to me during a reception or a visit to a scientific institution, would say something like "Thank you for what you wrote in *Sky and Telescope*," and quickly disappear.

In 1983 I was asked to be a Jury Member of the "Sakharov Hearings" that are held every three years in various cities in order to support Sakharov and other dissidents in the Soviet Union and Poland. This time the Hearings were held in Lisbon. Portugal also had an interest in discussions of freedom and dissidence, because it had gotten rid of an oppressive regime just a few years before. As guests of the Portuguese government we were gathered with about a dozen people who had recently come from the Soviet Union and Poland and who presented their testimony, prompted in part by questions from Jury Members. I was surprised to hear of the extensive misuse of psychiatry in the Soviet Union. How can medical doctors, and a country that claims to be civilized, torture healthy dissidents in psychiatric wards? Another revelation at the hearings came from a man in a wheelchair. Indeed, Jo-Ann and I had not seen wheelchairs in the tourist cities of Moscow, Leningrad, Minsk, Kiev, Alma-Ata, or Tashkent. The Workers' Paradise apparently finds it an embarrassment to have them around. A translation of parts of the testimony of Valery Fefelov follows.[7]

One day Foreman Sergeyev sent me to work on an electrical transmission line, not knowing that it carried 10 kilovolts. I climbed out onto a support, received a powerful electric shock, and fell to the ground. Due to a broken spine from the fall I have never again stood on my legs. The nine months spent in the hospital come back to haunt me even now. The medical care was bad, at times insufferable. I would wait for hours for the orderlies or nurses. As it turned out there was no specific kind of treatment carried out for my back or spinal column injury.

...

That's pretty much how I got my wheelchair. It was given to me only after I wrote a petition to the Central Committee of the CPSU Politburo and stated that until I got a wheelchair I did not consider myself a citizen of the USSR, since the state organ in the form of the Social Security Department of the city of Vladimir, by refusing to provide me with a wheelchair, by that very act, sanctioned the violation of my rights as a citizen to transportation as written in the Constitution of the USSR.

Constantly communicating with the handicapped I saw that my fate was not exceptional. It was one example of the general, ugly picture of the invalid's plight as a whole. Every commonplace trifle became a problem that required a great deal of nerves and time, mounds of paper, and communication with various departments to resolve. Even with all that, in most instances it was unfruitful. I realized that invalids in the USSR are the most numerous, underprivileged, and oppressed group of the country's population. Invalids, especially with terrible injuries and motor-function handicaps, due to their own physical shortcomings and extremely low pension, are deserted and limited in everything: in transportation and work, in medical aid and special treatment, in food and clothing, and in education and leisure. In the USSR it is almost impossible to meet invalids on the street. Many simply cannot leave home. The pensions are so small that invalids cannot buy themselves relatively good clothing, and are forced to deny themselves even normal food. When they appear on the streets in their old, strange clothing in awkward wheelchairs, their appearance in no way conforms to the desires of the Soviet authorities. For them it is more comfortable not to see invalids, to shut themselves off from them artificially with the established barriers already enumerated above. The situation of the handicapped compelled to live in state invalid homes is even more difficult (the USSR has 1500 invalid homes). Cast off by the power of the administration of these institutions, invalids drag out a miserable existence there: it is forbidden for them to start families and have children, there usually are no doctors in such homes, there is constantly a shortage of nurses, the libraries are poor, films extremely rare, and they do not even give any idea of what is going on in the world. For example, in the invalid home of Donetsk they do not give the invalids trousers, dresses, and other clothing. And so, in order to go out in wheelchairs, the paralyzed invalids are forced to wrap themselves in blankets. If work exists in an invalid home, then the choice is extremely limited, and this work pays poorly and is not counted in the work record. And of course

> these homes are filthy, unsanitary, and overcrowded. The service staff steal. In several homes the lives of invalids are so difficult and meaningless that suicides occur.
>
> . . .
>
> The reaction of the authorities to the formation of our [invalids' rights] group was almost instantaneous: on the third day after our announcement of the formation of an Initiative Group at a press conference in Moscow, a large pit was dug in front of my garage with a power shovel, which meant I was put under house arrest. Shortly afterward a delegation of workers from the KGB Oblast Administration appeared, headed by the chief of the investigation department, Colonel Shibayev. This visit was repeated a month later. Each time there were intimidations, threats to take away my pension, apartment, and so on, right up to confinement in prison. There were also in my apartment such "guests" from Vladimir as Procurator Obrazstov, the chairman of the People's Court, and Luchkov, manager of the department of Social Security. Next followed searches and the confiscation of all materials in which were only the smallest references to invalids, the letters of invalids with accounts of their bitter fate, bulletins, etc. (In my apartment alone five such searches were carried out.)
>
> . . .
>
> Just after the creation of our Initiative Group for the Defense of Invalids' Rights, we learned even more dreadful details about the authorities' deplorable attitude toward invalids. We learned of the existence in the Soviet Union of special concentration camps for invalids, directly from those who had served sentences in such camps. The prisoner's life, which is extremely difficult even for the healthy, becomes completely unbearable for invalids. Often it is harsher, since the invalids cannot stand against the will of the administration. The conditions in such camps are so difficult that invalids simply cannot survive them. Death discharges are rather frequent.

The repressiveness of the regime can be seen anywhere. When Jo-Ann, as a young girl, came with me to the Soviet Union, we were readily invited into private homes; the regime was apparently less critical when such a little foreigner was entertained. Otherwise, too many visits from foreigners might result in a refusal when one made an application for a larger apartment for a growing family. Why an application was refused, or who refused it, remain a State secret, according to our Russian hosts. A senior astronomer made it clear, in 1984 still, that he had to be careful with our discussions because they might be overheard—in his home! Other than those ever-present concerns and precautions, we had a good time because the people were loving, freely giving bear-hugs and celebrating with vodka, and the older children and family members had fun trying to speak English with Jo-Ann. The homes looked poorly made in our eyes, but such material matters appeared to be rapidly improving in the Soviet Union. It remained a haunting place, where I felt the KGB omnipresent, and government integrity absent. What can one make of it, for instance, when some 100 kilometers outside of Moscow and Leningrad there was near-starvation in the villages in winter, as we were told, because of poor distribution systems? Pan Am and other airlines may now fly the Delhi–Frankfurt route over the Soviet Union. On a clear winter day I saw the isolation of these villages, for there was white snow all around them, with the roads not plowed through.

How strange also that a great country could be so petty. Why take the trouble to put cockroaches in a mail package for Sakharov in Gorky?[8] Why not let whoever wants to emigrate go? This same regime makes the argument that matters of human rights are internal affairs, and it wishes to be a respected member of the international community!

Cronid Lubarsky used to say that the Reagan administration was the first that seemed to realize that the only way to deal with the Soviet government is from a position of strength. Cronid was enthusiastic, however, about Gorbachev's policy of *glasnost*, or openness, and he judged it to be without precedent in Soviet history. When we talked about it in 1987, the release of political prisoners had begun and no new arrests had been made, some private enterprise had been introduced, and a small amount of foreign capital admitted. Cronid believes that the reforms were not planned in advance by Gorbachev, who came to power as a dedicated Communist, but is also—a rare case!—an honest man. Gorbachev may have thought it possible to change things with only minor reforms, but even the slightest relief of ideological pressure led to such a flood of new ideas and criticism that he met strong resistance from the party bureaucracy and was forced to look for supporters in other circles, in mass media and among the intellectuals. That was clever, but it put him beyond the point of no return. He then practically repeated many of the ideas proclaimed by such dissidents as Sakharov, Orlov, and Lev Timofe'ev. His grip on the situation was not very strong as yet in 1987; he could still be toppled as Khrushchev was in 1964.

Cronid hopes the people of the Soviet Union, particularly the dissidents, can help Gorbachev: The government made a call for democratization and the people should act as human beings with a free will, public initiatives, and a real social life. In the Soviet Union, however, the people have lost the skills of free behavior in the past seventy years, and they may not believe that the call for democracy is more than a trap for new repression. Cronid thought that justifiable; it is a possibility! The inertia of the people may be the worst adversary of Gorbachev, worse than party bureaucracy. We must therefore help and urge them to meet the challenge. People and governments of the West should be cautious and combine support of Gorbachev with demands for real detente. Cronid believes that the West can expect from Gorbachev an ideology of "partner," as a replacement for the previous ideology of "enemy."

What about the danger of nuclear war between the US and the USSR? I learned a little about the effects of nuclear weapons when the Pioneer travels took me to Japan. First I gave a command performance in Tokyo for His Imperial Highness Crown Prince Akihito and his eldest son, Prince Naruhito; an interpreter from the Japanese Space Agency was also attending. The presentation included the showing of slides and the film of the passage underneath Saturn's rings, shown in a simple room of the palace. For foreign audiences I try to hold my enthusiasm down and speak distinctly, but here it was not necessary and we were soon continuing in English only. Both princes are academically trained, the father in biology and the son as a historian, and the questions and answers continued for an hour or so at a scientific level.

Takeshi Sato, the Director of the Planetarium in Hiroshima, became a close friend in Japan; we call him Ken-san affectionately. He took Jo-Ann and me to the A-Bomb Museum, a pilgrimage that will lead one irresistibly into the Peace Movement! The Museum shows pictures of the devastation, and wax displays of what actually happened to the people. It makes manifest the finality of cell destruction by radiation. Skin and flesh came drooping off the bones.[9] The museum is so impressive that even old soldiers come to tears. Even though we may be familiar with the destruction of cities during World War II, the devastating effect of radiation is new to us. The A-Bomb Museum is the strongest of deterrents to nuclear war. Such a disastrous demise must not happen to humankind.

If only more people, especially those in leading roles, could see the museum and the devastation it documents, nuclear arms might be eliminated. A United Nations report estimated that already in 1981 there were more than 40,000 nuclear bombs and warheads, equivalent to a million Hiroshima disasters. That is equivalent to more than two tons of TNT for every person on Earth. The study also warned that a bomb with the power of that used at Hiroshima can now be made small enough to be carried in a suitcase. Biological or chemical warfare would be even easier to initiate. Accidents are bound to happen. Some of them may occur where many people will be affected, even more than in Hiroshima and Nagasaki. An accident, a major one in a big city, will surely make the world rise in conclusive protest. A sad prediction for the future! Or could we be smarter? Can we eliminate the dangers before such an accident occurs? It can be done, just as the use of gas warfare was avoided after World War I.

For Americans visiting the A-Bomb Museum there is a feeling of guilt as well: America brought this horror into the world. We have that drooping flesh on our conscience. America therefore has a moral obligation to lead the world toward a future without nuclear and biological warfare. The United States may have a special aptitude for global concerns anyway, as it is a nation of immigrants from all over the world.

This reasoning brings us to an intricate predicament: we must confront totalitarianism with strength, but, at the same time, prevent global warfare. Some challenge! But is it more intricate than life itself, for which we must make daily decisions between give and take and balances of good and bad?

The Boston Study Group has studied the dilemma. The group consists of six scientists and experts on military affairs; the physicist Philip Morrison is one of them. The group first made an overview of arms and the expenditures for them, and then took into account the state of the world and major problem areas, as well as the technologies of the great powers. In 1979 they published *The Price of Defense*, and an update in 1982, *Winding Down: The Price of Defense*.[10] They concluded that the US and the USSR can significantly reduce military expenditures while still maintaining a safe defense, and they followed through by making specific recommendations.

For further guidance we have the Center for Defense Information.[11] It is controlled by a group of retired senior officers of the US Navy and other people with expertise in industrial and military matters. CDI seems to provide an antidote to propaganda from the powerful forces, in Russia and the US both, that Nehru blamed for World War I and that Eisenhower warned us of in his Presidential Farewell Address[12]:

> In the councils of government, we must guard against the acquisition of unwarranted influence, whether it is sought or unsought, by the military–industrial complex. The potential for the disastrous rise of misplaced power exists, and will persist. We must never let the weight of this combination endanger our liberties or democratic process.

CDI publishes regularly the *Defense Monitor* to report on situations; for example, early in 1987, "The Unraveling of Nuclear Arms Treaties: Another Step Toward Nuclear War." On each issue the credo of CDI is published:

> The Center for Defense Information supports an effective defense. It opposes excessive expenditures for weapons and policies that increase the danger of nuclear war. CDI believes that strong social, economic, and political structures contribute equally to the national security and are essential to the strength and welfare of our country.

I would like to have added the word "scientific" to that last sentence. It seems critically important for our safety that the non-communist world remain strong in basic science and engineering.

In Moscow in December 1979 a high-level Soviet scientist remarked what I remember as follows: "Aircraft carriers we could take out in the first half hour of World War III. Your science and technology, however, would make us think twice. You went to the Moon, where we could not. Most impressive were the organization and management to pull the capabilities in industry and universities together toward that accomplishment. You explore the outer parts of the solar system through extensive usage of new techniques and miniaturization. The United States is a powerful country." In other countries too, including the Third World, the American successes in basic science and the space program seem to make a strong impression, and they therefore have a deterrent value as well. This deterrent value is in addition to the pragmatic need of basic science for the invention and development of new weapons and defensive counter measures. It is therefore only prudent to reverse the present neglect and to amply support free research at universities and for space experiments, taking the funds from defense budgets if necessary.

How can we improve human rights while at the same time preventing a major war? What action regarding the world's problems is feasible? What can an individual citizen do?

In the first place, we have brains to invent new solutions. Arthur C. Clarke who invented the concept of communications satellites, now proposes what he calls the Peacesat system. He discusses our predicament in his lecture on "Star Wars and Star Peace"[13]:

> The real problem is not military hardware, but human software—though the right kind of hardware can certainly help. A stable peace will never be possible without mutual trust; without that, all agreements and treaties are worse than useless, because they obscure the real issues.
>
> Yet trust cannot be blind; it must be based on past experience—and even then may require constant testing. This is true of individuals, and even more so of sovereign states, whose governments and policies may change overnight.
>
> The greatest enemy of Trust is Fear, and it makes little difference whether that fear is baseless or well-founded. It is not paranoia but prudence that compels military planners to assume a "Worst Case" scenario when they are ignorant of a potential enemy's capabilities. That ignorance, and the fear it generates, can be dispelled only by accurate and timely information. It follows therefore—almost like a theorem in mathematics—that the only road to lasting Peace is through Truth.
>
> A classic example of this is the infamous 'missile gap' debate which dominated the Kennedy–Nixon campaign of 1960. After the initial shock of Sputnik, which opened the Space Age in October 1957, there was a tendency in the United States to exaggerate all Russian accomplishments in this area. Propagandists ably assisted by the US military–industrial complex claimed that the USSR was far in advance in the deployment of ICBM's—so the United States must start a crash program to overcome this 'enormous' lead.
>
> Well, the missile gap was a total illusion, destroyed when the new American reconnaissance satellites revealed the truth about Soviet rocket deployment. President Johnson later remarked that its reconnaissance satellites had saved the United States *many times* the entire cost of the space programme, by making it unnecessary to build the counterforce originally planned. I would like to quote his exact words, which should be inscribed in letters of gold above the doors of the Pentagon—and the Kremlin:

> "We were doing things we didn't need to do; we were building things we didn't need to build; we were harboring fears we didn't need to harbor."
>
> . . .
>
> In 1978 the Government of France, in a rather untypical fit of global responsibility, made a dramatic suggestion. It might be a good idea, President Giscard d'Estaing suggested, if there was an *international* body doing for the whole world what the Americans and Russians were selfishly doing for themselves. Such an International Monitoring Satellite Agency could verify arms control agreements, check border violations, and defuse crisis situations, by acting as a watchdog for the world.
>
> Establishing what I like to call "Peacesats" would present major political, administrative, and financial challenges—but the reward might be nothing less than the salvation of mankind.

Clarke did not discuss how the Peace Satellites would be organized or who would manage them. Even if they are not established, however, the information available to all is increasing, and this can only improve the international situation. As for the political power of individual people, there are examples of properly directed political campaigns that made a difference, even in totalitarian countries, such as those of Sakharov. The US is famous for them.[14] It is the nation where the people fired, or prevented from taking office, many crooks and otherwise unsuitable officeholders—from judges and governors to senators and a President.

A fine example of a specifically focused action concerned a group of Pentecostal Christians, who, seeking to escape Soviet religious persecution, lived for some five years in the US Embassy in Moscow. The Society of Americans for Vashchenko Emigration conducted a campaign of letter writing so effective that the Vashchenkos were allowed to emigrate. SAVE was cheerfully dissolved with a closing note "Please do not send us any money."

A larger campaign, called S.O.S., Save Our Scientists, is under the direction of Professor Israel Halperin in Canada.[15] It began in 1981–84 as a letter-writing campaign to liberate Professor J. L. Massera from prison in Uruguay. In 1984–86 the efforts turned to freeing Drs. Anatoly Shcharansky and Yuri Orlov from the Soviet Union, with a special action in 1986 for Orlov. The organization has grown into a world-wide network of scientists who write letters to officials in the country in question and who translate and distribute and finance the campaign's literature. The efforts are supported by more than a hundred Nobel prize winners along with thousands of other people, and the organization now feels strong enough to take on the cause of not only scientists but all victims in Chile.

Larger still is Amnesty International,[16] which chooses half a dozen human rights cases per month and tells its members how and where to write in detail, including titles and salutations and instructions to use a positive style. Amnesty International points out, and Cronid Lubarsky confirms, that even the first few letters make a difference in the treatment of a prisoner, that a large number of letters yields a definite improvement, and that tens of thousands of letters usually set the prisoner free. An important aspect of AI is that it concerns itself with human rights wherever they are violated in the world.

Amnesty International, Greenpeace,[17] which engages in direct action for peace and the environment, and Common Cause,[18] whose concerns range from nuclear proliferation to environmental issues, are the three groups closest to my heart. There are other organizations too and, in fact, it sometimes seems that there are too many; I will always check that there is a list of appropriate people endorsing a new mailing, for otherwise it may be a rip-off for financial contributions. I have so far counted twenty-nine such organizations, some of them most appealing, but specious, having accomplished nothing over the years.

A novel suggestion came from a professor in Japan, namely, a chain letter for scientists to declare themselves for peace. This campaign apparently continues—how could it be stopped?—and tens of thousands of letters must have been written. A chain letter can be powerful because it is a personal communication instead of a printed appeal from some organization, and because it reaches vast numbers of people. One could use such a letter to bring about a demonstration by hundreds of thousands of marchers for peace, reminiscent of the ones Martin Luther King, Jr., and his associates organized for human rights of black people. In Washington, D.C., there could be a gathering at the Lincoln Memorial and a march past the White House and the Soviet Embassy. A possible text of the letter could follow the Japanese example:

> Dear ((addressee)):
>
> This is a chain letter to promote a demonstration on ((date)) in Washington, D.C. It is for a Comprehensive Test Ban Treaty and Reduction of Strategic Weapons.
>
> If you intend to come, notify ((sponsoring organization)).
>
> Please join this antinuclear chain reaction by writing this letter to ten people.
>
> Sincerely yours,
> ((signature))

If the letter were issued by the major peace organizations it would start with at least 10,000. One week's time is probably needed for each link of the chain, for receipt and mailing it again. Asking each recipient to send the letter to ten friends seems reasonable, but we know from similar experience that only 20% would actually continue the chain. Even so, one can see that in ten weeks the number of letters written would be ten million! Soon thereafter it would be well beyond the saturation point, that is with more than one letter in the mailbox of each American family that may be in favor of such action. Only a few percent of the letter recipients would come to march, but that still would be a large number. The effect on makers of policy can be judged by what King accomplished, and by the effects of the great demonstrations against the Vietnam War by hundreds of thousands of people. In a democracy such as the US this is the way to express strong sentiments and to swing the voters of this country who seem to have a close balance, believing in a strong defense while also having a concern for global wars. Similar demonstrations could be promoted by chain letters within other countries. It would be a great day for peace in the world.

Taking part in such a demonstration is a stirring experience, as I learned on February 5, 1987. For me the participation did not come in a casual or emotional manner. It did not come easy. I have a military background. My experiences in the Soviet Union led me to mistrust their system, and Cronid Lubarsky had convinced me that it is necessary to be strong when dealing with the Soviet government, that we must remain on guard or suffer the indignities of Budapest, Prague, and Warsaw. I see it not, however, as a question of being rather dead than red; the challenge of our era is to survive as well as to be free, and the world is making considerable progress toward that intricate goal. In this complex political civilization an interplay of strength and negotiation is required, as the former US Ambassador to the USSR, George McKennan, implied in 1981:

> If we insist on demonizing the Soviet leaders—on viewing them as total and incorrigible enemies, consumed only by their fear or hatred of us and dedicated to nothing other than our destruction—that, in the end, is the way we shall assuredly have them, if for no other reason than that our view of them allows for nothing else, either for us or for them.

It is sad that in the United States there seems to be a rise in religious fanaticism that considers communists merely as heathens—although Jesus Christ was a true communist. As a result, ironically, hatred is spread and the risk of global warfare is increased.

Now is the time to act in negotiation toward a decrease in the disastrous armory. February 5 was the day for me to be counted. On that day an underground nuclear test was scheduled. I knew from the Center for Defense Information that the absence of arms control made war more likely, weakened our ability to resolve differences between the US and the USSR, increased the burden of military spending, and even reduced the opportunity for the US to monitor the size and composition of Soviet nuclear forces. Not listening to their peers at CDI, the American military leaders had prevailed to continue testing at the Nuclear Test Site in Nevada, even though the Soviet Union was still observing a moratorium unilaterally, already for eighteen months. The US was also deliberately violating SALT II agreements for limitation of forces, wasting our taxpayers' monies into a ruinous deficit, not merely doing research for its "Star Wars" Strategic Defense Initiative or catching up with the Russian initiatives in these fields, but increasing an already excessive overkill with yet more nuclear, chemical, and biological weapons. Weapons industries and their stockholders prospered, but poor people went hungry, students lost jobs, science budgets eroded, some of our telescopes had to be closed, and the United States was no longer the leader in space research. Instead of peace keepers, we Americans—with the drooping flesh at Hiroshima and Nagasaki on our conscience—appeared to be warmongers, exporting arms wherever there are markets and profits.

The United States had a leading role in the drafting of the Nuremberg Principles, which were adopted by the General Assembly of the United Nations soon after World War II. Principle VIa states that it is a crime against peace to prepare for war in violation of international treaties, assurances, or agreements. But on February 5, 1987, another nuclear test would be made in Nevada. Actually, the Department of Defense, when it learned of the planned demonstration, advanced the test by three days. How about that, for showing respect for the opinions of citizens who, after all, pay for the tests with their taxes? I had enough. Having been on black lists of the Sicherheits Polizei and possibly the KGB, now was the time to get onto the one of an administration that showed callous disregard for democracy and nuclear holocaust.[19]

So I went to Las Vegas with young people who were eager to vote with their feet for their own future. We admired the Greenpeace people, some of whom had made similar protests near Leningrad, and who now demonstrated by flying their balloon and banners right into the Nevada Test Site. They were arrested by sheriffs who reminded me of law-and-order Dutch policemen arresting the Jews, because the law told them to do so. But then I realized that this was a New World where democracy ultimately prevails, for the sheriffs recognized our right to demonstrate and even showed a considerable amount of sympathy. The sheriffs did not know what to do with that Greenpeace balloon and gondola lying within the Test Site desert. So they declared it an emergency landing—some emergency!—and took off the handcuffs of the Greenpeace crew and set them free to pack up their equipment. That was, of course, too easy, so they later walked back in with the rest of us, and then were handcuffed again, herded into the police bus and formally booked at the nearest District Attorney's Office. Amy Carter, the former President's daughter, was doing similar things on the East coast. We all got jail sentences, but these were eventually dropped or overruled.

Back in the Nevada desert we walked in the footsteps of Mahatma Gandhi: it was peaceful and forceful, victorious both. A policeman grabbed me by the arm. I said that it was not necessary, at which he held harder . . . then, remembering Gandhi, I told him "I want to shake your hand," at which he smiled and let go of my arm. We shook hands and became friendly right there. It was a bright day in that desert near Las Vegas. We cheered a spry old woman, a "Grandmother for Peace" who also had come to walk into that Nevada Test Site. She too had enough of vested interests and wanted her offspring to know she was glad to be handcuffed and jailed if her government would not listen to reason and only to demonstration. And we followed her in, 438 of us, holding up our handcuffed hands with fingers spread in a V for victory, to the cheers of our young ones, and on that spirited day we saw the Sun shining in a thousand eyes.

15. The Children's Ventures

In our planning of parenthood we wanted to have two small families of two children in the beginning and two children near the end of our childbearing years. The first two were Neil,[1] born in 1952, and George in 1956. Jo-Ann was then born in 1968, but after that the doctor in charge advised against having more children. Adopting one never occurred to us at that time.

Neil almost perished at an early age on a Sunday morning when he came along to the office. While Pop was concentrating on his science the little one proceeded to pull out drawers of filing cabinets. One of these responded by tumbling forward which was exciting to the little tyke because at that moment all of the four or five drawers came out in a great rattle. The problem was that on top of this cabinet stood Pop's new photometer, an affair of aluminum and steel weighing some forty pounds, which crashed behind the little one. This story could not be told for years because inside the photometer was a brand new photomultiplier tube. In those days this type of detector for the light from dim astronomical objects was best made by hand, by the great Professor André Lallemand at the Paris Observatory. I had to be honest with him, and he was a good sport in writing back that if I could get myself to the Paris Observatory, he would provide another one, free of charge. These detectors were so special that each one was given the name of a French girl, and I flew to Paris to pick up Marianne.

The children were deliberately trained in dangerous living. If a warning about a hot stove was not heeded, they found out the hot way. They would sit in my lap to steer the car at a young age. I allowed Neil to hitchhike at the age of 15. The Flagstaff police picked him up, called home to check if that was all right, and Liedeke replied: "No! But his father set him up to it!" At which they let him go, after an interesting night in jail. Already when they were very young, Neil and George would tag along to meetings and observatories. Trips to Europe were great opportunities to teach them hitchhiking as a sport. It has happened that we stood on one side of a highway, got tired of it, and switched to the other side; that was a great way to travel, not knowing where we'd end the day. One of their special memories is of an audience with Pope Paul VI in 1966 in which the Holy Father put his hands on the two little heads and said "Good boys." This made it impossible for a while to correct them about anything as they would wave a finger at me and say "Good boy!" Not that there was much correcting to be done, because we were close as a group of friends, although Liedeke and I remained parents as well and were not called by our first names. In turn, the children were treated as important people even when very young. There was a fine balance between freedom and rights of the parents. There was no hitting. Except once. It must have been a long day in the office and George must have been a nuisance at the dinner table. Out came father's hand and whacked the four-year-old on his head. Shock in the family, and instant regret. The little one got red in the face and looked furious and searched for what to do next and finally stammered: "You just wait till I'm sixteen!"

The 1960s were fascinating times, as the two young ones questioned our Calvinist morals. Liedeke worked for a Ph.D. in French Literature that had a difficult title: "The Concept of Justice in the Fiction of Albert Camus." When little sister was born in 1968, I therefore wanted to name her "Alberta," but this was vetoed, and she became Jo-Ann after Liedeke's mother—an exceptional person, teacher of French, widowed soon after the birth of Liedeke, her thirteenth child.

That dissertation on the existentialist Camus brought up such questions as "Why are we here?" "What do we live for?" and simpler ones such as "Why can't I sleep with a girl if I like her?" Shocking! But it had a nice ring to it too.

The 1960s brought us an awareness of the environment and a concern for ecology. All over the world there seemed to be a rebellion in progress, and a searching for Truth by the young. Khrushchev and Kennedy had almost blown the world apart in 1962, but later in the sixties man was going to the Moon to look back at Earth, the whole Earth. In awe we would ask "How can we protect it?"

The 1960s also brought the agony of Vietnam, scrutiny of that war, and action to stop it, and we were proud of the United States where the will of the people prevailed, especially the young.

I tried not to bring my science home too much, and I claim no credit or blame for the fact that both Neil and George completed Ph.D.s at the California Institute of Technology. Neil became an astrophysicist at NASA's Goddard Space Flight Center in Maryland and is married to Ellen Williams, also from Caltech and now a physicist at the University of Maryland. They had a little daughter in 1985, our first grandchild, but lost it after three months to crib death, Sudden Infant Death Syndrome. As many as one out of 500 babies die this way, but the medical world still refers to it as a "syndrome," the word used when one can identify a pattern but the cause and prevention are still a mystery. It was a shocking distress and puzzle. They overcame their grief, clinging closely together, and another offspring came in 1987. George is a geologist at the University of Arizona, married to Jennifer O'Brien, who is a geology student he met in the field in Canada and it was love at first sight, with a grandchild for us soon in the making. Every summer, back they go, the three O'Gehrels to explore Alaskan and Canadian fossils and plate tectonics.

Jo-Ann began her travels with Pop when she was six, and she initiated that: Every evening we sit around together before dinner making a toast to some event of the day, and one day I had described a Russian invitation and how I would stop over in Holland. Holland was a Magic Place of Relatives. While her friends had uncles and aunts and doting grandmothers in Gila Bend or Casa Grande or even Tucson, poor Jo-Ann had none, but she had always been told that there was a large family in "Holland." So she said quietly "I am coming with you." Knowing the determination in the line of women she springs from, I took it seriously, and for years our colleagues would welcome this astronomer with his little pal. It was such a delight to travel with a little one! For instance on the official occasion of a Pioneer presentation in Iceland we were proceeding toward the lecture hall in formal attire and stately step, first the President of the Academy of Sciences and me, followed by the Secretary of the Academy and Jo-Ann. This was probably too stiff for the child, for suddenly she challenges us: "I'll race you to the door!"... And the Icelanders, fathers too, immediately took off with her, coattails flying!

In our travels we have a tradition that is spreading among friends and other travelers. We step into airplanes with the right foot first. It is probably a carryover from the paratroopers' superstition, for whom it was a fine way to break the tension, but it has a deeper background. If a bride should enter her new home with the left foot, it would bring bad luck and she is therefore carried in to avoid the dreadful mistake; this custom of carrying the bride across her doorstep goes back into Greco–Roman history. Just why left is inferior to right is not clear, especially because our hearts are on the left. In India there is a clear preference for the right hand, and a temple or ashram is entered with the right foot. But why were we taught, in pre-War Europe, to give something to someone only with the right hand? In any case, a right foot first into airplanes has a new meaning as it is a reminder of the Miracle. Actually two miracles: the first one is that this machine will fly; a fully loaded Boeing 747 weighs some 356 tons, yet it flies, beautifully. The deeper wonder is that this little person is privileged to make this trip and meet other people elsewhere. On the practical side, our right-foot custom also helped in preventing complaints over delays or discomforts, for Pop's Miracle is still there! And so we have seen our children through long days of exhausting travel—Ahmedabad–Tucson can take forty hours—never complaining about hassles or tiredness.

With nose by the window! Always looking out at so much to see. We always ask for window seats away from the wing on the side where the sun is not. Not that the wings are uninteresting. The motion of the flaps emphasizes how important and controlling is the area of a wing. The flexing of that huge structure during clear-air turbulence or when flying through a storm, is most amazing. The aileron steering flaps are in continuous operation during flight and one can feel the instant reaction in the rolling of the plane. If one can see an engine during landing, the moment of reverse thrust is quite funny as the engine's parts seem to fly in odd directions to obstruct and bend forwards the streaming exhaust, while large air brakes stand straight up from the wings, which have been extended to maximum size for the landing. During the flight, one can see the opposition effect, the same as for asteroids (Chapter 11), here seen as a bright patch on the ground or clouds around the shadow of the plane (that is, in the direction exactly opposite the Sun). The effect also appears as a bright flashing when the patch moves over reflector paints that are used on highway signs and over glass windows in homes and cars. It does not appear on snow or ice because the crystals scatter light sideways; when the plane flies through rain with its lights on, there are bright streaks, but through snow there are bright flashes, again because of the sideways scattering by the crystals.[2]

In the rear of the plane, behind the engines, one can see the beginning of condensation trails, or contrails, which are caused by soot particles from the engines on which water droplets condense when the humidity is high where we are flying. Watching the formation of the contrails we are amazed at the speed of the plane, almost 600 miles per hour, 1000 kilometers/hour, and we should then also realize that the temperature is extreme, about fifty below zero. That contrail stripe behind the plane casts a shadow down below to follow us, which we can see even if we are too high up to see the shadow of the plane. We can, however, usually see the bright opposition patch at the front of the contrail's shadow.

The shapes of clouds, the thunderstorms with lightning, the Moon and bright planets, and a setting or rising Sun, all add fascinating extras to looking at mountains and deserts and oceans with icebergs, and trying to figure out just exactly where we are. Sometimes at sunset or sunrise we can see the shadow of the Earth, a wide dark band above and along the horizon opposite to where the Sun is just below the horizon; a purple hue lies above the shadow, where the Sun is shining through the upper atmosphere in its setting or rising reddishness. There may be wide rings around Moon or Sun, caused by ice crystals in thin clouds above us, or there may be showing bright rainbowy patches called "dogs"; on rare occasions the dogs, and even fully round rainbows caused by droplets, can be seen on clouds below the plane. On certain routes, Northern Lights can be seen if one looks for them carefully, screening the cabin lights away with a cloth or magazine. Such aurorae appear in a circle centered on the magnetic North Pole in northeastern Canada, so a flight from Iceland to Chicago brought us through the area of greatest activity, with a display of colorful curtains that waved along and came and went within minutes.

We like to meet people in the cabin and discuss the variety of books that are read on long flights, and we observe pilots too; especially Jo-Ann is good at being invited into their cockpit. American airline companies have prohibited passengers in the cockpits since about 1982, without exception they say. It makes me wonder when there are no exceptions to rules, and in this case particularly so, since the pilots seem to enjoy explaining their computers and procedures in reply to serious questions, and it seems to challenge them too, making them more aware of what is seen through the eyes of their passenger cargo. For us it is spectacular to learn how this machine flies. They are usually three: Captain in the left chair, copilot right, and behind him the flight engineer sitting sideways in front of instrument panels to monitor engines and fuel flow. Behind the Captain there are two "jump seats" and that is where we sit, dead-still during take-off and landing because the recorders have a twenty-minute memory, and if something should go wrong the investigators should not hear any chatter or distraction. The taxiing is strange compared to riding in a taxi, much more remote, knowing the enormous power this behemoth will soon roar out, while now it is merely strolling like a leopard on a leash. There is a tiny steering wheel in the Captain's left hand with which he controls the nose wheel, and thereby the directions we move into—without great precision it seems. The crew is busy with check lists for all moving parts of the machinery and safety lights and landing lights and interlocks of doors and hatches and communications with the tower for instructions for after take-off and regarding the take-off itself and telling cabin crew and passengers to sit down and put their seat-belts on. One is made aware of their responsibility for the life and well-being of the hundreds of people that follow behind. Finally comes the roar, and the surge of power and the thumping of wheels on that runway, faster and faster, committed to take-off, and still checking and watching and commanding and confirming, and then pulling the steering-wheel column back to bring the nose up, and up she roars, and soars in a flight of liberation from the gravity of Mother Earth.

But the controlling continues with reports to checkpoints and resetting the miraculous computers that now guide the flying machine. The computer system of a 747 costs a million dollars and is worth all of that, for it integrates and takes into account all data on velocity and acceleration, course and curves of plane and winds in order to state the position with a precision that is usually within a mile even after hours of flying through darkness and varying weather. On a route from Copenhagen to Los Angeles there may be as many as twenty checkpoints at which the crew resets the computers on the basis of radio signals from below.

During the flight it is rather quiet in the cockpit as far as the sounds of the engine are concerned, for they are a good distance away and behind;

up in this nose cone so far forward in space a wavy vibration is felt. The view is much better than from the little windows in the cabin, but still there are obscured directions such as straight up or down and backward. Looking out does not seem to be important to the flyers anymore. Their concern seems to lie more in the operations and in completing the flight on schedule than in looking out at the wonders of the universe, for which they sit on such an exceptional perch. It is rare that crew members know or care about Venus' bright light in the sky or about other astronomical or meteorological phenomena, and that may be why many of them can believe in Unidentified Flying Objects coming from outer space. They do have a radar view of the ground, and they use it diligently when flying near Kamchatka or other Soviet territory and when storms are ahead in order to find openings with less turmoil and turbulence. I saw the power of radar on a flight from Tokyo to San Francisco. The pilot warned me to pull the jumpseat's harness tight and on the intercom he told the passengers and flight personnel to buckle their seat belts as well. Nothing but clouds lay ahead, as an enormous storm system stretched out for hundreds of miles and way above our altitude ceiling. The flight was beyond its point of no return, so they could not go back, and the storm looked much more severe than their flight plan had predicted. So they turned off the automatic pilot and took over by hand, looking with radar for openings. Once they had made a choice of direction, they were committed to it, since an airplane cannot stop or go backwards.[3] After a while they got the hang of it and they enjoyed flying their enormous bird, which felt so large and yet seemed so small among the giant cumulus clouds lit up by lightning flashes all around us. How I wished then that one of the children was along for the ride!

When Neil was still a student he once wrote a letter to me asking for advice. Fatherly Advice! He wanted to accomplish something difficult in order to prove to himself that he could do it. The choice had already been narrowed down to two ventures, and Pop was invited to comment on each. The first was to row across the Atlantic, alone, and the other was to hitchhike around the world. From my own youth and reactions I had constructed the Pig Theory for Raising Children: a pig will go only into directions where he is not pulled. So I replied with a long letter praising the merits of rowing; at the end there was a catch, the titles of two monotonous books of rowing across the Atlantic. So he hitchhiked around the world with George in 1973. The only concession to their parents was to send a postcard each day, wherever possible, so that I might come and find them in case they disappeared.

To Neil, however, that trip was not difficult and he always dreams of new adventures, so next came a plan to bike the 2825 miles from Mazatlan to Panama. His general route followed the Pan American Highway down the Pacific coast of Mexico, inland to Mexico City, and south through Oaxaca and through each of the Central American countries, ending at the Panama Canal. The mid-1970s was one of the periods when such a trip through Central America was not made impossible by warfare. He started out Saturday 24 May 1975, biking with George and their friend John. Incidentally, they had to pay for trips like this themselves with monies earned by working in our yard, or babysitting with little sister, or painting stripes in parking lots during the nights, so it was poor man's travel. The following are practically unedited extracts from the diary Neil kept during the trip.

24 May

. . . near Zapopan, our first accident! John got sideswiped by a semi, breaking two spokes and tearing one of his paniers; luckily he was unhurt. We were very efficient, and repaired everything in half an hour. The event just points out the road and traffic conditions we've been encountering all along. So many times we were forced off the road when two trucks meet each other beside us.

Coming into Guadalajara was particularly hairy because it was late in the day, the traffic was heavy, and the road was under construction. We were beat, having ridden 137 km. The hotel we got was right downtown, and consequently cost 115 pesos (12 pesos = $1.00) even though it was rather shabby.

25 May: Guadalajara

. . .

Monday we were planning to leave but both John and I were still feeling pretty lousy. Lying around the hotel room we started talking about trip objectives, and John came up with the shocker that he was bummed out and wanted to go home from Mexico City. I guess George may be feeling the same way, in which case I'll be left holding the bag. I definitely want to continue, even though it looks like I won't be able to make it all the way to Panama City: at home we had estimated a daily distance of 80–90 miles, but it's turning out to be more like 50–60 miles. We'll have to see what happens as Mexico City is still about a week away.

28 May

I don't know how many times we were honked into the gravel. This was, of course, irritating, but it was better than the trucks that wouldn't honk. You'd hear a roar in your ear and look beside you to see the front ram-bars of a semi about a foot away from your leg, moving 100 km/hr.

That evening we sat around and talked. We came to the realization that if things like this would continue to occur, there would be a rather large probability of some serious mishap during the course of the trip (we estimated about 10%). I've thought a lot about this and want to say just once for the record that I accept this. If something happens to me I don't want people to think I got in over my head; it's all a well thought-through risk.

Three other things happened during the day to help make it a complete bummer. First of all we broke the number one rule of budget traveling and forgot to inquire at lunch how much the meal we ordered would cost. When it came we realized we were in for a big bill as it was very fancy, but we were still taken back by the 160 peso debt ($5.50 apiece)! Immediately after the meal, one of my paniers bounced too hard and got caught in the back wheel. The damage was a bent panier plate that I merely bent back. It remained wrinkled, however. Finally, in the same village a dog nearly bit George and me. Had he succeeded we would have had to cancel the rest of the trip and submit to the rabies treatment. Along the way we have seen plenty of dogs, but this was the first one to attack. In fact, the others all act extremely docile and even frightened. I think this is because they're so abused down here.

4–8 June

As I parted company with George Wednesday morning, I felt a funny combination of sorrow and scare. Only once before, in India, had I been struck this way when George and I separated for a day.

Things got off to a good start Wednesday I was allowed on the toll road to Cuautlo. For the first 1½ hours the going was all uphill, climbing out of the Mexico City valley. Then for the longest time it was down. I stayed in Izucar de Matamoros (100 miles) the first day. The next was again almost 100 to Huajuapan. I've been in the hills the last two days; a little up and down but also cool.

. . .

Oaxaca is smaller and dirtier than Mexico City or Guadalajara, but it is much cozier and more active. Saturday was the weekly street market and it was amazing in size. Also during the whole weekend, and particularly in the evening, the center square was jammed with people. Sunday there was even a band concert there. Because of all this, I suppose, there are a lot of tourists in Oaxaca. In fact, upon seeing a "gringo" (Yankee) elsewhere I always stopped for a chat, but here we just pass each other by.

. . .

June 9–13

One of the most hell-bent weeks I've ever spent.

Monday: Oaxaca–Marylu, 130 miles
Tuesday: Marylu–Juchitan, 50 miles
Wednesday: Juchitan–Tonala, 100 miles
Thursday: Tonala–Huixtla, 115 miles
Friday: Huixtla–Coatepeque, Guat. 75 miles

The first day was up in the mountains. I planned to spend the night somewhere about halfway to Tehuantepec, but there simply were no hotels. As the Sun started going down I still had 50 miles to go. A dash for it was attempted but after 20 miles it got dark, and in a frantic search I found a restaurant with hammocks outside. I was allowed to sleep in one. This town of Marylu is close to sea level at the beginning of a coastal stretch. It's hot and humid here and everyone sleeps in hammocks.

The second day I didn't feel very well so I made it a short day to Juchitan. It was my first experience in a jungle (on a bike) and it was amazing. Completely green everywhere, with constant bird and cricket-type noises issuing forth. Also extremely humid: I was covered with sweat to the point of having it fill my eyes so I couldn't see. To escape this, a trip to the ocean 15 km from Jachitan was planned; what a mistake! The entire way was rough dirt road. The tires and wheels certainly took a beating. Also, there wasn't really a beach, but just a fishing community. The water felt good anyway.

The trick to cycling in the heat and humidity here is to leave very early in the morning and break from 12:30–2:00. The extended break was not practical always because of approaching rains, but for the entire week and probably the rest of the trip the early schedule was used.

Something pretty disgusting is the number of dead animals on and beside the road; dogs, possum, cows, horses. This has been the case all through Mexico, but here in the jungle each carcass is surrounded by vultures pecking at it.

Thursday the plan was to make it to Tapachula, but the most amazing storm came up and stopped me 25 miles short in the town of Huixtla. It was more violent than any I've ever seen in Tucson. The locals say it happens everyday now until August. From experience through the weekend I think they may be right. The lights and water of the whole town were cut off and all traffic was forced to stop. This business with the lights seems to happen quite frequently in the smaller towns. I guess these summer monsoons are going to be really something else! It would be physically impossible to ride in one.

June 13–15

Friday: Huixtla–Coatepeque, 75 miles

Saturday: Coatepeque–Escuintla, 100 miles

Sunday: Escuintla–Sonsonate, El Sal., 112 miles

The whole day Friday I had my fingers crossed about getting into Guatemala. I had no visa and heard that one could get them only in Mexico City. But it occurred to me that somebody would have told me if one was needed. Well I bombed on up to the border and, being Friday the 13th, a visa was necessary. Fortunately, however, the place to get one was 12 miles back in Tapachula (I was crossing at Talisman Bridge). I left the bike, thumbed in, and not only got a Guatemala visa but one for El Salvador too. The entire crossing ended up taking 3 hours.

My first impression of Guatemala was very favorable. The road was nice, tractors in the fields instead of oxen, most of the people speaking some English and the Indians in the small towns seem to be left more to themselves in dress and religion than in Mexico. In fact, I had a most amazing thing happen. My bike broke down and just by chance, after fiddling with everything, in despair, found the problem. It was a clogged-up wheel in the back derailer. It wasn't at first obvious how to fix it, so imagine my surprise, after hardly ever seeing a ten-speed through all of Mexico, when a kid from a nearby Pueblo came running up with a spare he happened to have! After seeing his I was proudly able to fix my own. Adding this to a flat I fixed near Juchitan, I'm getting to be quite a bike man. That night in Coatepeque everyone was very friendly and I even found a Chinese restaurant that served a huge meal for a Quetzal ($1).

Some new things about Guatemala became evident. There is a large difference between the rich and very poor people. The beautiful ranches are Haciendas owned by a few very rich people. They are worked by poor laborers. Although Mexico appears much worse off, this is not actually the case since they have at least divided the Haciendas among the peasants.

I ran into one of those amazing thunder storms coming to Escuintla, but it cleared soon enough to allow me to complete the trip. Escuintla itself is a dirty town, but surrounded by beautiful scenery: three huge volcanic cinder cones one of which still belches lots of smoke. In the evening I watched a volleyball tournament. There is a lot of interest in sports here, mainly soccer and volleyball. Most of the participants seem to be from the small wealthy group. Also in Escuintla I made what later turned out to be the wise move of obtaining a visa for Honduras.

Sunday saw a most delightful downhill ride into El Salvador, crossing at Pijije. I make no excuses for skipping Guatemala City; I'm simply tired and want to get the trip over with.

June 16–18
Monday: Sonsonate–San Salvador, 40 miles
Tuesday: San Salvador
Wednesday: San Salvador–San Miguel, 90 miles

El Salvador is the most densely populated country in the Americas, and it's obvious everywhere. Even in the country, the roadside is lined with people walking, riding horseback or driving a cart. The buses that come by are packed to the point of overflowing. Since there is only so much resources and money to go around in a small country like this, the overpopulation causes great poverty. The streets of the larger towns are populated by a very low-class group; something like the untouchables in India. They don't work and have to beg for food, they live in the streets and are consequently filthy, and they let their hair and beards grow. I got the same lousy feeling I had in India.

Sunday I crossed over without any problems and biked till sunset to make Sonsonate. There I spent $5 on a hotel with a pool and air conditioned rooms. What a splurge! I even went out and ate my favorite, Chinese food.

Monday I cruised into San Salvador and discovered another touch of comfort and home, a McDonald restaurant. In the afternoon I tried getting visas without luck. This is when I found out how fortunate it was that I had gotten a Honduras visa in Guatemala; El Salvador and Honduras are in the middle of a border dispute and severed relations. The border is still open, however, and I found out, in fact, when I got there that the Honduran visa wasn't essential. The reason for the border dispute, which began in 1969, was El Salvador's desire to increase its territory to relieve the over-population. The Hondurans quickly rose to the aid of their small military and managed to keep their territory (partly thanks to United Nations' intervention).

Tuesday I got the rest of my visas and roamed the city a bit. What a terrible place! Lots of beggars and a tremendous number of military police. I've never seen a country with so much military everywhere.

Wednesday I biked to San Miguel and found a motel part way on the road through Santa Rosa to the border (Ruta Militar). It's shorter and, I understand, better than the Pan American route. I also met two Canadian and a single Dutch hitchhiker. They told me that thumbing is quite possible through Central America.

June 19: San Miguel–Choluteca, 90 miles

I was told Honduras was very poor and generally unfriendly, but compared to El Salvador it was a delight. There's plenty of room to move around, and the poverty is therefore more rural and not so drastic as in El Salvador. As for the unfriendliness, I got quite the opposite impression. The border is the only one that doesn't require a visa. I also had a couple of very pleasant encounters. In a small cafe a man speaking perfect English asked me to join him and a friend at his table and told me he'd cover me for anything I wanted. He then told lots of interesting things about Honduras. The main activities are lumbering, bananas, and mining. In the banana and mining areas private US firms are in charge and are exploiting the country. He didn't seem bitter, but just told facts. There was also a story about Pentagon encouragement of the El Salvador border attack in order to force the Hondurans to develop their armed forces.

In Choluteca itself, I met an American Navy retiree who owned a Texaco station, "Gringo Jim." He was a dumbhead, though very friendly, and only seemed to bring home the point that he had chosen Honduras to settle in because of its friendly atmosphere (I'm sure its cheap prices were also an influence).

One thing that's becoming a problem lately is my appetite. I just can't seem to find enough to eat. Thank heavens this isn't India, and good food is fairly easy to find. I discover, in fact, that cuisine is getting more like in the States the further I get from Mexico.

June 20–22: Nicaragua

Friday: Choluteca–Leon, 109 miles
Saturday: Leon–Granada, 70 miles
Sunday: Granada–Libria, C.R., 111 miles

Neil in Costa Rica.

The cycling problem I encountered in Nicaragua was a strong headwind. Particularly on Friday, I was slowed enough to prevent reaching Managua as planned. During my stay that night in Leon I read certain things about Managua that made me decide not to stay there at all: It was hit by an earthquake in 1972, and accommodations are still hard to come by.

On Saturday, I stopped a couple of hours in Managua and was truly shocked. The center of the city is still in total waste! My expectations for some reason was that it would all be cleared up and at least partly rebuilt. The epicenter of the quake was right in the center of town and really destroyed everything. It was like walking through a strange ghost city with no people around and huge fallen and burnt buildings everywhere. In the meantime the city has shifted its growth and population centers to the outskirts of old Managua along the main highways. The plan is to rebuild the center eventually.

Nicaragua in general is more beautiful and also wealthier than the other Central American countries I've seen. Its landscape is dotted with volcanoes and there are two huge lakes, Lago de Managua and Lago de Nicaragua. Saturday night I stayed in Granada on Lago de Nicaragua. It was a nice place and my room only cost $1.00.

Unfortunately Sunday was clouded over and very windy. My ride along the lake into Costa Rica which I expected to be so beautiful was only miserable.

June 23–27: Costa Rica
Monday: Liberia–San Ramon, 90 miles
Tuesday: San Ramon–San Jose, 50 miles
Wednesday: San Jose–Cartago, 14 miles

Thursday: Cartago–San Isidro, 70 miles
Friday: San Isidro–Villa Neilly, 120 miles

I've been hearing all along how much I'll like the people in Costa Rica, but the first couple of days it was the scenery that impressed me the most. The northern part of the country is a wild bushland with a backbone of volcanoes running down the middle. Recently, one of the volcanoes, Arenal, has been active, causing some destruction and a lot of excitement. This area is sparsely populated and is largely used for grazing or timber, or simply not used. In fact, the only part of Costa Rica with a sizeable population is the central plateau, actually a huge valley in the central mountain range.

The riding in Costa Rica was made difficult by several factors: although the road was mostly brand-new and in good condition, there were some very bad stretches including a big-rock gravel part near the Nicaraguan border. Needless to say, the second cycling challenge was the mountains. The highway rises from sea level in Puntarenas to 4000 feet in San Jose to 11,000 feet where it crosses the Continental Divide south of Cartago by the Montana del Muerte (Mountain of Death). It's the highest point of the Panamerican Highway.

I got caught short on Monday because of the unexpected steep grade up to the Central Plateau. After huffing all afternoon I didn't have the energy or time to make it into San Jose, and therefore stopped in San Ramon. I'm glad I did: San Ramon is a beautiful little town. Being tucked away in the mountains, and built around a big central cathedral, it looked like a town in Europe. The people are mostly white skinned throughout Costa Rica and well educated, so the atmosphere was even European. The reason for the white skin is that diseases brought in by the Spaniards killed the Indians off completely; the reason for the education is that since 1948 military spending has been cut to zero to finance better schools. I can now see why tourists like the atmosphere and why so many Americans choose to retire here.

In San Jose there is quite a lot of English spoken thanks to tourism. I found an English bookstore and read the *Tico Times* and *San Jose News*, two English newspapers (Costa Ricans are called "Ticos"). Concerning newspapers, I had my picture and story taken for a possible article in *La Nacion*, the main paper in San Jose. I think I'll see if there's similar interest in Panama City.

Tuesday night I went to a jazz club and listened to a lousy group. It was an exciting time nonetheless, because I met Roger and Linda, a husband and wife team also cycling. They call themselves "International Odyssey" and are going from Chicago through South America! They go slowly and even take trains and rides occasionally but are seeing the countries thoroughly. He's easy-going; she is of Dutch Indonesian background. We got to be great friends. Wednesday I cycled to Cartago to get ready for the big pass, and ran into them at their hotel by the highway. Thursday, as I was starting out on the big climb, I again ran into them on a sightseeing trip.

The climb to 11,000 feet was certainly no easy deal, and took a good 5½ hours, mostly standing on the pedals. I was disappointed, however, to find nothing at the top; no marker, no snow, no panorama, and it wasn't even cold. The way down made up for it, though, by giving me one of the biggest thrills and scares of my life! Most of the way had been in the clouds and I had no way of knowing that the cloud cover on the back side was concealing a big thunderhead. The way down was much steeper than the other side, and, at first, I had great fun racing away. Then the rain started, and it poured. There being no place to stop, I had to suffer it out and get off the mountain before night fell. It was very cold and I found it hard to control the bike. On some slopes it was necessary to apply both brakes full, together with my foot pressed hard against the back wheel, to keep from running away. I pulled into San Isidro del General completely soaked and exhausted. I consequently didn't make it into Panama on Friday as planned, but had to stop in Villa Neilly on the Costa Rican side.

June 28–July 1

Saturday: Villa Neilly–La Lajas, 90 miles
Sunday: La Lajas–Aguadulce, 100 miles
Monday: Aguadulce–Panama, 120 miles

I haven't said much about the cycling itself. Everyday starts at the first hint of dawn or usually at the first rooster crow. I eat sandwiches made the day before for breakfast. Once on the bike it's just pedaling. My seven enemies are wind, rain, sun, mountains, bad roads, poor health, and equipment trouble. The rule of thumb seems to be that a day never goes by without having to battle it out with at least one of these. On the other hand they seldom hit more than a couple at a time. My hands now have nice calluses but my rear is a constant bother. The bike requires constant fixing and cleaning. In Panama I broke gear cables. Everything is easily done though. The only permanent damage is a badly dinged back wheel.

I thought Panama would be an easy country. On the map the road stays close to the ocean, making it therefore flat and cool. Also it is recently paved in most parts making the pavement smooth. However, it turned out to be just about the hardest country. The most irritating thing was that the road is over small steep rolling hills the whole way. I hate them more than mountains. Also, although the temperature may be a little lower in Panama, the humidity is extremely high; almost to the unbearable point. Because of these factors, I needed refresher stops more often than before.

Here Panama was throwing another hurdle at me. Panama is much less developed than other Central American countries. It has its highway and its canal and that is about it. The rest is mostly jungle that the Indians don't even go into. Even along the road, the villages are far between and facilities are limited. The first day I got caught in a rainstorm in the late afternoon. I was near the town of Remedios, but a little doubtful that there would be a place to stay. Just as it started to rain and I started to panic, a miracle happened: a little hotel appeared in the distance along the roadway. It was one of the only roadside motels in all of Panama!

The fates had saved me, but as always happens, I had to pay back with interest. The whole day Monday, everything went against me: wind, rain, hills, road conditions. In fact, the road got worse the closer I got to Panama City until in the end I almost had to walk the bike. Around midday I started to go into a frenzy, I just couldn't see another day of it. The whole afternoon I fought that road with every ounce of strength I had. There were no stops, and looking back I can't even remember what the countryside looked like. Everything was concentrated on the victory of Panama City. And then suddenly it was over and I was riding over the bridge crossing the canal and entering the city. I think I was too exhausted to appreciate that it was finally finished.

The second son, George, made an important discovery as part of his geology dissertation at Caltech.[4] His professor had suggested mapping and rock sampling of the Alaskan islands that lie to the west of British Columbia—geologically called the Alexander terrane—to see if these islands had drifted from where southern California is now. The trip would have taken millions of years, with the islands carried on a plate of the Earth's crust, as it shifted on the surface with respect to others. The mapping had to be done carefully, and George spent five summers, four months each, on an island just west of Ketchikan in southern Alaska. The weather is generally quite rainy there, especially in August, and while there are no people living on the islands, there are dangerous black bears, so that he always had to be armed and ready to fire. One day he was walking through a thick part of the forest over a fallen tree and happened to slip sideways into thick growth. On top of a big one! The bear must have been equally surprised because he or she took off into the distance never to look around or stop, leaving an equally shaken geologist behind. Another student elsewhere was not so fortunate and lost both her arms to one of those bears.

We were at the time getting a little concerned about George as he seemed to become more and more of a hermit, although in rugged health. The food during the long camping was mostly C-rations and fish caught right there, occasionally augmented with other fare on a rare trip to Ketchikan or through sympathetic delivery by a local fisherman. It was an accomplishment to survive on that lonely rain forested island.

He had the rock samples transported back to Caltech to analyze their age, composition, and structure. The maps and rock characteristics were compared with what was known for the geological units in California. But George did not find the predicted fit. The search was then on to find the original site of these islands, and eventually he found a match with what is now an area off the east coast of Australia!

Moving at the typical rate for continental drifting, about 4 centimeters/year, the Alexander terrane would have taken some 300 million years for that journey.

The discovery of these large-scale motions on the Earth's crust has revolutionized our understanding of many geological features. The plate that is carrying India is moving northward, and much of it has already been pushed underneath the plate carrying China, thereby actually lifting up the entire Himalaya mountain range together with Tibet. The North America Plate, carrying Arizona and most of the rest of the continent, is moving also at about 4 centimeters/year nearly due west; the Pacific Plate, which carries some of the coastal areas of California, is moving mostly north; the relative motion results in a forceful rubbing of the plates along the San Andreas fault, with severe earthquakes as a result. Off the west coast of Oregon and Washington states, a small plate moves due east, colliding with and being pushed underneath the North America Plate. By being forced down it also heats up and produces volcanic material, which may erupt as an active volcano in Oregon and Washington. That is why a few years ago Mount Saint Helens blew its top, and why the people in Seattle are similarly in danger from an eruption of nearby Mount Rainier.

The oldest rocks on George's islands, dating from 600 to 530 million years ago, were volcanic, suggesting that the terrane was born perhaps as an island arc, like the islands of Japan or Indonesia. Then came a period of mountain building, from 530 to 490 million years ago, followed by a second period of volcanic-arc evolution, from 490 to 425 million years, and mountain building again from 425 to 400 million years. For about 25 million years the mountains shed debris, and after that came a long time of quiescence, from 375 to 250 million years ago.

To pinpoint the origin of the Alexander terrane, George compared it with what is known about the geology elsewhere along the rim of the paleo-Pacific. An excellent match emerged: rocks that formed offshore of eastern Australia showed periods of volcanic activity interleaved with two periods of mountain building at the same times as the ones in the Alexander terrane.

Another match supported the first: magnetic fields locked into the rocks of the Alexander terrane showed a history that allowed a determination of latitudes. The latitude had originally been believed as being of the Northern Hemisphere; George now hypothesized that they might really be southern. On that new assumption, the latitude of the Alexander terrane proved to match that of eastern Australia for the period from 450 to 375 million years ago.

Then he found something else remarkable: The magnetic record for the Alexander terrane indicated that 375 million years ago it moved somewhat to the north, while Australia at that time was moving south. At that same time the geological histories also diverged: the Alexander terrane entered a long period of quiescence, whereas eastern Australia continued to be tectonically active. Apparently Australia and the fragment had rifted apart. From 375 to 225 million years ago the Terrane maintained a latitude that was roughly equatorial, but it was not stationary in the west–east direction. Records of fossils suggest what may have been going on. A layer that is about 225 million years old shows fossils hitherto found only in Peru. Evidently the Alexander terrane had moved eastward across the equatorial Pacific basin and had come up against rocks that are now a part of South America.

The magnetic record further suggests that by 200 million years ago the terrane was moving northward. The rocks deposited on it 160 million years ago resemble those of California, and George concluded that the fragment then was a part of the western side of a system of faults much like the San Andreas. By about 100 million years ago the terrane had arrived at Alaska and welded to the western edge of North America.

The movements of large and small plates of our planet's outer shell are probably driven by heat convecting up from the molten core of the Earth. The plate movements control the landscape, climate, and habitat around us, but they in turn are the result of that convection from the interior parts of the Earth. Thus we can see everywhere in our delicate world the connections with global and planetary aspects and with the formation of our solar system over 4 billion (4,000,000,000) years ago.

Research and Reality

16. Asteroid Impacts and Planet Formation

On a clear, dark night the Milky Way shows not only thousands of stars, but also vast clouds of dust and gas. This is an important astronomical observation anyone can make with the unaided eye in open country on a clear, moonless night, by lying down and looking carefully up at the Milky Way. There are bright areas that are caused by millions of stars; we see them individually as single stars in large telescopes, but the unaided eye cannot resolve them, and it therefore sees only the combined effect as a shine of illumination. In this luminous belt there are, however, also dark patches. They are not clouds in the Earth's atmosphere because on the following night they are still there, in the same place. (Actually, they do move among the stars, but their distances are so great that the motions are imperceptible, except over thousands of years.) The dark clouds are made of dust particles, each grain so small—a hundred-thousandth of an inch—that if you had one on your hand your eyes could not resolve it. By the millions, however, in clouds many times the size of the solar system, the grains can obscure the light from stars farther away in the Milky Way enough to cause a dark patch. This dust and the gas that subsists between the stars are the material from which stars and planets form.

Current theory postulates that star formation begins when these clouds of gas and dust increase in density, perhaps as the result of the shock from a supernova, the forceful fireworks that concludes the life of a giant star. A supernova produces two shock waves: one comes from the radiation and travels with the speed of light, the other one, traveling slower, is gas expelled by the exploding star; both shocks compress the interstellar gas and dust. In the regions where the clouds are compressed, the gas molecules weld themselves onto the dust particles, and the dust particles weld together also, in their now more frequent collisions, clumping together into larger objects like fluffy dirty snowballs.

Where the mass concentration of the interstellar cloud is large enough, gravitation begins to play a role. Local clouds of dust clusters and snowballs begin to attract each other gravitationally and stick together in more massive objects. We call these the "planetesimals": sandy or snowy objects that can be as big as rocks or mountains, up to tens of kilometers in size. The gas cloud also begins to contract under the influence of gravity, and if the mass of the cloud is large enough, most of it can be compressed enough by its own gravity to form a star. The leftover material—gas, dust, and planetesimals—continues to collide and crash together, eventually accumulating into planets wherever there is enough material.

We can observe them today, those planetesimal building blocks of the solar system, in the form of comets and asteroids. The only major changes they have undergone in the time since their formation have been due to collisions and evaporation of volatile substances. In the nebula from which the solar system formed there must have been a great deal of colliding, crushing, and breaking up, as well as accumulation. In the solar system there is now still a region of collisions, namely within the asteroid belt that lies between the orbits of Mars and Jupiter. The belt contains a million objects larger than a kilometer in diameter, and an ever increasing number toward smaller meteorite sizes.

Asteroids are, on the surface or throughout, probably conglomerated piles of rubble, of material like that of the rocks on Earth, that formed in the inner part of the solar system, where the Sun was intense enough to evaporate the gases within the planetesimals. Comets are more like large dirty snowballs, filled with ices and volatile substances, that formed farther away from the Sun, where it was cooler because the Sun was distant, so the dusty rubble included gases, ices, and other volatile substances. The cometary cores are held together by their own weak gravitation and, much more tightly, by "vacuum welding," a sticking of grains that is known to occur in a vacuum. Sunlight may have evaporated the outer snow layer of the cometary core, such that the objects may now be dust-colored dark instead of white. There are an estimated 10^{13} of them, with nearly circular orbits in a huge but tenuous spherical cloud at the outskirts of the solar system. This is the Oort Cloud, named after Professor Oort of Leiden University. Comets may leave the Oort cloud and come through the inner parts of the solar system when their orbits are disturbed by a massive object, such as a yet-unknown companion of the Sun, or another planet, or by a passing star or a cloud of gas and dust in interstellar space.

The asteroids had charmed me already in student days. Studying them involves some of the excitement of a hunt. (Although I had learned to detest hunting in New Guinea when hearing the wounded cry; once it was one of our men, shot instead of the pig in the bush at night!) Comets and asteroids have to be chased and carefully followed, or one loses

Phobos, one of the satellites of Mars. Asteroids may look something like this.

them. Once the observer has pinned them down carefully, their orbital parameters are computed and published so that they can be observed and studied again later. Paul Herget used to make those calculations for us at the Cincinnati Observatory; now it is Brian Marsden who heads the Minor Planet Center for that discipline, in Cambridge, Massachusetts.

The importance of asteroids and comets as leftovers from the early stages of formation of planets and stars did not become clear until the 1960s; by studying them we are looking back in time. Since our starting days at the Yerkes Observatory when there were few people interested in asteroids, this discipline has expanded into a wide area of exploration. By the early 1980s, a listing of those involved in studies of asteroids showed nearly 200 scientists.[1] They were engaged in telescopic surveys, orbit determinations, occultation observations, photometry, radiometry, spectrometry, interferometry, polarimetry, mass determination, radar and radio observations, planning of missions and mining in the future, studies of origin and evolution, and interrelations of asteroids with comets and meteorites.

Shapes and rotations of asteroids are easy to observe, and that type of study was in full swing already in the 1950s, but it took astronomers until the late 1960s to determine the composition of these distant objects. Different materials have different colors, so careful measurements of the brightness of an asteroid at selected wavelengths should give a hint of its composition. I tried to use the spectrograph at the McDonald 82 inch for this in 1958, but not with enough determination. The effects would probably have been too subtle for that instrument and for most of the asteroids anyway. Tom McCord pioneered that work with a photoelectric photometer when he was at the Massachusetts Institute of Technology; fortunately one of his first objects, the asteroid Vesta, shows a very clear color-dependent reflectivity.

For years I had a dream to build a telescope dedicated to the small bodies in the solar system. There was so much to be done, and the planetesimals are like fugitives, always moving and changing their phase angle. If one misses an observation at a critical phase, one is anxious to catch up as soon as possible. But the observing schedules for large telescopes are usually set months in advance and they cannot accommodate such needs. In 1970 we therefore designed a 72-inch "Asteroid Telescope," and the engineer Max Kaufman, of Phoenix, constructed a working model.[2] The size had been set through the advice and support of Aden Meinel, who did that sort of thing for colleagues in addition to the work he did as director, successively, of the Kitt Peak National Observatory, the Steward Observatory, and the Optical Sciences Center of the University of Arizona. Aden gave us a 72-inch lightweight mirror, a leftover from some spy-in-the-sky program. Jack Frecker took it to the Corning Glass Works in New York State where it was heat-treated in an oven at a temperature that would make the quartz just fluid enough to make it sag in the approximate shape for our telescope, giving it a focal length of 194 inches. This means that the mirror has a "speed," like that of a camera lens, of *f*/2.7.

We came close to putting it all together: We would figure the mirror at the Optical Sciences Center; for a building we would use a surplus radar tower on Mount Lemmon north of Tucson, and the dome had already been granted by President Harvill of the university. Still needed was a relatively small NASA grant for the steel structure of the telescope and for the completion of the project by the same crew that had proven its competence during the Polariscope ballooning. The money was available, and colleagues in New York State would participate. However, there were questions regarding previous grants to the laboratory for construction projects on the mountain, such as a Schmidt telescope that had been slow in coming into operation. Could I and my crew guarantee prompt and proper execution of the ambitious telescope plan? I was unable to resolve the quandary with Director Kuiper, and the money was not granted.

The years 1970–77 were frustrating times within a divided Lunar and Planetary Laboratory; it is too recent to write a history now. The contrasts were great: one day I would spend attempting to see the director, or even his business manager, and in despair writing a memo instead; the next day I would be with Charlie Hall's people seeing new data coming in from the Pioneers. One week would have nothing but miserable meetings within the faculty; the following week I'd be lecturing the Sri Lanka. These were times of learning solar system facts and human nature both.

Asteroids

Catalog number	Name	Perihelion (AU)	Aphelion (AU)	Inclination (degrees)	Diameter (kilometers)	Type of composition
1566	Icarus	0.2	2.0	23	2	Unclassifiable
1864	Daedalus	0.6	2.4	22	3	Silicaceous
1862	Apollo	0.7	2.3	6	2	"Apollo"
2062	Aten	0.8	1.1	19	1	Silicaceous
1620	Geographos	0.8	1.7	13	2	Silicaceous
1627	Ivar	1.1	2.6	8	8	Silicaceous
1221	Amor	1.1	2.8	12	1	?
434	Hungaria	1.8	2.1	23	12	Enstatite
944	Hidalgo	2.0	9.7	42	29	Dark
4	Vesta	2.2	2.6	7	555	"Vesta"
29	Amphitrite	2.4	2.7	6	199	Silicaceous
1	Ceres	2.6	3.0	11	933	Subclass of carbonaceous
153	Hilda	3.4	4.6	8	224	Pseudometal
279	Thule	4.1	4.4	2	131	Dark
624	Hektor	5.0	5.3	18	150×300	Dark
2060	Chiron	8.5	18.9	7	140	Subclass of carbonaceous

The Astronomical Unit (AU) is the mean distance of the Earth from the Sun.

Eventually I loaned the lightweight mirror blank to the Multi-Mirror Telescope, because we were anxious to start with the optical work for it while five more of Meinel's blanks were being sagged at Corning in New York. The MMT is located on Mount Hopkins, south of Tucson, and operated jointly by the Smithsonian Astrophysical Observatory and the Steward Observatory of the University of Arizona. Those astronomers had made a novel design: the telescope has six *f*/2.7 mirrors on a single mounting; the mirrors therefore are aimed at the same point in the sky and their images are combined at a single joint focus, giving the MMT the light-gathering power of a large telescope but with the structure and cost of a smaller one. That was an imaginative project; it yielded a working astronomical instrument and engineering experience with a new concept for future, larger telescopes, and it has become a productive tool for extragalactic astronomy.

Nonetheless, whenever and wherever I could, I was still trying to survey the solar system and to pin down the characteristics of my asteroids.

The nomenclature can be confusing. A small pebble entering the Earth's atmosphere seen as a "shooting star" is called a meteor. When the rock is large enough to survive the flight through the atmosphere without burning up, what is left is called a meteorite. The small objects involved in the early stages of solar system formation are called planetesimals. The larger objects are referred to as asteroids or minor planets. The table above lists a variety of objects that are all called asteroids. The first column gives the number in the catalog of minor planets that is published every year by the Institute for Theoretical Astronomy in Leningrad; the International Astronomical Union endorses that catalog at its triennial meetings. (During the IAU meeting of 1958 we saw the computations in Leningrad being done with abacus and logarithm tables, but now it is computerized.) The asteroid's name appears next in the table. The name is usually chosen by the discoverer and the result is rather chaotic, with names of cats, dogs, and mistresses along with nobler and mythological designations. I proposed in 1971 that the IAU would ennoble the naming, and an International Committee was duly appointed. Cats and friends are still being honored, but human-rights advocates are now vetoed as being too political. Asteroid Sakharov was possible because it was named before the committee existed; Galya and Lubarsky were allowed because there was no mention of their eponyms' accomplishments; but Lech Walesa was "impossible."

Next in the table come the parameters that describe the orbit. The asteroid's perihelion and aphelion are its closest and farthest distances from the Sun. The inclination is the angle between the orbital plane and the "ecliptic plane" where the major planets move. The last column gives information on the composition of the asteroid's surface.

Geographos was discovered on a photographic plate of the National Geographic–Palomar Sky Survey mentioned in Chapter 10. Vesta, Amphitrite, and Ceres move in the asteroid belt proper, which lies between about 2.2 and 3.3 AU. The others are more unusual asteroids, at an exceptional distance or with highly elliptical orbits, decircularized by massive Jupiter, which always lurks beyond the asteroid belt, controlling the motions of the rubble. Most of the "near-Earth asteroids," which occasionally cross or come close to Earth's orbit, still have aphelia in the asteroid belt; they probably are fragments from collisions within the belt; examples are Icarus, Daedalus, Apollo, Ivar, and Amor.

I can look at that table with personal pride, and embarrassment as well. I discovered Daedalus and many other asteroids. But not Chiron. For the observing run in which it was found, Charles Kowal had kindly suggested that we coordinate our programs by observing separate parts of the sky. It was a reasonable and efficient proposal, but for some stupidly competitive reason I waved it aside. Nature made the final judgment: Charlie found Chiron, while I failed to recognize its images on my plates. If we had not had Charlie as a good observer, this exceptional object would have remained unknown for many years.

About one-third of the comets are discovered by amateur astronomers, scanning the sky with large field glasses and recognizing the new moving object among the thousands of nebulosities familiar to them. Human vanity may be a stimulus for searching for comets, as they will be named by the discoverer; astrophysicists therefore inspect their photographic plates carefully for a possible discovery of a cometary object, their namesake. They do this in addition to their normal study of stars and galaxies, and it overrides the disdain that some of them may have for an object as simple and nearby as a comet.

Comets discovered in the 1970s

Name of comet	Aphelion (AU)	Perihelion (AU)	Inclination (degrees)	Description
Gehrels	large	3.3	176	Diffuse with condensation; 10 arcmin tail.
Kowal	large	2.3	6	Diffuse. No condensation. No tail.
P/Gehrels 1	9.0	2.9	10	Diffuse with slight condensation. No tail.
P/van Houten	8.6	4.0	7	Asteroidal motion, but hazy appearance.
P/Kowal 1	7.6	4.7	4	Diffuse with some condensation; 2 arcmin tail.
P/Swift-Gehrels	7.4	1.4	9	30 arcsec in diam. with sharp condensation. No tail.
P/Kowal 2	5.8	1.5	16	Diffuse with slight condensation. No tail.
P/Gehrels 2	5.6	2.4	7	Diffuse; 2 arcmin tail.
P/Gehrels 3	4.6	3.4	1	Nearly stellar. Slightly diffuse.

In the early 1970s, I made a wager with a colleague that I could go to Palomar five times and discover five new comets. It was to be done with an old technique, called the blink comparator, newly applied for this purpose: one looks alternately at two photographic plates taken at different times—"blinking" between the two plates—such that a moving object can easily be seen among the fixed stars. The method is a powerful way to find a comet so faint that it does not show an active tail but only a fuzzy nebulosity moving among the stars; there are many nebulae in the skies, but this one moves. The essence of the wager concerned the composition of comets and the distance from the Sun at which they would start to evaporate material; if volatile methane and ammonia were abundant, the comets could be seen much farther from the Sun than if they were mostly water ice. The table above lists the comets found with the blink comparator, including my five; for honesty I should admit that I went to Palomar two more times but without finding a comet.

Comets that return periodically are designated in the table with a letter P before the name. But when a comet's aphelion is large, as for the first two in the table, it may take so long to return that we omit the letter P. The sixth line in the table is for a comet that had originally been found by Lewis Swift in 1899, but was then lost. Now Brian Marsden has recognized that it was the same comet as the one I found. Recognizing comets and asteroids from their orbital characteristics is quite an art, but Brian is one of the few experts who has not only sophisticated computer programs but also a personal familiarity with these fugitive objects, knowing them by name and a number and history of discovery and quality of orbit determination.

Even though we cannot resolve asteroid surfaces from Earth and they are faint because of their great distances, we can measure color, brightness, and polarization, and how these characteristics vary with time. Such probing of the distant objects is called "remote sensing." A few comets and asteroids come close enough to be observed with radar telescopes; the largest one, near Arecibo, Puerto Rico, has a reflector dish with a diameter of 305 meters.

The determination of surface compositions has progressed especially rapidly. The compositions are being determined by comparing asteroid spectra with spectra of meteoritic samples measured in the laboratory. The same photometer as is used on the telescope to observe sunlight reflected by asteroids is operated in the laboratory to observe lamplight, or a sunbeam mirrored in, reflected by the samples. The instrument usually allows one to vary the phase angle, the angle between incoming and reflected light, in the laboratory. Asteroids can generally be classified as either dark and carbonaceous or light and silicaceous, but there are also types that show the presence of metallic matter. Carbonaceous asteroids have characteristics similar to meteorites containing carbon, while the silicaceous ones are more like sand or rock containing silicates, although they may also have a small amount of iron and magnesium. A search is on to find specific "parent bodies"—asteroids and for some perhaps the Moon or Mars—from which the meteorites fragmented.

The fact that different asteroids have different spectra is of fundamental importance. They apparently are not all covered with the same interplanetary dust collected over the years. Instead, material liberated from an asteroid in a collision settles gravitationally upon its own surface like dust after a storm. Why do asteroids differ in composition? The explanation is still incomplete. One possible answer is that asteroids are pieces of larger "differentiated" bodies that were broken up in collisions. A clear example of a differentiated planetary body is the Earth. It heated during its formation when planetesimals, attracted by the increasing gravitation, collided with the growing Earth, heating it by their impacts. At the end of the bombardment, when the region around Earth's orbit finally cleared, the estimated temperature was about 40,000 degrees centrigrade (Celsius). The Earth was then a molten globe of lava, within which the heavier substances settled by their weight toward the center: the Earth obtained a heavy metallic core. The outer regions were left with the lighter materials, the silicates that became continents and ultimately sand and soil, after the heat radiated away and the outer mantle of the Earth solidified and later eroded. When in the crowded asteroid belt similarly differentiated objects were broken up by collisions, the fragments from the mantle became the sandy silicaceous meteorites and asteroids, some of them perhaps carbonaceous, while the fragments from the center are now seen as metallic meteorites and asteroids.

The present mass in the asteroid belt is only one seventeen-hundredths of that of the Earth. This is a small amount of material for the large space between Mars and Jupiter; elsewhere in the solar system the

Asteroid impacts on Earth

Diameter (kilometers)	Number of objects	Impact probability (years between impacts)	Impact energy ("Hiroshimas")
10	10	10^8	10^9
1	1,000	10^6	10^6
0.1	100,000	10^4	10^3

masses of the planets show that there originally were greater concentrations in the nebula from which the solar system formed. Since the early days of the formation, Jupiter's gravity has jostled, swept up or thrown out of that region most of the early planetesimals, such that a major planet could not form there for lack of material. What we now have left over from the formation stage, in the large gap between the orbits of Mars and Jupiter, is a small mass of mostly collisional fragments and rubble piles reconfigured by their own gravitation after the collisions.

What about the asteroids closer in, near or even within the orbit of Mars? The average velocity of the belt asteroids with respect to each other is substantial: about 5 kilometers per second. This differential velocity implies that the collisions are forceful, with fragments flying off in all directions. Jupiter's gravity may then further perturb the orbits of fragments so they become more elliptical and move into the inner parts of the solar system. They then occasionally come close to Mars, Earth, the Moon, and Venus, perhaps even Mercury, and their gravitation may further affect the orbits. The ultimate fate is a collision with a planet—being "swept up" by that planet—or ejection from the solar system.

Our knowledge of the statistics of objects that cross the orbit of Mars comes from telescopic searches and counts of craters on the Moon, but a precise determination is lacking for the smallest, faintest objects. There probably are at least 1000 having diameters of 1 kilometer that occasionally come close to the Earth. This includes old cometary cores that have passed by the Sun often and close enough to have had their volatiles exhausted so that they now are nearly indistinguishable from the asteroids. The table above gives an estimate of the statistics of Earth-approaching asteroids and cometary cores, in a simplified form with approximate numbers.

Objects larger than 10 kilometers do not seem to occur close to the Earth; apparently the fragmentation in the asteroid belt is such that the large ones survive the collisions with their orbits pretty nearly unchanged. We are fortunate that those orbits in the asteroid belt are stable, that they stay where they are and not hit the Earth, for their diameters range up to 1000 kilometers! Such large objects may have hit the Earth during its formation, but if that had happened in more recent times, life would have been obliterated. Even at 10 kilometers there may be only ten or so that cross the Earth's orbit, and only once in about 100 million years, on average, will one of these hit the Earth, according to various astronomers's estimates.

Meteor Crater, Arizona. D. J. Roddy, USGS

The energy of the collision, in the last column, comes from the kinetic energy of the asteroid, $\frac{1}{2}mv^2$. The velocity v of a near-Earth asteroid with respect to the Earth is typically 20 kilometers/second; the masses m in the table are taken from what we know for the meteorites, namely that they have a density of about 3 grams per cubic centimeter. The impacts have energies that are hard to image. To compare them with better-known disasters, the last column gives the energy in terms of the Hiroshima A-bomb explosion of August 6, 1945, which had an equivalent of 13,000 tons of TNT, or 5×10^{20} ergs. Expressing the energy in Hiroshimas is not done callously, as we do not forget the suffering there. The asteroid impacts have much larger energies, but they do not have the nuclear-radiation effects of atomic bombs.

Meteor Crater in Arizona demonstrates that the Earth is still being bombarded by asteroids. It is near enough to Tucson that we include it on tours for our visitors, especially those from abroad: from Tucson north to the Meteor Crater, then on to the lava fields and historic Indian dwellings near Flagstaff, the Grand Canyon of course, and return through canyon landscapes near Sedona or via the Petrified Forest. The object that made the Meteor Crater had a diameter on the order of 30 meters and was metallic; the iron fragments from that asteroid are called Canyon Diablo meteorites, after a nearby canyon. The impact occurred some 40,000 years ago. It released about 10^{23} ergs of kinetic energy, 200 times that at Hiroshima, and left a crater 1.2 kilometers in diameter and 167 meters deep (that is, below the crest of the rim, which rises about 50 meters above the surrounding plains).

More serious was the much greater impact with a 10-kilometer object some 65 million years ago, at the end of what geologists call the Cretaceous Era and the beginning of the Tertiary. The names of the eras refer to layers of sediments, whose sequences and times have been determined by geologists. The boundary between the Cretaceous and the Tertiary sediments is clearly seen because it involved the demise of about 60% of the animal species that lived during the Cretaceous; fewer but different species flourished during the Tertiary. Evolution took a turn from dinosaurs to the smaller mammals, and, eventually, people. The Darwinian gradual evolution through survival of the fittest apparently has had a few impacting discontinuities.

The impact probably caused this massive extinction. Another one appears to have occurred 34 million years ago, causing the Eocene–Oligocene extinction. There still is some debate whether or not the dinosaurs were eliminated by the Cretaceous–Tertiary impact. Were they already on their way out through other causes? That collisions of the Earth with comets and asteroids can cause major extinctions is not doubted, however; it had been pointed out already in 1973 by Harold Urey.

That there was an impact at the time of the Cretaceous–Tertiary extinction is seen from the high abundances of heavy elements, such as iridium, found in the 1- to 2-centimeter-thick boundary layer between the deposits of the Cretaceous and the Tertiary periods. That iridium content was first measured by Luis and Walter Alvarez, father and son, and their team at the University of California at Berkeley in 1980. They found that the iridium abundance in the boundary layer is similar to that in meteorites and unlike that at the surface of the Earth. Like other heavy elements, iridium sank into the core as the molten Earth differentiated. Other evidence confirms that an asteroid or comet hit the Earth at the time of the extinction associated with the Cretaceous–Tertiary boundary. A single large crater has not been found, but it may have disappeared due to plate-tectonic movements. It is possible that there was not a single asteroid or comet but a group or a shower of them, either landing at the same time or spread over a longer period. In any case, the cloud of asteroidal matter and of surface soil or water ejected by the impact(s) was massive: It settled as the 1–2-centimeter clay layer over the Earth. The dust cloud in the atmosphere must have been thick enough to block out most of the sunlight at the surface. It stopped the weather patterns and rain-making that are due to differences of solar radiation on land and oceans. For lack of sunshine there was no photosynthesis, and most plants including algae died, depriving fish and terrestrial animals of food. The Earth was, for a while, not green. There may have been huge firestorms, as soot has been found in the clay layer. The effects of such an impact with an energy in the millions of Hiroshimas are hard to imagine. The studies that showed how the impact could have eliminated much of life on Earth also made the effects of global nuclear warfare more clear, and the frightening concept of a "nuclear winter" emerged.

We must conclude from the table that such disastrous impacts will occur again. This should not be dismissed from our concerns because of the small probabilities, as some of my colleagues argue, because the small chance is countered by the terrible consequences. Another collision with a 10-kilometer asteroid would eliminate human society; this is certain from the studies that were made for the Cretaceous–Tertiary event. An impact can, however, be avoided by changing the orbital velocity of a menacing asteroid. Preliminary studies of this problem have been made for the US Department of Defense, and they indicate that with enough advance warning, the present technology of spacecraft would be sufficient to do the job. It would take years of preparation and execution, but the same years would also be required to determine the orbit with sufficient precision. One would also want to make physical studies to ascertain the strength of the asteroid before using powerful engines on its surface; breaking it up would aggravate its effects, like turning a bullet into buckshot.

The table was filled in under the assumption that with each factor of 10 decrease in size, the number of objects increases by a factor of 100. This is true in the asteroid belt, but for Earth-approaching fragments that scaling factor is not yet known; the size distribution is among the parameters to be determined by a telescope that would be dedicated to surveying for comets and asteroids.

An asteroid telescope would seem indispensable to discover near-Earth asteroids, to learn their orbits and their characteristics. For me the astrophysical study of small primordial bodies in the solar system is the most important, but there are other interests too: concern for the hazards of future impacts, and also some preparation for future missions to these objects, with perhaps even the possibility of mining them. Studies of asteroid-mining technology are already under way at a few universities in the form of research on how to separate platinum, for instance, from the other metals on the surfaces in space.

After the failure of the plans for a 72-inch Asteroid Telescope, the dream of a dedicated facility was rekindled in the late 1970s on Palomar Mountain. A few of us had begun to penetrate that Holy of Holies For Extragalactic Studies with some solar-system observations, but we could not get enough telescope time on the Big Schmidt to suit our needs. Eleanor Helin ("Glo") would come up that mountain with excitement and persistence in the search for near-Earth asteroids that she had initiated on a smaller Schmidt together with Gene Shoemaker.[3] She was one of the first women astronomers tolerated at the Observatory and in the Monastery, although it was still considered necessary to equip her with a strong stick to block the door of her room. Charles Kowal was effectively using the Big Schmidt to discover Chiron and a thirteenth satellite of Jupiter. I was doing my own wild things in discovery of comets and asteroids described in Chapter 11. We wanted so much to do even more!

There was another force driving us toward a special telescope, the dream of Krzysztof Serkowski. Serkowski was born in 1931 in Poland, where he had lived through German occupation and the communist regime. He had decided at a young age to be either a biologist or an astronomer. Later, when he found science more lively in the West, he came to Flagstaff, Arizona, where he was known as one of the best astronomers in polarization studies of stars. Still later he had a research position in Australia. On a memorable day in March 1970, I thought it would be nice if Kris would join us in our polarimetry. Formalities were simple in those days. I called him up in Canberra, had a difficult connection, but could shout "Kris, we have an opening. Are you interested?". . . "Yes.". . . "When could you come?". . . "In three weeks.". . . Salary was not discussed, that would have been considered inelegant. He came on time and became a research professor at the University of Arizona. With Jack Frecker he modernized our Minipol photopolarimeter, but his dream was to discover planets of other stars.[4]

> Those ethical and aesthetic values which are common to most human cultures are in agreement with the notion that geological and biological evolution, progressing toward a greater variety and an increasingly sophisticated, seemingly purposeful structure, which means toward an increasing information content, is progressing toward a better and more beautiful world. Increasing intelligence and consciousness is the most characteristic feature of biological evolution. . . . If our moral principles are universally valid, we may expect that civilizations which survived the technological crisis are beaming signals toward us which may save mankind from self-destruction. Wouldn't it be worthwhile to spend one cent a day of our tax money per person in an effort to detect those signals? Such expenditure, essential for having good chances of detecting the signals, is not likely to be approved by the Congress until we know that the planets exist outside of our solar system and until we know which types of stars have planets. Therefore, we should not ignore the possibility that searching for extrasolar planets may be more important to the survival of mankind than any other human activity.

The techniques for discovering faint companions of bright stars are not simple. A future telescope in space may perhaps make a high-resolution picture of such a star and its companion, but Earth-based detection has to be indirect. A planet orbiting a star also attracts the star, such that both orbit about their center of mass. The star's orbit is very small, because its mass is so much greater than the planet's. But if one looks at the system from the plane of the orbit, and if one has a precise enough spectrometer one could observe the changes in velocity of the star as it moves to and fro in its small orbit, by observing the Doppler shift in the light it emits. That was Serkowski's idea. It required a spectrometer with much greater sensitivity than had been built before in order to measure such small "radial velocities." He also needed lots of time with the same telescope, without equipment changes, to observe the small effects carefully on several selected stars over the years such orbits would take; Jupiter, for instance, takes 12 years to orbit the Sun.

Serkowski did not worry where the funding for the telescope would come from. Gentlemen do not concern themselves with money. Ideas matter, and once these are in place, financing follows. It is true. He did not succeed in completing the project, however. During our meetings I noticed a quivering of some of his muscles. We all have such troubles when we are overworked; it is a warning to take a day off. But his were getting worse. So he came in to tell me, rather casually, that he had amyotrophic lateral sclerosis and that the doctor had predicted that he would live only three more years. He actually lived eight more years. His mind seemed brighter—it is a known effect of that illness. We also found out what a remarkable human spirit Kris had. He showed us how to live and how to die graciously. He never complained. His only remorse was that his illness caused problems for family and friends. He had a remarkable gift to uplift his visitors even when, in the end, he could hardly move or speak.

One day, David Black came to Tucson to present a lecture and to look in on Serkowski's project, for which he was the monitor in its NASA funding. He also was a special friend of Serkowski's, so he wished to visit him. I warned him it would be difficult: he would have to put his ear inches of Kris's mouth to hear a few words that might still be heaved out. Kris, however, had asked his son to put a microphone near his mouth and for half an hour he gave a lecture, reviewing the state of the art for the detection of planets of other stars, describing possible causes of systematic error, and endorsing the crew members who had taken over in his place. His style was positive and we felt cheered as we left his home, thinking that he might go on yet for a long time. Inside the house, Kris went to sleep but did not wake up. He had died within and for astronomy, just the way he wanted.

At the memorial service, a friend of Kris's reminded us of Galileo's wish that the Inquisitors on their way to heaven should come by Jupiter to see its satellites revolving. The wish was now expressed that Kris on his way to heaven might pass by a star to see its planets revolving. His project was taken over by Bob McMillan, who succeeded in building and operating the proper instrumentation.

By 1980, McMillan and I had consolidated our plans for a dedicated telescope in order to study comets and asteroids, and planets of other stars. Peter Strittmatter, director of the University of Arizona's Steward Observatory, made us a splendid offer. About 70 kilometers west of Tucson is located the Kitt Peak National Observatory and on that mountain the University of Arizona also has a station. Its 0.9-meter (36-inch) telescope had not been used during the previous two or more years, and we could have it and its dome for our programs. At first we wanted to replace the 36-inch with a copy of the Big Schmidt, as we had been dreaming and scheming to do on Palomar Mountain. Its design could be reproduced, a 72-inch mirror would again be available, and we were even offered a 48-inch corrector lens—the original lens from the Schmidt telescope in Australia, which had received a new and improved corrector.

However, at this time there were also rapid developments in electronic detectors, and we had to consider the possibility that these could replace, and even improve on, the photographic plates. After considerable study we opted for the new electronics on the basis of faith in the future. The decision was made to go for the most modern technique, the charge-coupled device, the CCD, and use the 36-inch Steward telescope.

Fast-moving objects, such as near-Earth asteroids, make trails on photographic plates; one cannot command the telescope to follow them the way one can for stars or planets, because the motions are fast and unknown. A longer exposure causes only a longer trail instead of a brighter image on the same spot, so it is of no help in locating smaller or faster objects. For observing these bodies one therefore prefers a more sensitive detector, one that uses fewer photons to make an image—has a higher quantum efficiency, as we say—and this is what the CCD does.

The electronics option had the advantage of computer processing of the data, instead of the slow darkroom procedures, followed by examination and measurement, required for the photographic plates. Still, the plates at Palomar are large, 35 centimeters square, while the CCDs are small, on the order of 1.2 centimeters, and so they would cover a much smaller area of the sky, a factor of 850 times smaller, assuming the same telescope characteristics. But they are about thiry times more sensitive and they can observe over a range of wavelengths three times greater than can the photographic plates—a gain of a factor 90. So if CCD detectors can be made larger than about 35 centimeters divided by the square root of 90, that is, larger than 3.7 centimeters on a side, they will be more effective than plates. Attempts to make CCDs as large as 5.5 centimeters are already being made in the electronics industry. One can also put several CCD in the focal plane and that is beginning to be done, the state of the art in 1988.

In the meantime, we have already made a device with a small CCD, to develop new techniques for scanning of objects in space, such as spacecraft debris, and also for astrometry, the measurement of positions.

Charge-coupled device is, for an astronomer, a sophisticated name for a simple gadget (my friends in engineering say, it is a simple name for a sophisticated gadget). Our first one has 512 rows, and each row has 320 picture elements, "pixels." The pixel, made of a silicon oxide and other materials, is sensitive to the incident light and converts it into an electric charge by the photoelectric effect. The pixel also acts as a capacitor to hold that charge until the astronomer commands it to be moved. At that command the pixels are coupled together (which is how the gadget got its name) and the charges move from one pixel to the next, in the 512 direction to the end, and from an end register they are shifted off into a computer.

Astronomical observatories generally use their CCD detectors in the "stare mode," as one does a picture camera; the shutter is opened for an exposure and at the end of the exposure, the charges accumulated in the pixels are quickly transferred to the ends of the rows and shifted into the computer, which then produces a picture of it, provided it is programmed to do so. Our use of the CCD is in the "scan mode" and I therefore refer to the new techniques as "scannerscopy," with "-scopy" pronounced as in "spectroscopy." We transfer the pixels' charges from row to row slowly and continuously at a rate that exactly matches the rate at which the telescope scans the sky. By turning the drive of the telescope off, we let the Earth's rotation scan the CCD over the field that may contain a comet or asteroid, and also stars of known position; it is then straightforward to determine the differences in coordinates. The disturbing effects of the Earth's atmosphere are minimal, since the telescope was always aimed through the same amount of air. Within the computer it is as if a slowly moving filmstrip is being received and stored for processing. The shutter is, of course, held open during the scan, which may last 30 minutes. For hunting asteroids we then close the shutter, reset the telescope to the previous starting position, and repeat the same scan. The computer superposes the scans and subtracts them, canceling and ignoring the images of stars and galaxies, which occur in exactly the same place. However, the objects that moved during the 30-minute interval are not subtracted, and so they show up in the difference image. That is how we find comets or asteroids, because they move with respect to the stars. To make the process more certain of success, we use a third scan to confirm the discoveries.

By 1981 we had a detailed proposal ready and we sent it to NASA, where it was extensively reviewed. Funding for a new 72-inch telescope, a "scannerscope," was not available, but generous support was granted to develop the new CCD techniques on the old 36-inch, which we then proudly named "The Spacewatch Telescope." It had been built about 1919 by the Warner and Swasey Company in Cleveland, Ohio, as "The Great American Telescope." It was not as large as those built for Hale around that same time, but it was the first to be made entirely within the United States, including the mirror, while the optics for Hale's telescope had come from Europe (except for the 200-inch, later). That 36-inch mirror is still there. It is the only glass mirror on Kitt Peak, because later reflector were judged to need less expansive but more expensive materials to avoid distortion of the image when changes in temperature occur, such as Hale had seen at the Hooker Telescope (Chapter 10).

The 36-inch had been planned by Andrew Douglass (1867–1962), who was the only astronomer at the University of Arizona at the time. He found Mrs. Lavinia Steward willing to pay for the telescope and dome in Tucson, so it became the Steward Observatory. For nearly 40 years it was in operation on the campus of the University of Arizona, even though Douglass had become involved in other studies—he became famous for counting tree rings to determine ages and dates for climatic events on Earth, as each ring is one year in the growth of the tree. He was succeeded at the observatory by Edwin Carpenter (1898–1963) who in the 1950s proposed to the University that the telescope be moved to a darker site because the light pollution from new homes and buildings surrounding the dome made work on faint objects impossible.

In 1958 the newly founded US national observatory decided to locate on Kitt Peak, which lies within the reservation of the Tohono O'Odham, the People of the Desert, previously called Papago ("bean eater") Indians. The Tribal Council was reluctant to have anything happen to their mountain, because parts of it are sacred in the Indian tradition. Meinel, the first director of the national observatory, conferred with Carpenter and they invited the council to look through the Steward reflector at celestial objects. As a result, the council gave the astronomers—whom they called "hegam mo ge ce, e cew wu : pui," "them with the long eyes," the colon representing the eyes—permission to locate their telescopes on Kitt Peak. It was agreed that jobs would be made available for Indian people to participate in the work on the mountain, and that their woven baskets would be sold in a Visitors Center on the summit. A part of the mountain was set aside for Carpenter's 36-inch and for other telescopes of the University in the future. Carpenter and Walter Fitch designed a great dome and selected the site with the best local conditions on that mountain. Sadly, however, Carpenter died of a heart attack just before his telescope was ready for operation, so he never got to use it up there.

The optical arrangement for the 36-inch is one that was first described by Isaac Newton and it is therefore called a Newtonian reflector; the primary 36-inch mirror, *f*/5, sends light up to a flat mirror that brings the beam out near the top of the tube of the tube high in the dome. That focus is reached with a platform that moves up and down in the dome's slit, which is opened with a large sliding door whenever the observations are to be made. The Newtonian platform is supported by a high steel structure inside the dome, at the top of which there are a ladder and doors that lead to a pulpit outside at the very apex of the dome. This may be the only dome in the world having such a high place for night-time inspections of the skies. Astronomers need to know what is happening outside, as clouds and winds can restrict the observations. At most observatories one uses a suspended walkway around the dome called a "catwalk." From the pulpit of the 36-inch, however, one can look all over the sky with one sweep of the head. It is a magnificent place from which to feel on top of the world and in touch with the heavens.

When new instrumentation is brought to the mountain, the principal investigator usually comes too, and then, in addition to the regular preparation for the night and checking the sky frequently, there is great activity of adjusting equipment and fixing unexpected problems—often with the help of a colleague or engineer who also came, or who is consulted by phone. Decisions and adjustments and a first inspection of the results have to be made during the night, and some of this continues

NOAO photo by A. Paulsen

Telescope domes on Kitt Peak, Arizona, as seen from the open dome of the 4-meter reflector. The Spacewatch telescope is the first round dome on the right.

during the daytime which does not improve the chance of having a good sleep. It is a special life, attractive to many and not at all to some (a theoretician might say: attractive to a few and not at all to most).

The trend of the past two decades has been to provide for the astronomer, and his night assistant on larger telescopes, a "warm room," where one can wear lighter clothes than a parka or electric suit. Developments in remote control and electronic data acquisition make this possible. In the future, astronomers may not even have to come to the mountain anymore. A few are already beginning to observe from home or office, communicating via telephone or microwave link with a computer and night assistant in the "warm room."

For the transmission of Spacewatch data we have a microwave link that was made by Marcus Perry with the advice of other experts; it runs from Kitt Peak to a hill on the west side of Tucson called Tumamoc, and from there on to our computer in the Space Sciences Building. Its carrier wave has been running successfully for years through weather good and bad, and even lightning storms. We actually still record many of the

The Spacewatch Telescope. The stairs lead to the viewing pulpit at the top of the dome.

Mark Sennet /People Weekly, ©Time Inc.

data on magnetic tape and ask the Kitt Peak bus driver to take the tapes to Tucson, as many as fifteen large reels. Sometimes that was a bit heavy and there were delays and protests, so I once strutted into the office of Kitt Peak's business manager, Buddy Powell, saying that maybe a letter of complaint was in order. But he is a wise man, and he said I should instead write a letter of thanks, for the transporting that had already been done. Thus we became friends, the bus crew and I, and every morning when I am in Tucson I bike by the bus just before it leaves for Kitt Peak in order to wave a cheer and a yogi blessing to Mary Trujillo the driver, Gloria Allen, the secretary, and several other friends. That bus has never had an accident.

We are in good hands on Kitt Peak. There is an excellent kitchen to provide the finest of foods for the astronomers and their assistants, day and night. Lodging is provided for daytime sleeping, which is not easy. The lodge for the Arizona astronomers is a steel-beamed structure, which is the worst for soundproofing, so one can find out who are the real observers, who use not only their eyes at night but also their ears in the daytime to hear their sleep-disturbing racket on those steel steps and the doors that are slamming shut behind them.

For observing faint objects a dark sky is needed, so that our observations of comets and asteroids are made during the half of each month centered on new moon. During the moonlit half of the month, the secondary mirror of the Spacewatch Telescope is turned away from the CCD to another port, where an optical fiber leads the light into the radial-velocity spectrometer for the detection of planets of other stars. It will be the most exciting discovery made with the old telescope when it finds an indication of another planet elsewhere from observations made over the years. McMillan is the principal investigator in charge of this program, working on it with Peter Smith, Marcus Perry, graduate student Bill Merline, and undergraduate Toni Moore; he is also co-principal investigator on the CCD work. The two programs thereby have an integration and mutual support, and they are both related philosophically to planet formation and the origins of life.

What a privilege to develop something new! To do it together with a competent crew: McMillan of course, but others too. The extensive computer programming for the CCD scanning was done by Jim Scotti, a young genius at that sort of thing. Additional help for special CCD tasks is given by Marcus Perry and Lyn Doose. Jack Frecker designed and built nearly all of our equipment, using the philosophy that any operator mistake that can be made will be made; his products therefore are astronomer-proof, which is an order of magnitude tougher than foolproof. Frecker has worked with me for 26 years, as has our secretary, Shirley Marinus. Some of our team members work on both projects, while still others help with specialized tasks, and we are supported in astrometry by Brian Marsden. It is a smoothly running team that works together in a friendly and efficient manner, sharing the excitement of observing the universe.

17. Teachers and Textbooks

It is not sufficient to merely discover a comet or asteroid. One must determine its orbit so that it can be found and studied later in detail. We use the Spacewatch Telescope for determining orbits not only for our own discoveries but also for some of those made by other astronomers elsewhere. To do this we determine a precise position as soon as possible after the discovery, and again a few nights later. The orbit parameters are made more precise by further observations over the following months and years.

The rotation of an asteroid produces periodic variation of the reflected sunlight, as we saw in Chapter 11, and its lightcurve gives an indication of how fast it is spinning in space and how irregular is its shape. The techniques for the analysis of asteroid lightcurves have been improved since those early days at Yerkes. In Tucson, four high-school teachers have been active in this field. The first was Larry Dunlap, who came into my office in 1966 to ask for part-time employment. He said it would be good for his teaching if he could tell the students about advanced research. I was impressed but taken by surprise, so to test his determination I asked him to come back in three months. He did return, and his work was such a success that soon other high-school teachers, Ron Taylor, Bob Sather, and Carl Vesely came to be trained by Dunlap in intricate routines for data processing and analysis.[1] The teachers' daily sessions with large classes of restless teenagers have trained them in patience and persistence, making them especially suited for work in research. I like to call them my "Junketeers," since some astronomers disdainfully call asteroids the junk or vermin in the sky ("minor planets are minor to me" is another witty quip). They do not understand as yet that asteroids occur everywhere in the universe, that their stars, too, were formed from them.

The work of the teachers was meticulous, and they needed only on-the-job training. Each first studied published papers to see the whole process. The new man would then be assigned tasks with an experienced team member, who would provide supervision as needed. They wrote their first preparations, painstakingly as teachers will do, in bright red ink on white paper, but these could not be read, for at night observatories use red lights, because these spoil our dark adaptation the least.

The teachers quickly became independent, selecting the objects to be observed, making the observations and processing the data, and finally publishing the results. They have published thirty-five papers in various books or periodicals, about half of them by Taylor who is continuing with one or two per year, mostly in *Icarus*, the international journal of solar system studies.

The dedication of *Icarus* was written by Sir Arthur Eddington; it is printed in each issue:

> In ancient days two aviators procured to themselves wings. Daedalus flew safely through the middle air and was duly honored on his landing. Icarus soared upwards to the sun till the wax melted which bound his wings and his flight ended in fiasco. . . . The classical authorities tell us, of course, that he was only "doing a stunt"; but I prefer to think of him as the man who brought to light a serious constructional defect in the flying-machines of his day.
>
> So, too, in science. Cautious Daedalus will apply his theories where he feels confident they will safely go; but by his excess of caution their hidden weaknesses remain undiscovered. Icarus will strain his theories to the breaking-point till the weak joints gape. For the mere adventure? Perhaps partly, that is human nature. But if he is destined not yet to reach the Sun and solve finally the riddle of its construction, we may at least hope to learn from his journey some hints to build a better machine.

The high school teachers have used telescopes ranging in size from 41 to 229 centimeters (16 to 90 inches) at the observatories near Tucson, at the Lowell Observatory in Flagstaff, Arizona, at the Cerro Tololo Interamerican Observatory in Chile, and at the Mauna Kea Observatory in Hawaii. They have had their interesting times, and nightmares too. One of them dreamed I had sent him to the Moon to observe with a large telescope, but he had come back with nothing to show, as it had been too cloudy.

One of the jobs Dunlap worked on was to improve the knowledge of the shape and the period of rotation for asteroid Hektor. After my South African observations, I had made two more at different aspects, that is, at times when Hektor occurred at different places with respect to Earth in its orbit about the Sun, namely in 1965 with the 84-inch reflector on Kitt Peak and in 1967 with the 61 inch in the Santa Catalina Mountains north of Tucson. However, while Dunlap was working on the analysis during his school's Christmas vacation in 1967 he needed urgently one more observation to determine properly the rotation and the peculiar elongated shape of Hektor. We sent a telegram to the Cerro Tololo observatory in Chile asking for one clear night. Their schedule had already been set, but the observatory was so new at assigning nights to outsiders that the 31st of April was included. The rescheduling was worked out with a cheer, I went to Chile to get Dunlap his data, and he established that the period was 6 hours, 55 minutes, and 21.115 seconds.

Such a refined analysis of observations made from various aspects not only allows a more precise determination of the rotation period, it can also give us an idea of how the axis of rotation is oriented in space, and an insight, a visual model, of that physical object. Think again of the lightcurve of the book we looked at in Chapter 11, and consider how it would be different if the rotation axis has different orientations—pointing at you (flat lightcurve), or pointing perpendicular to your line of sight (maximal variation). When an asteroid's rotation axis is perpendicular to the plane of its orbit, and if the orbit is not tilted with respect to the ecliptic, the lightcurve appears be the same from whatever direction we may be looking at the asteroid. If, on the other hand, the rotation axis lies in the plane of the orbit, there will be times when the full lightcurve is observed and times when it is flat. The intermediate cases are more intricate to analyze, and that is where the high-school teachers made themselves experts.

Taylor became an international authority in that discipline. When a senior astronomer from the Peoples Republic of China was invited to visit the Minor Planet Center in Cambridge, Massachusetts, he accepted with the wish that he might stop in Arizona to have a visit with "Dr. Taylor." He then came, accompanied by a younger associate who is applying Taylor's techniques at the Purple Mountain Observatory, near Nanjing.

Larry Dunlap's work earned him a Master's degree in astronomy at the University of Arizona, even though he had one already in education. His thesis was *A Model Study of Asteroids*.

Bob Sather invented a new method to determine the orientation of rotation axes in space from precise photometry over a range of phase angles.

Carl Vesely became the assistant of George Van Biesbroeck during his last years. Van Biesbroeck had worked full days in the office, until close to his death when he was ninety-four years old, and yet he left an unpublished legacy of his photographic positions of comets. Vesely worked on that after Van Biesbroeck's passing; four papers based on these data were published in the *Astronomical Journal* by Van Biesbroeck posthumously, with Vesely and Brian Marsden as coauthors.

I am close to teachers in my family also, because Liedeke is one, and so is our "manna-daughter," Jasna. Jasna is actually the daughter of Yugoslavian astronomer Leo Randič, but to us she came as manna from heaven unexpectedly in 1980, to live with us as a family member while getting an advanced degree and teaching junior classes in English as a

Second Language at the University of Arizona. Liedeke teaches French and Art History in an advanced high school that was started in 1977 for students who are especially motivated or have pushy parents. The level lies between that of the normal high school and college. The students are being prepared to take the national advanced-placement examinations, which give them college credit if they pass, so that they will not have to take certain first-year courses in college. The art-history program is especially exciting because each summer Liedeke takes a group of about thirty-five students and six teachers on a four-week tour through Europe. In the evenings they go to shows and discos, of course, but they also seriously study art treasures; college credit is given after completion of a trip report. The group therefore visits the inside of a cathedral, but also goes onto its roof to inspect the buttresses that hold up the high walls of these great edifices.

In addition to the observing and studying, the fundraising and administrative work I do, there is also some organizing and editing of books in the planetary sciences. These are intended to serve as textbooks for advanced graduate students. Unlike high school or college texts, which are usually best written by a single author in order to have uniformity of presentation, it seems less feasible for an advanced textbook to be produced by a single author. The results of new research are best reported by the people actually working in the field, so a text at the highest level must be a collection of such reports if it is to be up to date. Such textbooks did not seem to exist in planetary sciences when we first looked into the situation during the early 1970s. We therefore developed a new type of advanced source and textbook by adopting and modifying the strengths of two types of books—conference proceedings and compendia—that existed at that time.

Conference proceedings are usually published quickly after the workers in a given area have gotten together, with only minimal editing. That approach encourages contributions from many researchers, and thus achieves a broad, democratic survey of current topics and viewpoints in the field. It can be a fast procedure, with the results of the conference available in printed form within months. It does not, however, usually allow the time for close editing of contributions, for authors' revisions in response to new data, nor even for including what was learned from the conference itself. Proceedings usually do not have full and balanced treatment of a topic, and they rarely include a general review of past work. They therefore tend to be too imperfectly organized, repetitious in some topics and lacking in others, for use as standard texts. They also tend to be expensive and are not generally available in science libraries.

Compendia are organized by an individual editor and usually contain contributions from that editor and a few of his colleagues. Gerard Kuiper and Barbara Middlehurst organized several thoroughly edited books in this manner, but it was hard to get the extensive work for good chapters done by single researchers, who were the best in the field and therefore busy; one of their planned books did not even make it into print because of the delays and frustration. A compendium lacks the stimulus of an "open forum" conference and the reactions to different results and opinions, so we thought this type of book would not be suitable for the rapidly developing fields in planetary sciences. Also, a compendium tends to be rather autocratic, as the editor chooses authors who are often his associates, which does not engender breadth and objectivity in representing diverse views.

About 1970, I wanted to switch away from polarimetry to new studies of asteroids. The proper procedure seemed to create a book on polarimetry to summarize the field, and a book on the physical studies of minor planets to open a new one. Hannes Alfvén came into the act with his contagious enthusiasm, especially buoyed by his newly awarded Nobel prize in physics. He was interested in asteroids, so we changed the order, but both books were published in due time. Other books in the planetary sciences followed in what became known as the Space Science Series of the University of Arizona Press.[2]

These are our new kind of textbooks. I am the General Editor for the series, and my task is to plan the topics and supervise the books as they are being produced. The primary goal for each book is to enhance a discipline or to open up a new one. Each book is stimulated by a large, international, and open conference, where the topic is thoroughly discussed and chapters are aired; later, the submitted chapters are thoroughly processed into carefully structured books.

Topics may be chosen because of new developments in a field, or because a difficulty has arisen or been resolved, or because of a spaceflight that just happened—as in the case of the Pioneer and Voyager flights past Saturn, after which we made the book "Saturn." The aim is to eventually cover all major aspects of advanced research in the planetary sciences, and to update each some seven years later, with another editor to obtain a new approach. The initial suggestion for a book and conference may come in discussions with prospective editors or with any of our colleagues. We have learned to examine the qualifications of potential editors for a new book carefully because the success of the venture depends greatly on them in the planning, drive, and scientific evaluations during the process. We give the editors considerable help, however, and the actual editing is mostly done by Mildred Matthews and her assistant, Melanie Magisos.

At about the time I was switching my research interests and hoping to produce those first books, Mildred Shapley Matthews came to Tucson. Mildred is the daughter of Harlow Shapley, and she brought to our laboratory a long tradition in astronomy. Shapley was a renowned astronomer, particularly because of his studies of the Milky Way and because he was the Director of Harvard College Observatory during the early part of the twentieth century. It was Shapley who located the solar system at the periphery of the Galaxy, rather than at its center, just as Copernicus had removed the Earth from the center of the solar system.

Shapley also discovered an asteroid, which he named Mildred, after his eldest child and only daughter, but it was not followed sufficiently, and it is one of the three asteroids that remain lost.

Shapley's autobiography[3] ends with the following anecdote.

> Not so long ago, when I was going down a street in Harvard Square I met a dean, who stopped me and said, in effect, "It is your turn to entertain the Exam Club. Your paper is slated for next Monday."
>
> "Oh, no," I said, "things are not going well. I have nothing to contribute. I don't know anything. Why do you say it is my turn?"
>
> "Because it is your turn. All you have to do now is to give me the title of what you are going to talk about and I shall send out the notices. That will give you time to do some useful thinking."
>
> "All right. I will do it since I must do it. And the title of my talk can well be: 'The Scientific Blunders I Have Made.' "
>
> "Oh, no," said the dean. "Not that; it is only to be a one-hour program."

The Shapley home was a lively environment, and Mildred met many students and astronomers then, and again during her career, working as an assistant and editor.

One of the first steps in putting together a conference and textbook is to send an "idea letter" to all known researchers in the field; we ask their opinions and suggestions for the organization of the proposed project, and for the best time and place of the meeting. If, and only if, the response shows strong support do we proceed with scheduling. All colleagues who responded strongly to the idea letter, whether critical or enthusiastic, are invited to serve on a large international Organizing Committee. This committee never meets, but the membership of perhaps thirty people yields a willing and knowledgeable constituency for consultation by phone and for helpful refereeing of manuscripts. The book's editors, with advice from committee members supervise all preparations, including the invitations and information circulars to interested colleagues, the selection of topics and authors, and the allotment of time for presentations. An invitation is, however, not required to attend the meeting, and contributions to the book may be made by people unable to come. Past conferences have been major events: for instance, 183 people, 38 from outside the United States, attended "Protostars and Planets" in 1978.

When there are several volunteers for a chapter, we make an "arranged marriage." At first we received strong reactions to the joining of authors: "Don't you think I am good enough to take care of that topic alone?" Soon, however, the people realized that they needed to provide only a part of the text while still being involved in the whole chapter, and these "shotgun weddings" are now favorably regarded. Only rarely does a prima donna insist on doing it alone, and we now mostly have the opposite problem of too many wanting to see their name listed as authors, rather than leaving the discussion of their work to a few selected reviewers. On the other hand, we sometimes ask additional experts and encourage them to participate when their contribution is essential. The authors interact extensively to work out new treatments; they sometimes get together even before our meeting. In that sense, the books are study projects to integrate the topic or discipline that is selected for each book. Authors are discouraged from publishing elsewhere any text to be included in the books, or from using previously published writings, so that the books are largely composed of new material. For presentation at the meeting we also encourage contributed papers on subtopics, and these are usually published in a special issue of *Icarus* or other appropriate journals; their up-to-date science thereby becomes known to the chapter authors and is included in the book, as are the results of the discussions at the conference. The meeting can be held anywhere in the world; it is in Tucson only if our colleagues indicate such a preference. We prefer to hold these study sessions as quietly as possible, without reporters, so that the participants can concentrate on science and feel free to express themselves.

As soon as a Table of Contents has been agreed upon, the chapter authors are asked to submit outlines, and these are circulated among all authors to facilitate cross-references and minimize duplication. Authors are asked to submit preliminary versions of chapters at the time of the conference; subsequently, however, they are expected to make revisions to assimilate material from other researchers and further progress in their own work, and to incorporate referees' comments. All chapters are thoroughly and critically read, usually by two helpful and knowledgeable colleagues, and these sometimes suggest extensive revisions. Once the manuscripts are approved, they are edited to form a concise text; we make an effort to make the style even and readable—we have rewritten a few chapters. The finished material is typeset as soon as possible, and, finally, an index, a glossary, and cross-references are added. The printing is done by photo-offset and a special illustration is provided for the cover. Bill Hartmann is an excellent painter and planetary scientist both, and he has provided some of the covers, including the one for this book.

The table on the next page shows the estimated expenditures in time and money required to produce a 1100-page book in the series. Each book represents a venture of half a million dollars! We have received steady support to pay for the editing and typesetting from NASA and other donors. The University of Arizona Press supports all aspects of the production. It is thereby possible to have the price of an 1100-page book still less than $50, so that students may be able to buy it; without that support the number of printed books would be too low for an affordable edition. They are specialized volumes, for a limited readership, so that usually not more than 1500 copies are printed.

There is a perennial tradeoff between quick publication and delayed delivery by busy authors of major, time-consuming chapters. We aim to

Book efforts

Activity	Time (man hours)	Cost
Writing	4,000	$170,000
Meeting	4,000	$190,000
Processing	3,000	$120,000
Publishing	1,000	$40,000
Total	12,000	$520,000

never hold up the publication of a book for a late author; in one case we have actually deleted a chapter to prevent a delay in publication. The interval between the meeting and the appearance of the printed volume is ideally less than a year; for some, however, it has taken longer because the editors were not effective enough.

These procedures have been developed with the help of the editors of the books; Donald Hunten, Clark Chapman, Joe Burns, and others have made major suggestions for improvements. Through 1987, we have interacted with some 860 colleagues, half of them more than once in different books. We have had interesting times seeing a variety of personalities, authors, and editors at work. They are, most of them, hardworking and cooperative, some prima donnas, and a few procrastinators. I learned long ago not to be the one to argue with colleagues, but to leave that to Mildred and Melanie. If an author is late with a manuscript I might telephone him and say "you so and so, you are late again!" and that would not get us anywhere, while Mildred and Melanie have the required tact and effectiveness.

My favorite story to illustrate Mildred's ability concerns my attempt in 1973 to obtain permission for a visit to Irian Jaya, my old stomping ground. Since I had to be in Washington, D.C., for a meeting at NASA Headquarters, I tried the Indonesian Embassy there. The answer was a firm no, as that part of Indonesia was tightly closed to visitors because of internal strife. Somehow I thought of giving Mildred a call in Tucson. A few hours later, I was called out of the NASA meeting to go immediately to the Indonesian Embassy to have tea with the Ambassador, who then invited me to give a lecture at Irian Jaya's University. I have been a good sport in not asking Mildred just what she did or said.

She has become a legend among space scientists not only because she is descended from Harlow Shapley and is a slavedriver with charm, but also because of her total dedication to these books: everywhere she goes she brings manuscripts to edit, even on vacation in Italy or in intermissions of the opera.

Would this new approach to making advanced textbooks work for other sciences as well? We tried to expand into astrophysics, but the leading scientists did not seem anxious to spend their time on writing such textbooks. Perhaps our situation is different because planetary sciences involve a wide range of disciplines (geology, meteoritics, astronomy, physics, meteorology) and planetary scientists have had interactions over the years with NASA and its aerospace contractors, so that they may be more used to communicating and explaining to newcomers and outsiders.

We have also tried a start a series of books and meetings on global aspects of the future of humanity in 1973, but the war in the Middle East, followed by the oil crisis, disrupted our international contacts and cooperations of that time. But at least our planetary sciences series appears to be successful and stimulating, and is making good text and source books available on nearly all aspects of the solar system. This should be accomplished by about 1992, affording an almost complete review of the solar system and a documentation of the phenomenal growth of planetary science. After 1992, and some of this has been started now, we plan to update the major topics and reviews of the planets with a new book every six to ten years. We hope that the series will have a long-range effect on the quality of planetary research.

18. Fun, Funding, and Human Dynamics

Tucson's school district is on an early schedule, so we get up early and I am on my bike to the office, usually by six o'clock when the streets are still quiet, with only a milkman in his truck and a paper boy on his bike, still dark in the winter but bright day in summer, and glorious sunrises in between. Exhilarating to start on a ten-speed bike! Having the day ahead, with first some yoga, and a quiet time for study in an office that has cloth hangings and souvenirs from India and elsewhere. Later come the mail, visitors, and telephone calls. Weekends and holidays are spent similarly. The vacation breaks are the monthly observing on Kitt Peak and the yearly stays in India. With some work at home in the evenings, it adds up to more than 4000 hours per year, of which I spend about one-third on research, a third away at the telescope or elsewhere, and the remainder on maintenance and administration. Long hours may not mean anything, of course, because some people are more efficient than others. Sometimes I sit and think, and sometimes I just sit. My scientific papers and this book may require as many as eight rewritings, while some people can make do with only one or two—they probably are the ones with superior power of concentration.

My workplace is in the Space Sciences Building, which we designed in the 1960s with high office doors as reminders to the occupants to stand tall in science. I will describe in this chapter my daily work in that building, which is truly different from that on the Mountain.

I was hired at the University of Arizona in 1961 by Richard Kassander, who was interested in my program of scientific ballooning, and by Gerard Kuiper, who had been my professor at the Yerkes Observatory. Kuiper had had hard times at Yerkes, and when his directorship was terminated on 4 January 1960, he reacted by asking Aden Meinel to speak on his behalf to President Richard Harvill of the University of Arizona, which Aden did. Harvill was a leader who recognized that with the selection of Kitt Peak as the site for the US national observatory, Arizona would become a center for astronomy. Kuiper soon came on a visit to Tucson, bought a house, and was offered a position at the University (in that order).

Our assignment was to build telescopes, to develop new techniques and research projects, and to get the University of Arizona involved in space programs. In 1961 we were still in the Institute of Atmospheric Physics under Director Kassander, but in a few years we had our own Lunar and Planetary Laboratory, a dream we had already had for Sul Ross College near the McDonald Observatory in Texas. Kuiper had problems there too,[1] but it also was difficult to have personnel come and live in the isolation of West Texas; we tried to hire at least one engineer, Ed Roland, who lived in Indianapolis and later did agree to come to Tucson.

The title, Lunar and Planetary Laboratory, was to indicate that work in planetary astronomy would be respectable, even though Kuiper, who was the director, and Harold Johnson, Elizabeth Roemer, and Frank Low who joined us soon, continued to work at least part-time in stellar astronomy, and we did that in our polarization program too. Most astronomers still had a disdain for objects as nearby as the Moon or minor planets. All the way until the end of the 1970s, there was little contact between astronomers studying star formation and planetary scientists, who study the origin of one stellar system in depth. In 1978, I had an opportunity to bring the groups together to study protostars and planets, and that conference and the resulting book have turned into a regular dialogue by that name.

Too many discoveries had been made in polarimetry for me to continue that work alone, so eventually we established a Polariscope Group, with graduate student David Coffeen writing a dissertation on Venus and Ben Zellner one on interstellar grains. Later came Ed Tedesco, Johan Degewij, Jon Gradie, and David Tholen with dissertations on asteroids. George Coyne, who was then a young Jesuit astronomer, and Krzysztof Serkowski, Andrzej Kruszewski, and Stephen Shawl worked on stars; Tomasko had several students with dissertations on light scattered within planetary atmospheres, and we helped Joe Veverka to get started on asteroids and Santiago Tapia on galactic objects. Others followed and worked with these, which made me a grandfather in this field we had called photopolarimetry.

The University of Arizona could claim the motto *Praesidium Libertatis* as well as Leiden University. There was no publish-or-perish pressure, no need to write papers until they were ready. For publishing the data on asteroid Hektor, we could afford to wait until the observations of 1958–68 were completed. The greatest laggard was a paper on Mercury, for which the observations were completed by 1970 but laid aside until 1986, because we could not interpret the data. In 1986, a conference on Mercury clarified the situation, and Rob Landau was able to understand the puzzling observations in terms of variations in reflectivity over the surface. On the other hand, our interpretations of data on the Moon—in terms of electrically charged particles hopping around on the surface!—looked intuitively wrong even to me, let alone my peers, but that is what the data indicated, and that is what we published, like it or not (two decades later I was told that at least some of it might be correct). In another case I failed to let observation overrule intuition: In the 1970s on my Palomar plates I noticed a few nebulae that were square in shape, but I did not report them because they were so unusual. I now understand that they do exist and that their square shapes have been interpreted.

In my job I am free to leave and return, free to spend concentrated periods of study without interruption, and free to choose research topics or get involved in any issue of my liking. I have never forgotten Chandrasekhar's example of switching to a new topic every nine years or so. Most of my colleagues do not agree that is a good idea, and they stay in a field to become its experts. I realize the risk of not completing a topic of study in the period of nine years, not exhausting it, as Chandra usually does with a definitive book written on the subject. If the topic is not exhausted, there may be a shallowness to the next effort, which may be merely an addition while one still sometimes returns to the previous work.

There is always stress of wanting to do better in science, with advance planning of the next days, weeks, and years of work. I love this pressured life. Always more to do and new ideas to test. A favorite game is to leave problems written out on the desk at the end of the day; a clear statement of a problem invariably brings an answer the following morning. A side effect is a racing mind, in the shower or wherever, or sometimes waking up in the midst of the night. New solutions must quickly be jotted down or they may not come back for weeks. On the other hand, some problems need to be massaged, elaborated on sheets of paper, discussed with others or read about in journals, sometimes laid aside in frustration, while at other times the answer comes through unexpectedly from some intuitive combination. Unhappy is the occasional day when nothing seems to move; this stymied agony is miserable; at such times I cannot even bring up the willpower to go hike the mountains.

At the university, even senior people like myself are evaluated each year. If a select committee of the laboratory considers the performance insufficient, that opinion is clearly relayed, and the dissatisfaction may be expressed in a lack of salary increase, or in the allocation of adverse office space. Sometimes these committees have expressed dissatisfaction over my not teaching many, if any, classroom courses. But it is difficult to combine teaching regular hours on the campus with observing on a distant mountain—sometimes as many as fifteen nights per month. I could have a much easier life by occasionally using existing equipment at the telescope rather than conducting projects to develop new techniques. Actually, I do quite a bit of lecturing, especially in India and often to senior people, who in turn lecture at their own institutions.

The teaching load in our Department of Planetary Sciences is light because, with a few exceptions, only graduate courses are taught. I have helped to teach "The Universe and Humanity; Origin and Destiny" for undergraduate students; this course does not have a prescribed syllabus and leaves its instructors free to roam over a nearly limitless range of topics. At the end of the course, the students evaluate it, anonymously, and I have enjoyed reading remarks such as "Now I understand our human predicament better," or "This man has changed my life!"

Our research is paid for by federal funding from agencies such as the National Aeronautics and Space Administration, and the National Science Foundation. There also are military sources, since the US Navy, Army, and Air Force have offices that support research in astronomy and planetary sciences because it may have a military application, such as orbit determination of artificial satellites. Encouraged by the views of Arthur C. Clarke (see Chapter 14), I am glad to provide information, because more knowledge, spread widely, can only increase the chances for peace.

Sometimes the military supports basic science without a military application. I have asked the monitors for their views on that, and they answered that it is important for the United States to have a strong base of fundamental science; I have also heard that such support gives the agency contacts in universities and other laboratories that may be useful should other needs arise. The first federal support of astronomy at an American university was in 1946 by the Office of Naval Research; it helped me get started in polarimetry. In 1974 I had an opportunity to dedicate our polarization book in the Space Science Series to the monitors of our grants and contracts; to be specific, Jean R. Streeter was singled out for the friendly professional manner in which she monitored the astronomy program at ONR (1950–70). On one of her visits, about 1964, she noticed that we were still using slide rules, mechanical desk calculators, and tables for logarithms and trigonometry; she offered to pay for our first electronic calculator, a large contraption on its own stand by the desk, but it could be taught a simple program to execute.

In 1983 NASA began to run into financial restrictions for planetary research at universities, and we had to try our hands at raising funds from private sources. In despair we tried crazy schemes such as placing an advertisement in *The Wall Street Journal*, and writing to Elizabeth Taylor my admiration and the offer to name an asteroid after her. Eventually, a constituency was developed of about 60 donors for whom we publish occasionally *The Spacewatch Report*. The Spacewatch search for impacting asteroids has popular appeal because of the connections with extinction and evolution. Donations are mostly in the $10–$100 range, a few as much as $1000. Barney Oliver gave us $103,800 as well as good technical advice. The Callahan Mining Company keeps sending a yearly contribution, even though it has met with hard times in the mining industry. And then there is Greg Goebel, an engineer with Hewlett–Packard; he and his company have donated equipment. All gifts are tax deductible and administered by the University of Arizona. We have raised about 300,000 dollars, including the value of donated equipment, and this manner of support is continuing.[2] Spacewatch thereby belongs to many people who follow its progress and results.

Much of my "maintenance and administration" time is for such fund-raising and the writing of proposals and reports to the supporting agencies. We are envious of the scientists in West Germany at the Max Planck Institutes and those in India at research organizations such as the Physical Research Laboratory. Once these people have passed a trial period, they do not have to concern themselves with where their salary and research support come from, and for the rest of their lives there is less pressure on them than there is on us to perform and raise funds.

On the other hand, writing a proposal is not a bad chore, as it requires us to carefully think through the planned work and to review what has been done. A cover page shows the topic, the proposers, and the necessary signatures. An abstract, a budget summary, and a table of contents provide a little more detail. Then we come to the background, objectives, and approach of the proposed research. The proposal closes off with descriptions of personnel, facilities and budgets, and attachments, such as the boastful listings of the accomplishments and publications of the investigators, and reprints or preprints of papers in the area of the proposed investigation. The thick package is nicely bound, with perhaps a picture of the research instrument or facility on its cover.

At the Lunar and Planetary Laboratory—and I am also associated with the Steward Observatory—I work within a group of about twenty people who share their interests in science and in the use of our Spacewatch computer and other facilities. The ensemble is vaguely subdivided into three groups headed by Research Professor Martin Tomasko, Senior Research Associate Bob McMillan, and myself. Of the twenty I am the only one whose salary has been supported over the years by the State of Arizona. Of late, Tomasko also receives his salary from the State—this is called "hard money"; "soft money" comes from grants and contracts. Our salaries are for the academic year of nine months plus one month vacation; we may write proposals to appropriate agencies for work to be done during the three summer months. The salaries of the other people and the cost of all travel, equipment, operations, secretarial help, and "overhead" have to be obtained from such grants and contracts.

We write perhaps twenty proposals a year to a variety of agencies and their subdivisions; about fifteen are successful. Between 1961 and 1988 we raised a total of at least eleven million dollars in this way; that includes two space projects we executed in cooperation with the Optical Sciences Center. From these funds, the University collects an overhead, which rose gradually from 25% in the 1960s to the present 47%; this goes for expenses ranging from utilities to library facilities.

The funding agency sends the proposals out to referees, sometimes as many as twelve (!), who are expected to be expert, objective, helpful, and prompt in their replies. Occasionally they are anonymous, competitive, and excessively slow. The uselessness of extensive review of proposals has been documented in an article in *Science*.[3] Other scientists, such as Ernst Öpik,[4] have also protested against anonymous refereeing, arguing that a critic can say what he wants but remains unaccountable, while the author of the paper or proposal has no chance to reply or to remove a referee's misunderstanding before the damage caused by a negative opinion is done. Especially if the review is critical it should not be anonymous. The greatest objection to having many referees is that, by definition, out of a plural evaluation comes a common denominator, mediocre science. Indeed, the referees may not have heard of the new imaginative idea, and they usually do not have the time to study what is being proposed, so they reject it as unlikely to be successful. Some of the best science projects have been judged by referees to be impossible. We are keenly aware of that: In addition to Serkowski's instrument, Polariscope, Multi-Mirror Telescope, and Spacewatch were judged, at an early date, to be technically impossible. We therefore tell our students that an evaluation of "impossible" may be the best indication that you are far ahead of the heap. Explore the idea with zest! I learned that philosophy first in the Soviet Union, with its even more extreme bureaucracy: what they say is "impossible," those are the interesting ideas to pursue.

Monitoring grants and contracts cannot be a challenging job any more, when the decisions on a program are relegated to the votes of referees and committees, and the best people may no longer want to be monitors. A competent grant monitor should not need more than an occasional consultation with an expert on a specialized topic. The monitor usually has had previous experience with the proposed research and the proposing scientists, and knows the investigators and their students from visits in the field; he or she has learned who would be best able to perform a new project.

Edwin Snider is one such monitor for the National Geographic Society. We have known him a long time, for he was the one to take on Serkowski's search for planets of other stars when the other agencies' referees thought it premature. Later he supported a search with our Spacewatch equipment for satellites of asteroids: I had great hopes for discovering a new class of even smaller objects orbiting around the already small minor planets. We had already chosen a name for them: minor satellites. It seemed reasonable to expect that in the dust cloud of the early solar system some small planetesimals might have found stable orbits around larger ones. But we discovered no such pairs and I wrote my report in disgust, not understanding why we had failed; this was my final report, we had nothing further to say, and I could not write a meaningful paper in a scientific journal. But Snider knew that if I would just calm down and work out a paper, reporting everything we had done in detail, surely some further clarification would emerge. And indeed, so forced, in his gentle and gentlemanly manner, we gained insight as the work progressed. With the help of Jack Drummond we comprehended that the many collisions in the asteroid belt knock any satellites out of stable orbits, so that they could not stay bound to their primary asteroids. Jack found densities and collision rates in the literature to make the estimates more precise, and we produced a useful result. It was just a matter of putting in more work, as Snider had known from his previous interactions with projects and people.

In the wealthy days of the Kennedy and Johnson administrations, federal money was freely available for research. I remember one of our proposals for which the budget already was inflated, but a telephone call came from Washington, telling me to airmail a new budget with every item multiplied by a factor of 1.5. When the Steward Observatory proposed to build a 60-inch telescope, it was suggested that they turn the "6" over, which they did.

The 1980s, by contrast, have brought science funding at universities to a sorry state, when young people have to face declining budgets, and when Star Wars and refurbishing old battleships are judged more profitable than promoting scientists or the space program. The effect is seen in a flight from science by bright young Americans. We have, for example, had feelers and an advertisement out to hire a young Ph.D. to enter a research career and lead us into new experiments of his or her own invention. We emphasized two things: that the salary standards at the University of Arizona are relatively low, and that the job requires much observing with telescopes on the mountain. Such a position has a starting salary of about $23,000 per twelve months; our highest paid researchers make somewhere near $65,000. No one has yet walked into our tall doors saying, "This is the job I must have." Instead, there have been hesitant phone calls with a first question, "How much is the salary?" or apologizing by saying that the present employment is ending and that the candidate now has to look for something else. Excitement and curiosity seem to have been replaced by fear of review.

Another aspect to the discouraging trend is the tendency to do only safe science. This was especially noticeable during the mid-1980s when many proposals were made for the Fad of the Day, Comet Halley. That was a safe topic because of the public attention that reappears with this comet once in seventy-six years. The resulting waste on Comet Halley makes one cry for our scientific conscience: For instance, some ten good position measurements per month would clearly have been plentiful, but there were months when more than 600 were published.

It is sometimes argued that the increasing emphasis on administrative procedures is an unavoidable trend of the times. But the trends take over only if we let them. If we have the will, we can cut out unnecessary rules, laws, forms, meetings, committees, procedures, regulations, refereeing, and administrative positions. In our daily lives we can point out excesses, challenge them, and promote awareness and elimination of unnecessary paperwork. Such a peaceful but powerful protest, if it were done by many people, could gradually reverse the increasing trend. I call it the Law of Human Dynamics that the drive toward human freedom forces us to diminish bureaucracy; the converse is also true.

I refer to this law in refusing to attend unnecessary faculty meetings, to serve on wasteful committees, and to write unneeded or anonymous reviews. It is fun, and a challenge anyway, to tackle the stifling bureaucracy. As astronomer Sidney Wolff has written[5]

> We must find ways to restore that willingness to take risks, to accept the possibility of failure as the price of being on the cutting edge of science to encourage the next generation of scientists to choose the more difficult course of action.

Planetary scientist Gene Levy once told an international audience how he had to talk for four hours to convince an influential member of a high-level advisory panel—a Nobel laureate—of the importance of asteroids, so that the panel would include a recommendation for studies of asteroids in its report. What if Levy had not been there?

Too many advisory committees have sprouted—for NSF, for NASA Headquarters, for the NASA centers, for individual programs—and the mavericks, the exceptional talents, the innovative researchers are rarely on them; they would not be chosen because they are not team players, or they would not accept because they prefer to do their research.

I believe that the underlying cause of the weakening of NASA after 1969 lies not just in a shortage of funds, but also in an affluence of words. In its early days NASA was ruled by strong administrators, as the Soviet space program is today. But then came the era of more reviews. Advisory panels have given uncertain or conflicting recommendations. Competing programs and missions put out illustrated and expensive brochures to influence the opinions of the panels and administrators.

An advisory committee discussed in 1972 how an existing Pioneer-12 spacecraft, modified for imaging, could make a flyby mission to several asteroids and a comet.[6] It would not have been as refined as the Soviet–French mission planned for the mid-1990s or the American Comet Rendezvous–Asteroid Flyby mission later, but it would have been done in 1974, a quarter century before CRAF, for an estimated $40 million. However, the Pioneer was hung in a museum in Washington, while subsequent study groups did not even refer to the earlier reports. All the panels, including the ones I was on in the 1970s, considered only the large, numbered asteroids, never even mentioning the hundreds of smaller ones that would come into the field of view of such a mission.

At the University of Arizona for some years now we have made futile attempts to give the secretaries and the telescope personnel a deserved raise. The raises have been repeatedly refused by administrators who ruled that their system and description of ranks does not allow it. The rank system has become so complex, especially for telescope operators, that it takes much time—wasted for research—to fill out the forms and to explain to those administrators the duties of our experienced secretaries, who after many years have become assistants in our research, and what it is we do on the mountain. Finally, we were told the "good news": an expensive consulting agency has been engaged to look into the situation. Anyone of us could, of course, solve the problem quickly and well (as I learned from Charlie Hall), by simplifying the system.[7] The consulting agency will, most likely, do the opposite. Its first act was to demand that every staff member of the university fill in a ten-page questionnaire. The painful irony is that these impediments to our research take time and resources away from our research. We need to apply the law of human dynamics.

University administrations evaluate us, but we do not evaluate them. Researchers build up the university's reputation and bring in millions of dollars in grants and contracts, but we have little say in deciding the amount or disposition of overhead taken from the funds, in deciding salaries or rank of our personnel, or in judging the performance of the administrators, the deans, the vice-presidents, or the president. I am not suggesting that we do that; I am suggesting that we all throw out more paperwork. The size of the administration at the University of Arizona steadily increases—four new associate vice-presidents are being installed—without a corresponding increase in students or faculty, and there is no opportunity to register dismay at the added paperwork the new administrators create.

At the Kitt Peak Observatory the director has to report in detail to a board consisting of about forty astronomers and administrators. Just getting them together in any one year costs the equivalent of a technical position! Directors have commented, even in writing, on the negative aspects of too much oversight; for at least one of the directors it was a major factor in his resignation. But the Board persists. In the mid-1980s the "good news" was to have an additional layer of administration, with a Super Director over and above the directors of Kitt Peak, Cerro Tololo, and the National Solar Observatory.[8]

The law of human dynamics implies that the oversight could be reduced. At the Yerkes Observatory, Director Otto Struve used to receive, once per three years, a conspicuous envelope from the administration of the University of Chicago. In it, he knew, were faculty ballots for the directorship. He would shove the envelope unopened into the garbage can, to save his faculty more time for research. This story was told to me with glee as an example of what a good director can do. As for Kitt Peak, it is certain that more astronomy could be done with the time and money saved from reduced bureaucracy. To astronomers the highest priority seems to be to again open the Kitt Peak telescopes that are now mostly closed during July and August. Such closures make a mockery of our competing for time on these telescopes, rushing to the mountain, not going to bed at 3 AM though we may be exhausted, and risking health problems or even death. It hurts to see those closed domes during some of the finest nights of the year.

And there are beautiful nights for observing! Southern Arizona has a good climate—of weather and people both—even if it is not the very best place for astronomical observing. It is not as free from clouds as the observatories in Chile or in Hawaii. Even so, with the Spacewatch Telescope we can observe almost 70% of the available time in a year, provided that some of it is done between the clouds. Harold Johnson has determined that the sky is totally clear, "photometric," about 25% of the time. The desert heat will bring winds that produce dusty air, particularly in the springtime. Some of our best nights actually occur during the monsoon of July and August when there is, on average, a cycle of some five days' building up of cumulus clouds over the hot desert and the mountains, five days of violent storms (but we can sometimes observe during the later parts of these nights), followed by five nights when it is remarkably clear because the dust had been rained out, and the seeing is good because the desert is wet and relatively cool, with a stable temperature gradient above it.

But there is a negative side, too. As cities grow and the desert is developed, the light pollution gets worse; when I return to the mountain each month I see new lights on the desert. Tucson is becoming brighter, partly because of the mistaken notion that more lights bring greater safety. Our neighborhood is one of the darkest, without street lights, even though it is near the geographical center of Tucson, while it is also one of the safest, with burglaries rare. We studied that situation in a neighborhood committee, prompted by the increasing pressure on gullible citizens from the utilities for more street lighting, a worldwide phenomenon. Our committee's conclusion was that

> . . . if people feel the need for more lights, it is best to have them right at the house, on front and back porches, illuminating the home and its windows. A light on the street may cause shadows of trees and bushes near the house, and these are ideal for a burglar to hide in; streetlights have a blinding effect for the people inside the home who may have heard something and wish to look out. But if light is shining near the house itself, the burglar has to look into it and *he* loses his sharp night vision.
>
> The blinding effect is also noticed for those of us who like to take walks, for which our neighborhood is so nice. Bright lights shining onto the street make the walks less enjoyable, make it harder to see the evening twilight and the starry Arizona skies.

Southern Arizona is a glorious place to live. It is great hiking country, with trails into the mountains in all directions. There are surprisingly few people on them, so close to a city of about half a million inhabitants. Within a ten-minute drive and a twenty-minute walk one can be almost alone among steep mountain walls and rugged canyons.

One night Jo-Ann and I slept on Picacho Peak, on the very top, which is so narrow that we were afraid that our roaming dog, Pong, might fall off. We have a favorite camping place in the Santa Catalina Mountains where she slept one night when she was only about two years old. When we were nicely tucked into our sleeping bags the kid talked about a bear above us and I thought that was amazing, for the ancients also had seen a bear in the Big Dipper area, but from what she said later it turned out to be the shapes of the branches in the trees that looked like a big bear. The child saw things nearby better than I did, while the majestic mountains in the distance seemed to make little impression.

The mountain canyons have enough water running during parts of the year to nourish brush, weeds, and cacti. In the foothills, the straight Saguaro cacti dominate; many over a century old and more than 10 meters tall. They are found mostly on the south slopes, as they are Sun seekers.

In the mountains there are volcanic outcroppings with round rocky bombs that must have come from deep down below, cooling rapidly when they penetrated the surface. Other mountain ridges were pushed and buckled against each other by the movements of the Earth. One can see gashes on the slopes that have been made by miners looking for copper and other minerals in years past.

What a life, so close to my Mountain with merely one hour driving over open roads, radar police the only hazard, and always something of interest! Along that road are crosses for Tohono O'Odham people who died in traffic accidents; some of them have a little oil flame at night, so relatives must come from afar to maintain them. There are cows and horses, rabbits and coyotes, and snakes and tarantulas near the desert road, while higher in the mountains one can encounter especially at night a coatimundi, bobcat or skunk, squirrels and raccoons, foxes and deer, field mice and javelinas, lizards such as the Gila monster (whose name makes it seem larger than it is), roadrunners and a great variety of other birds, and even a brown bear with her cub has been sighted. California woodpeckers stop by during their migration each spring, and they like to rest on the windowsills of our lodge on Kitt Peak. Just when the astronomer inside is nicely asleep after a long night and maybe snoring a little, the woodpecker will turn around and what does he see in the glass of the window? Another bird! Which then, of course, must be fiercely attacked, with hard woodpecker's beak against the glass. The other fights back, and so it continues, making the astronomer startle up and out of his sleep. This is made up for, however, by a close view of the fierce attack, because the bird sees only his reflection and not the astronomer's head. A small strip of anti-reflecting coating on the glass at a woodpecker's height has brought the sleep and peace back again, but it has ended the chance of seeing a unique attack.

Not only am I surrounded by this grand country of flat desert with mountains going up to 7000 feet (2100 meters), as Kitt Peak does, and a few even higher, but it is also a great region for astronomy. There are about twenty-two telescopes around Tucson and at least a hundred astronomers living in the city or visiting from everywhere, astronomy being an international trade. Here scientists and engineers are pushing the frontiers, experimenting with new types of optical and electronic equipment, some of it designed for spacecraft, to explore the stars, moons, planets, comets, nebulae, galaxies, and asteroids. We are asked sometimes if being an astronomer is difficult. It is not, no more than being a specialist in anything else. An auto mechanic who can fix automatic transmissions, for instance, has a difficult job that must require years to become expert in. As with any field, one grows into it gradually. As an old soldier I see a similarity of research projects to military campaigns: choosing the openings and approaches, advancing with all power and persistence, having a few breakthroughs among the standoffs and defeats. This is, of course, a general's or historian's perspective.[9] The death and suffering are fortunately missing from the academic world—and when I walk on this large campus with so many young students I hope we can keep it this way in the world at large, too, remembering Sep Postma and other young ones wasted in earlier wars.

A science writer once asked to what extent we researchers are stimulated to greater performance by working within a university of excellent reputation. The answer to that is a clear affirmative, for such a place makes fruitful cooperation possible. It was, for instance, the cooperation of the Steward Observatory with the Lunar and Planetary Laboratory that made the Spacewatch programs possible. There also is an emotional uplift from under the sunlit palms of a campus where 33,000 students are brought together with a professional faculty and first-rate facilities.

Yet the choices of methods and of topics of study are my sole responsibility. The environment that stimulates also brings a pressure to do better next time, by choosing more promising projects. The stress is usually good, and there is euphoria and excitement as well as disappointment, such that the job is a paid hobby of having fun with facts and thoughts. To suddenly see something new that I did not previously understand and that no one understood before is the greatest stimulation. There may be too many ideas, the execution may be incomplete or superficial, but that, in the end, is accepted too. It is a privilege to be an astronomer!

It is good, however, to get far away occasionally, to quiet down and see life and research more in perspective. My first experience of that was in France, where Audouin Dollfus invited us to spend the summer of 1977 at the Meudon Observatory outside Paris, with its research quarters in a forested park that used to belong to a palace. But as we arrived, what a shock to see, already posted, a public lecture by the visiting Great Professor, in French! I cannot do that. To solve the problem, we developed a method that works in any language. I lectured into a recorder in English, exactly the way I would have spoken, including requests for the next slide, etc. Audouin listened to that, and he spoke in French into his recorder, and his secretary typed it phonetically in capital letters, large enough to be read from a distance, grouping the words the way one would speak. All I needed to do was practice a few times, listening and reading simultaneously. It went so well that I tried it later also in Spanish, which I know hardly at all, in Caracas, Venezuela. The audience wondered about my Caracas accent—sure enough, for Humberto Campins had been my coach and he was born there. It would be fun to try it next in Russian or Gujarati.

The most popular Gehrels in Paris was Liedeke, because she really speaks French; she became the favorite, just as Jacqueline Kennedy had been. One Frenchman said to me in awe, "Elle le parle comme une Parisienne," "She speaks it as a Parisian," which is the highest accolade. Then he flirted with her, like a Frenchman should, and promised to let her into the Cave of Lascaux. To me that sounded like a boast, because only five specialists are allowed in per day. But the permission came promptly, and a few days later we stood in the fog of Southern France in front of a tall gate, duly impressed by the instructions which stated we could not tell anyone, before or after, nor speak inside the cave, and that we must be out within half an hour. One of the discoverers of the cave, Jacques Marsal, lives there and eventually he came out to let us in. But not Jo-Ann. Too young. Well. . . . A little while later, he growled that he would count her as an adult—quite an extrapolation for a nine-year-old. Maybe he had gauged our determination not to be split up, or, more likely, Jo-Ann's charm worked on him as well as it did on airline pilots who invite her into their cockpit. Inside was a miracle. Fifteen thousand years ago, some master, a Leonardo da Vinci, must have entered with his apprentices and laid out the colorful paintings of varieties and groupings of animals all over the walls and ceilings, so real and three-dimensional that the nine-year-old had to exclaim out loud and almost touch and ask questions, which the Frenchman could not resist answering, and we stayed in for an extra hour. Curiously, at about the same time of fifteen thousand years ago, people were making the same type of paintings in the Bhimbetka Caves near Bhopal in India.

My lectures usually involve only scientific problems, but once, in 1983, there was a moral problem as well. I was invited to give the yearly von Braun Lecture in Huntsville, Alabama, in memory of Wernher von Braun who did so much for the American space program in its early stages. At first I fired up a reply saying I could not do that, I would have to rise to a higher spiritual level of freedom from bias before I could honor von Braun. I rushed on my bike to deliver that letter to the Post Office . . . but did not mail it. For now if ever was the time to rise to that lofty forgiving level. My problem was explained at the end of the lecture:

> As considerate people and especially as scientists we try to be free of bias and prejudice. But none of us is truly free. We all start with our education and inheritance as background. Mine was in the Netherlands in World War II; I grew up in Resistance, escaped from the German occupation, was trained in England as a saboteur, and parachuted back into Holland. I have had close encounters with the Peenemünde rockets which have been described, for instance, by von Braun and Ordway in *History of Rocketry and Space Travel.*[10] Friends and family suffered terribly, and some of them perished. My brother died in one of the Nordhausen camps. How then could I

> speak in memory of one of the leaders of Peenemünde? Deep emotions had to be overcome.
>
> But I did overcome. And I could speak in memory of Wernher von Braun, because he was a dedicated pioneer and laid foundations for our space programs. It is our inheritance to stand on his shoulders as we continue to build a future of space exploration.

Had I really overcome? The original version of this talk said that I could speak in memory *and in honor* of Wernher von Braun.[11] Later, however, I learned that he had visited the underground factory of the Dora–Nordhausen concentration camps. Only in the later 1980s did I learn that his close associate Arthur Rudolph was the director for production of the V-2 rockets at Dora–Nordhausen. Rudolph came with von Braun to the US, but was deported in 1983 when it was revealed that he personally participated in the procurement of concentration-camp laborers with full knowledge of the grotesquely inhuman conditions. The prisoners were made to dig tunnels by hand, during long days, beatings and torture and hangings were common, and there was no medical treatment. There was, in fact, medical experimentation. No wonder it took us and the world so long to find out, for who could even believe at first such large-scale cruelty? An estimated 20,000 of the total 60,000 population died, which does not even take into account the ones who were murdered on the way to the camp, as was described in Chapter 5.

Even in Dora–Nordhausen the human spirit prevailed. The challenge for the prisoners was to sabotage the rockets. And they were successful; the launches in Holland were seen to fail frequently. In the Nordhausen factory, however, the sabotage had to be carried out while bodies of some who had been caught were hanging from the ceiling. All hangings were during roll call. The challenge then was to shout one last defiance. The hangmen reacted by gagging the convicts with a stick wired tightly back in their mouths. These little sticks are still seen in the museum at Dora.

Regarding the V-2 rockets' results, von Braun and Ordway write: "The V-2's took their toll. More than 1500 V-2's were reported to have landed in southern England or just off its shores, and they were responsible for more than 2500 deaths and great property damage." No mention of the 20,000 dead at Nordhausen. von Braun could write this? In 1975? In 1987 I made my pilgrimage to Dora.[12] I learned then that von Braun's own brother had a high position in that gruesome place. Why glorify von Braun and his henchmen, while to the best of my knowledge no book has been written about Hans Oster or Count Klaus von Stauffenberg and other German heroes who were willing to resist and sacrifice? The excuse is usually made that Hitler and the SS were in control and responsible. The critical question is whether Rudolph and the von Brauns tried to do anything to help the prisoners and to diminish the rocket barrage on England and Antwerp. And the answer is, no. Had I realized the situation more clearly, I would have refused to speak and recommended instead that the von Braun Lectures be discontinued.

Ernst Stuhlinger was my host in Huntsville. His own life story is interesting. Not wanting to be involved in nuclear research for the Third Reich, he let himself be sent as a private soldier to the Russian front. After some time, however, the system found out about his Ph.D. in physics, and he was then sent to von Braun at Peenemünde. Stuhlinger's view is that if I had been born in Germany and von Braun in the Netherlands, our roles would have been reversed. This may be true, especially if I had had the dream of rocketry as grandly as von Braun. Then again, there was some German Underground.[13] And would I have neglected in 1975 to mention the 20,000 deaths? The Hitler Youth in its training included killing animals with the bare hands. I hope that I could not have done that. And if I had been in von Braun's place, a visit to Dora would have turned me off. Still, "What if I had been a German?" remains a haunting thought. The moral of my von Braun story is that none of us can really be free of bias and background.

My travels to India had succeeded in getting my spirit raised to a high level, as the remainder of this book will try to describe. But in my condemnation of von Braun I have failed to be influenced by India enough to rise to a Nirvana remoteness from human strife.

The sojourns in France, Sri Lanka, and India did bring plans for new research. The development of the Spacewatch Telescope was using my full attention, but I wanted to switch to the outer solar system, our new frontier. Are the extinctions of species occurring periodically, as a few colleagues have suggested, and not at random as the table on page 191 implies? If so, what would cause such periodicity? Are there, well inside the Oort Cloud, other planets or "Chirons," or another cloud of comets? In fact, how well do we know the membership, the completion of the solar system?

Let us now look at those places for gaining perspective and for turning within: Sri Lanka and India.

INDIA AND BEYOND

People and Science

19. Sri Lanka's Telescope and Arthur C. Clarke

Independent Sri Lanka, which was called Ceylon while it was a British colony, is the island country south of India, a beautiful place of white sandy beaches with tall palm trees and cool ocean breezes. Deeper inland one can travel to mountains covered lushly green with tea bushes or rubber and banana trees, villages in between, people and animals everywhere, and children waving at tourists passing by on busy roads. The major exports are tea, gems, spices, and rubber and coconut products.

Colombo, the principal city, has wide streets except in the crowded old harbor district; in the days before airplanes that harbor was the main connection of the island with the outside world. A large grassy green lies along the beaches to the south of Colombo harbor, by the main road along the coast, and here people come to stroll and fly their kites; it is called "Galle Face," as the road leads to a small harbor town called Galle, where some descendants of Dutch settlers still are living as *Burghers*.

Some 74% of the sixteen million Lankans are Sinhalese who speak Sinhala and are primarily Buddhist, while 19% are Hindu Tamils from the south of India; some of these came to Sri Lanka many generations ago, others were brought over during the British raj to work on the tea and rubber plantations. Relations between Tamils and Sinhalese are hampered by differences in background, religion, poverty, and appearance, the Tamil people having a darker complexion. The Tamils are seeking some autonomy or possibly independence in a seaboard strip of land on the north and northeast of the island.

The island has had other names during its varied history, and it has seen many invasions. The first recorded history or legend begins about 2500 years ago when Prince Vijaya and 700 followers came from India and settled in the north. In modern times there were three periods of colonialism, each lasting about 150 years, first under the Portuguese, then the Dutch, and finally the British, who granted independence in 1948. An old Arab name for the country is Serendib, and I understand that the word Ceylon is from old Dutch maps, where the island was labeled *Zeil an*, which literally means "Sail onto." It was a stop for the sailing vessels during their global explorations.

The north, with the cities Anuradhapura and Polonnaruwa, is culturally of the greatest interest. There are Buddha statues several times larger than life and pyramid-sized earth-filled structures called stupas or dagobas, which have a square base symbolizing Earth and a hemisphere for a roof symbolizing Heaven. In many places one sees outlines or partial ruins of palaces, and more are being discovered and excavated every year. There also are great waterworks of canals and storage lakes called tanks that sustained a large population long ago; they are being restored and are again sustaining a growing population and new agriculture now.

When I traveled around the island in 1945 on my soldier's motorbike or hitchhiking with military vehicles, it was overgrown with thick jungles all the way to the east coast, but the increasing population has been cutting the forest back, so that the jungle remains only in a small area against the southwest side of the central plateau. Sri Lanka's mountains do not go high—the highest, Pidurutalagala, reaches 2524 meters (8281 feet)—but it is a delight to travel through these carpets of rich green growth on steeply undulating terrain that is watered aplenty by the monsoons.

Adam's Peak is a pointed mountain of 7360 feet on the southwest side of the central plateau, and a point of pilgrimage for people from all over Sri Lanka, young and old, and from all religions: Muslims believe that after having been expelled from Paradise, Adam stood there on one foot for a thousand years after which he was reunited with Eve on Mount Ararat. Hindus believe that Lord Shiva stepped off this peak into heaven; while the Portuguese thought it was the apostle Saint Thomas. One can indeed see what looks like a footprint in a rock at the top. The Father's Path is a hard and long way up from the west, while the Mother's Path is from the east. We climbed the latter under the guidance of Father Mervyn Fernando, Catholic priest, educator, and amateur astronomer, who taught us about his beloved Sri Lanka and its people with a fine-tuned feeling for its various cultures and religions. The path is lit with electric bulbs, such that from below this illuminated stairway up and up is a dream of Jacob's ladder ascending into Heaven. Stone steps have been arranged up the peak, for kings used to be carried up. By tradition, one climbs during the night to see the sunrise from the top. And what a sight the rising sun was from that holy mountain! It was a clear day and we saw the shadow of the peaked profile on the hazy jungle below. People were praying. The reward for the hard trip was this holiest moment in their lives. But not for long could they keep quiet, as they felt compelled to cheer and laugh and ring the bell; we rang it once, but Father Fernando five times, for this was his fifth visit. In the end, everyone skipped or stumbled back down that stony stairway through the monsoon-watered jungle, the fit and the crippled, the young and the old, and mothers with babes, all enriched and jubilating with happiness.

The lush green of the island is granted by two monsoons a year. In the summer the heating of the continental areas of Tibet and India causes the air above them to rise, as hot air will, and as it rises it pulls in cooler air from the south, from above the surface of the cooler ocean. As that very humid air is pulled over the mountains, the moisture condenses and comes to the ground as life-giving rain. However, the monsoon does not come straight from the south, there is a curious deviation, the Coriolis effect, named after the French physicist Gaspard de Coriolis who first described it. At the equator the circumference around the Earth's axis is greater than up north; as the Earth rotates to the east, the air surging from south to north overshoots the smaller circumference toward the east. The monsoon is thereby seen in Sri Lanka as going toward the northeast, coming from the southwest and it is therefore called the Southwest Monsoon. Because it takes time for the sun to heat up the continental landmass, the Southwest Monsoon is not centered on midsummer, June 21, but a month or two later; it is felt from about May through October.

The opposite mechanism works in winter, when the oceans are warmer than the land. During the winter the air above the Indian Ocean has a lower pressure than that over the continent of India and Tibet, and the air is then pulled from north to south. But again the Coriolis effect makes it deviate, and the winter monsoon is the Northeast Monsoon, in November through February. Both monsoons bring lots of rain to Sri Lanka, especially the Southwest Monsoon as it comes from the large Indian Ocean in summertime, but even the Northeast Monsoon comes from across the Bay of Bengal, still a large mass of water.

Although the frequent rains make Sri Lanka so green and lush, the clouds are not good for observing the stars and planets. The central mountain plateau does, however, have a few areas that are dry, as indicated by their sparser vegetation. Apparently, the clouds are emptied when they rise against the surrounding slopes.

I became enchanted by Ceylon in 1945 and kept in touch through mail and visits. As my interests turned to astronomy, the correspondence turned particularly on one topic: Paradise did not have a telescope!

In going to Sri Lanka I had to learn to deal with jet lag, caused by differences in time zones and the vagaries of travel, and I had some of the worst cases. Once there was a layover in Bangkok of several hours to catch a Lankan plane to Colombo, the airline had arranged for a rest in a downtown hotel, but when I returned to the airport nicely on time there was no plane as "the pilot had left early because he was ready and wanted to go home." That was unique and did not happen again after they reorganized their national airline.

In Arizona, we had the pleasure of a few Lankan visitors, and several of us in Tucson remember an impressive lecture by Vice-Chancellor L. Siriwardene regarding his country and his expectations for its future development. My astronomy lectures in Colombo and Peradenya always arouse a lively interest. Some of the young ones who asked questions years ago have ended up in the United States or elsewhere as scientists. That "brain drain" might be stopped if an observatory were built in Sri Lanka to stimulate both astronomy and the interaction between engineers and scientists. I have been a member of an Advisory Committee for Astronomy at the University of Sri Lanka and made field trips to look for sites for telescopes. We made several studies and wrote proposals for construction of an astronomical observatory on the central plateau in 1973, 1975, and again later. The country's only helicopter in 1978 was made available to us for a day to look for possible telescope sites, and with it we found several promising locations. Near Hanguranketa in the central highlands there is a near desert that should be further investigated for cloud statistics during the night. From correspondence and through the work of the advisory committee, the following information was gathered for astrophysics in Sri Lanka, as compared with Arizona, for example.[1]

The climate did not seem good enough for a large optical telescope. Data on night-time cloudiness were not available, but the days have about 70% of the sky covered on a yearly and national average. That is at least twice as cloudy as in Arizona, but there are variations that depend on the precise location. For teaching purposes a telescope of moderate size, say 30 inches in diameter, might be installed near a university, in addition to the 10- and 12-inch reflectors that exist in the Colombo area. Later, a telescope as large as a 70-inch might be considered. Radio astronomy is much less affected by clouds, and it therefore seemed more promising for Sri Lanka, especially if it could be done in cooperation with the existing radio observatory at Ootacamund ("Ooty") in south India. Theoretical studies were also recommended, although it was noted that in astrophysics, theory is never far removed from observations. The most interesting debate was on the question of how much should be done by Lankans and to what extent the support should be sought from elsewhere. We seemed to agree that the greatest desired benefit would be obtained if the people did everything themselves, except, perhaps, for the figuring of a large mirror, because the appropriate tooling and experience are difficult to come by. The Lankans' involvement in hardware and software would be good for training of students, and it might stimulate new industries in the country. Telescopes are computer controlled, as is much of the auxiliary instrumentation, and a telescope project would therefore require computer expertise and facilities. A telescope is not easy to install and put into operation; we had learned that lesson from a US-built 48 inch that had been sent to Hyderabad in India, where its operation had been delayed for the longest time. I had learned the same lesson in the engineering for the 28-

inch balloon-borne Polariscope. It may be faster to obtain ready-made equipment, but it is better in the end to have it built by a local crew because their expertise is vital during the operation: The people who built the equipment should be available when something goes wrong and needs to be fixed immediately.

We discussed various possibilities for research topics and for observations that could be made from Sri Lanka. The geographical location—latitude near 7° north—is ideal for observations of objects in the solar system because the ecliptic, the plane of planetary orbits, is never far from the zenith. The situation is better than at Tucson's latitude of 32°. Because the telescope would be of modest size, we suggested observing programs concentrating on bright planetary objects. The center of our Milky Way galaxy is, however, not far from the zenith either, and for giving students a broad training, one should also observe the stars and our galaxy. The Lankan telescope would primarily be a device for teaching and providing some experience with astronomical observing, after which graduate students would travel to larger facilities in better climates to carry out more extensive research. Japan, India, the Vatican, and the United States, among others, have offered cooperation and help in the training of students.

No telescope has as yet been built. It is one thing for administrators and committees to suggest a telescope, and another for someone to want to use it. The wish and drive should come from below, not from the top, and a telescope should be planned and built by its users. No Lankan astronomer-to-be showed up to push the project locally. And there were deeper reasons for the lack of success. Whenever a new party was voted into governmental power it had a tendency to undo what the previous government had brought about. Our telescope plan was no exception; the next government saw it as a white elephant and threw it out. The same thing happened to another set of elephants. Peradenya in the central highlands has famous botanical gardens, whose main entrance path used to be lined with topiary bushes carefully trimmed into the shapes of elephants. The elephant is, however, also the symbol of one of the political parties and when its opposition came into power these meticulously kept bushes, after all the painstaking care by their gardeners over many years, had to be cut off and thrown out.

There is an exodus of good people. For instance, a geologist whom I met first in 1951, when he was a student at the University of California at Los Angeles, returned to a promising position on the Peradenya campus. He was maneuvered out, however, because the people who had not been abroad resented the competition, or because he was Tamil, or both. He became a professor in an African country, always yearning to go back home. Nearly all the members of our Advisory Committee have gone abroad; Vice-Chancellor Siriwardene and a former campus president are working in foreign capitals. The reasons for leaving the country are often financial because the academic salaries are some of the lowest anywhere.

The problems of Sri Lanka may lie in part in the character of the people and in their philosophical background. Perhaps the Lankans are less inclined to science than Indians because Buddhism involves more passive contemplation than Hinduism. But the astronomical sciences are more philosophical than most other disciplines, so we seem to have a contradiction. Could it be that the inquiry into our origins has little appeal to Buddhists?

Several Lankans also expressed doubts about the practical value of astrophysics. Should a small and poor country not concentrate on practical science, on biology, agriculture, and water management for instance, rather than on astronomy with its expensive equipment and without a practical return for the poor? In India that issue was never in doubt as there was a firm conviction already in the 1930s that basic science is essential for any culture.

Eventually it will all fall into a place proper for Sri Lanka. The racial strife and bloodletting between Tamils and Sinhalese may also have a positive effect, as it sometimes has had elsewhere (Chapter 2 has an example). The brain drain may eventually be stopped by raising academic salaries to a respectable level. Sri Lanka now has an Institute of Fundamental Studies under the leadership of Cyril Ponnamperuma, who is also a distinguished chemistry professor at the University of Maryland. George Coyne, the director of the Vatican Observatory, is taking an interest in Sri Lanka and occasionally teaching astronomy there.

For me personally, the Lanka experience brought back the soldier's youthful introduction to other cultures and philosophies. And it has brought me many friendships: the Gunawardena family, Father Mervyn Fernando, Devendra Lal and others in India, and Arthur C. Clarke, the famous British science writer who fell under Sri Lanka's spell long ago.[2]

> It may well be that each of Ceylon's attractions is surpassed somewhere on Earth; Cambodia may have more impressive ruins, Tahiti lovelier beaches, Bali more beautiful landscapes (though I doubt it), Thailand more charming people (ditto). But I find it hard to believe that there is any country which scores so highly in *all* departments—which has so many advantages, and so few disadvantages, especially for the western visitor.
>
> Ceylon is also the right size; you can drive from one coast to the other in half a day, over roads which are usually good and often excellent, though sometimes afflicted with unusual hazards, such as elephants without rear-lights. Yet despite the fact that the island is only 140 miles across at its widest, it has two distinct climates. The central mountains (up to 7000 feet high) trap the monsoon rains alternately on the west and the east; unless you are very unlucky, you can always find good weather *somewhere* in Ceylon. And as the island is only a few degrees north of the Equator, winter never comes. . . .

> And always it is the same; the slender palm trees leaning over the white sand, the warm sun sparkling on the waves as they break on the inshore reef, the outrigger fishing boats drawn up high on the beach. This alone is real; the rest is but a dream from which I shall presently awake.
>
> I came to Ceylon in 1956, intending to stay for six months and to write one book about the exploration of the island's coastal waters. Fourteen years and twenty books later, I am still here, and hope to remain for the rest of my life.

Arthur did what I dreamed of in 1945, to settle some day in gentle Sri Lanka. He is still there, the Chancellor of one of the universities, writing prolifically about our future in space and many other topics, always with a philosophical perception. To date he has written more than thirty fiction and thirty non-fiction books. Arthur is a distinguished scientist; *Ascent to Orbit*[3] is a scientific autobiography that consists of some twenty impressive papers reprinted from technical journals.

Clarke came up with the idea of communications satellites that orbit the Earth at precisely the altitude at which they have the same angular speed as the Earth's surface. In these "geostationary" orbits they therefore remain above the same part of the Earth, which is convenient for the transmission of telephone and television. He published this remarkable suggestion in *Wireless World* in 1945, long before astronomers thought about space:

> It will be possible in a few more years to build radio-controlled rockets which can be steered into orbits beyond the limits of the atmosphere and be left to broadcast scientific information back to Earth. . . . There are an infinite number of possible orbits, circular and elliptical. One orbit, with a radius of 42,000 kilometers (22,300 miles) has a period of exactly 24 hours. A body in such an orbit, if its plane coincided with the Earth's equator, would revolve with the Earth and thus remain "stationary" above the same spot on the planet. It would remain fixed in the sky of a whole hemisphere and, unlike other heavenly bodies, would neither rise nor set.

And he went on to describe the possibilities in detail. Such communications satellites are now in common use.

India, for instance, has adopted the idea for bringing television to its villages—some of them so remote that the only source of electricity for the television set is a bicycle-driven generator contraption. The satellite transmissions provide entertainment, educational programs, and specialized instruction on farm methods and birth control. Clarke is therefore an honored man in India, where he went to lecture on several occasions and where he was a Sarabhai Professor in 1984. For Lankans he is their most famous citizen. For his many readers worldwide he brings happy hours with his fascinating books. For scientists he is one of the great writers, bringing out from our results a justification of our findings and our daily work.

20. Indian Leaders, Dying Young

Devendra Lal became a new friend quickly when we met at an international conference. He was born in Varanasi, called Benares by the British, where Buddha delivered his first sermon some 2500 years ago. Varanasi is also a pilgrim city for Hindus, for whom a bath in the River Ganga is sacred. Devendra is a lucky man, totally spoiled by Mrs. Lal, Aruna, a beautiful Gujarati who works full time at being his devoted assistant. They had decided to have no children because of their concern for the overpopulation in India. When we met, Lal was Director of the Physical Research Laboratory in Ahmedabad, a city in the western desert about one-third of the way from Bombay to Delhi in the state of Gujarat; it is not a large city by Indian standards, about three million people. Lal invited me to PRL. He claimed that India provided a climate of active accomplishment in science and that I was wasting my time in contemplative Buddhist Sri Lanka, where basic science did not seem to bloom and no telescope was being built. My hesitation out of loyalty to Sri Lanka ended when he mailed me a round-trip ticket and arranged my appointment as a Sarabhai Professor for 1978–79. That was a profound honor and a challenge.

Vikram Ambalal Sarabhai was the founder of PRL. He died suddenly in 1971 at the age of 52, and the Sarabhai Professorship is one of the ways in which his memory is honored. He was born on August 12 in 1919, a day that was celebrated in Ahmedabad in 1979 with the planting of trees and a symposium to honor promising scientists; I saw some of his previous associates having tears in their eyes when they spoke of Vikram on that occasion.

Young Vikram grew up in Ahmedabad in a powerful family that owned Calico, a prosperous textile mill. Vikram knew of India's proud tradition in physics, and he wanted to be a physicist too. Every student in physics learns of the Saha equation, Bose statistics, and Raman scattering; these are named after the Indian physicists Meghnad Saha (1893–1956), Satyendranath Bose (1894–1974), and Chandrasekhara Venkata Raman (1888–1970).[1] Raman, a Nobel Laureate in physics (1930), was the uncle of my Chicago professor Subrahmanyan Chandrasekhar, Nobel Laureate in physics (1983). Raman headed the Indian Institute of Science in Bangalore, which the British physicist Lord Rayleigh had helped to plan, and later the Raman Research Institute, also in Bangalore, which was founded by him. His son is the present director.

Sarabhai studied cosmic rays and made measurements at various latitudes in India. He obtained his Ph.D. degree in England, but he also worked at the Indian Institute of Science in Bangalore where Homi Jehangir Bhabha was one of the physics professors. It was an exciting time, for India was struggling for its independence, which came in 1947. Bhabha and other scientists were important advisers to Jawaharlal Nehru, the first prime minister: India was to become truly independent also in the basic and advanced sciences. The physicists were convinced that science was necessary for the development of India. Raman put it strongly[2]:

> Unless the real importance of pure science and its fundamental influence in the advancement of all knowledge are realized and acted upon, India cannot make headway in any direction and attain her place among the nations of the world.

The advice from the scientists was often prescient. I have seen a copy of a letter from Bhabha to Nehru describing a possible solution to India's poverty through the peaceful application of atomic energy. That letter was written in 1943, when nothing was known in India about the Manhattan Project, in which the Atomic Bomb was developed. But Bhabha and Sarabhai had studied in Cambridge, England, where the astrophysicist Arthur S. Eddington, had written, already in 1920[3]:

> If, indeed, the sub-atomic energy in the stars is being freely used to maintain their great furnaces, it seems to bring a little nearer to fulfillment our dream of controlling this latent power, for the well-being of the human race—or for its suicide.

After he returned to India, Bhabha went to Bombay to build the Tata Institute for Fundamental Research, one of the great institutions in science, where about fifty senior scholars work on problems ranging from the origin of the universe to commercial application of electronics equipment. It is housed in a majestic building on the seacoast with gardens and a walkway to the surf, but within sight of a shantytown that is a haunting reminder to work harder to help the poor. The Institute has spun off industrial companies whenever one of its scientists wished to strike out in a commercial venture to make use of some progress in, say, electronics. India suffered a terrible loss when Homi Bhabha died in a plane crash at the age of 56; he was on a Boeing 707 flying into Zurich when it crashed into Mont Blanc.

In the plans for independence, India was going to be blessed with centers for research and development all over the nation. Some privilege for India's scientists! Some obligation! It was done consciously, as Sarabhai explained[4]:

> The real social and economic fruits of technology go to those who apply them through understanding. Therefore, a significant number of citizens of every developing country must understand the ways of modern science and of the technology that flows from it. . . . An ability to question basic assumptions in any situation is fostered by probing the frontiers of science, whatever field one may be engaged in, whether it is Biology, Genetics, or Space Research. It is this ability, rather than an empirical hit-and-miss approach, which proves most effective in tackling the day-to-day problems of the world. It follows from this that countries have to provide facilities for its nationals to do front-rank research within the resources which are available. . . . Broad understanding of the physical and social environment in which man lives is the most urgent task which faces all humanity. . . . Lack of insight concerning the environment in which man operates has posed a problem at all times. . . . The consequences . . . are more serious to the security of the world than they were ever before. The task of promoting an understanding of science is of course at the core of the problem of education. . . . Acquisition of technology by itself does not contribute to this understanding.

The issue is important also for us today, to consider our involvement with matters other than those of our scientific pursuit. The Soviet physicist Peter Kapitza addressed this clearly[5]:

> The future of civilization depends on whether existing governments are able to provide solutions to global problems. . . . But, for this, problems must be expressed clearly and convincingly and widely discussed. This can be done mainly by scientists, since they can talk with sufficient authority on the possible solution of global problems for the benefit of mankind. Thus, we should not stand aside from the solution of such problems but realize their connection with our scientific work.

Professor P. R. Pisharoty, an associate of Sarabhai, argues that the general character of India's scientists should become broader and more inquisitive.[6] In India he sees a "lack of a scientific spirit and a scientific culture, even among the majority of our science graduates. Most of them are happy to be camp-followers of the scientists and technologists of the advanced countries. Until we generate a scientific temper and a scientific culture of creativity and innovation among a larger fraction of our Indian population, we are likely to continue to browse in the backyards of science and technology created in the advanced countries." Pisharoty therefore believes that centers where children learn to experiment are at least as important as advanced research institutes. In the Science Centre that Sarabhai founded in Ahmedabad, Pisharoty observed "young boys and girls busy with many activities, moving about freely, using their hands and heads, enquiring, questioning and learning with pleasure. It is a great joy to see their curiosity filled with happiness." Such thoughts are appropriate for science centers and planetariums all over the world. San Francisco, Hong Kong, and Hiroshima, for instance, have excellent centers where people may try for themselves an array of experiments. Professional scientists, however, sometimes show an apathy if not a disdain toward such places of popularization. An exception is in Bombay where members of the Tata Institute for Fundamental Research are involved in the Nehru Science Centre and its Planetarium.

Those who knew Vikram Sarabhai tell me that he was a remarkable person, a powerful organizer who succeeded with winning charm and lively interests. He was beloved by all. There must have been strong similarities between Sarabhai and George Ellery Hale in personality and in their accomplishments as founders of great institutions. Vikram's main concern was with people, and he was driven to help the poor of India. He could do that by giving them work and thereby raising their standards as well as those of the leaders in science and industry. He seems the personification of the theme of this book, to bring out the godliness within us. There are stories of how Vikram would gently talk to a lowly person who would never forget it.

Vikram drove himself too hard. In addition to keeping up his family obligations, he was the pioneer of pharmaceutical industry in India, he kept up his own research, he was the leader of a team of cosmic-ray physicists, and he became more and more involved in organizing new facilities. In addition to the Science Centre, he established in Ahmedabad the Textile Industrial Research Association, the Indian Institute of Management, the Physical Research Laboratory, and the Nehru Foundation for Development. After Bhabha's death he was asked to take over as Chairman of India's Atomic Energy Commission. He initiated India's space program and founded the Indian Space Research Organization; one of its four centers, in Trivandrum at the very southern tip of India, is now called the Vikram Sarabhai Space Centre. India's Rocket Launching Station is also located near Trivandrum; it is sponsored by the United Nations, and investigators from any country can launch rockets with scientific payloads there. In the end Vikram was so much at Trivandrum. . . and then he died there. He must have been reading a newspaper in his room; when he did not come out the next morning he was found that way, peacefully passed away because of a heart attack.

The cremation was done according to Indian tradition along Ahmedabad's Sabarmati River. Thousands of people came out from the city and from all over India. His son Kartikeya could not come, as it was too sudden, and he was away as a student in the United States. It therefore was 18-year-old daughter Mallika who lit the pyre, and that was probably the first time a girl had performed this act of reverence in India. The next generation sets the father's spirit free.

Too many of our Indian leaders have died young. Homi Bhabha and Vikram Sarabhai built institutions, and died prematurely. It happened again with the sudden death of Vainu Bappu (1927–82), senior astronomer in the development of modern astrophysics in India and President of the International Astronomical Union at the time of his death. His influence brought the astronomers to India in 1985, but he himself was not there, as he had already died of a heart attack just before the previous IAU Assembly in Patras, Greece. Bappu's work continues with a new 93-inch telescope at what is now named the Vainu Bappu Observatory at Kavalur, southeast of Bangalore, where he was the Director of the Indian Institute of Astrophysics. The telescope was officially opened by Prime Minister Rajiv Gandhi, who stayed over to be at an astronomical observatory for a night.

Rajiv is the grandson of Jawaharlal Nehru (1889–1964), who, before he became prime minister, spent years in British prisons because of his participation in the Independence movement. Nehru used that time to write a remarkable book. Remarkable also that the British allowed him to do so. The book is in the form of letters to his daughter. The full edition is *Glimpses of World History*,[7] and it is a rich source of information and interpretation. Nehru's daughter, Indira Gandhi (1917–84), later became prime minister herself. Nehru educated her in the arts and sciences as well as in politics, and she in turn seems to have done that for the present prime minister, her son Rajiv. India has been ruled by this dynasty, and there seem to be no prominent opposition candidates even though one hears and reads in the newspapers of strong criticism and sometimes even rumors of corruption at the highest level of government during the past decades.

People remember horrible events in the context of their own circumstances at the time they were told. Many of us remember exactly what they were doing when President Kennedy was shot—I was observing with a telescope in Texas. When Mrs. Gandhi was shot, I was at the Spacewatch Telescope on Kitt Peak about to start the work of the night. It was a dreadful night, and the following days I had a strange feeling of being close to crying, which I never do. A few days later, the memorial service of the India Club granted some relief for all those in Tucson connected with India. I told about a visit with her:

> Last July I had a brief meeting with the Prime Minister in her New Delhi residence to discuss the case of Soviet Academician Sakharov. I was shown to a small room, where most people are received. When you see pictures in *India News* of the Prime Minister with foreign dignitaries, there is always that sofa and her chair. Inside her chair is a little knob for a bell to bring in probably a secretary for dictation. So I sat there and wondered about the closeness, the informality of all this.

After some minutes she came in and I know her to be a reserved business-type person, so no handshake of course, simply the Namaste greetings. No chitchat. She sat down and quietly but charmingly looked at me and did not say a word. It was in perfectly good communication, as was all of the following interview. I knew that it was up to me to state my case, which was to bring to her attention the plight of Andrei Sakharov and the interest of Indian scientists in having him come to Bangalore and work on the proton decay experiment. This is an investigation carried out in the deep gold mines of southern India in order to determine the age of basic material, particularly the decay time of the proton, the nucleus of the hydrogen atom. The decay had been postulated by Sakharov.

She knew about proton decay and the case of Sakharov. I was watching her closely to see if I should explain the physics or to tell her more about Sakharov and Mrs. Bonner in Gorky. It was not necessary.

She had a few comments: It was difficult for her to do anything in the Soviet Union because there had been so much publicity about Sakharov. She made one comment that I still do not understand when she said that also the case of Afghanistan could have been resolved quietly at a high level a long time ago. Another comment was that she did not know who would respond; even if she would communicate with President Chernenko, it seemed to be not he but someone else who would reply. It was difficult to know who was in charge.

Her demeanor was friendly. She asked why my young daughter had not come, as was scheduled. She did say, "You have been here before," and I reminded her that it was to show pictures of Saturn. During the first flyby of the planet, pictures were made with a small spacecraft telescope of the University of Arizona. Soon after the death of her younger son Sanjay, it was felt that she needed some nice bright things to see and I therefore showed Saturn to her household in 1980. So I reminded her of Saturn and its rings and she smiled and charmingly said, "Oh, yes, that was nice."

Regarding Sakharov she did not speak further, nor did I ask her whether she would or would not do anything. I merely asked her to consider a probing at high level. I had the impression that something might be or was already being done. In fairness, I mentioned that Sakharov also in India would always be a dissident, as well as a far-sighted leader and possibly a personal friend of hers. She acknowledged these possibilities and we admired Sakharov to have a cause to live for, and to die for if necessary.

Indira Gandhi had a remarkable premonition of her own impending death. Found among her papers later was an undated note in her handwriting[8]:

> I have never felt less like dying and that calm and peace of mind is what prompts me to write what is in the nature of a will.
>
> If I die a violent death as some fear and a few are plotting, I know the violence will be in the thought and the action of the assassin, not in my dying—for no hate is dark enough to overshadow the extent of my love for my people and my country; no force is strong enough to divert me from my purpose and my endeavour to take this country forward.
>
> A poet has written of his "love"—"how can I feel humble with the wealth of you beside me!" I can say the same of India. I cannot understand how anyone can be an Indian and not be proud—the richness and infinite variety of our composite heritage, the magnificence of the people's spirit, equal to any disaster or burden, firm in their faith, gay spontaneously even in poverty and hardship.

The night before the fateful shooting in the garden of her New Delhi residence, she had said to a large crowd[9]: "I am not interested in a long life. I am not afraid of these things. I don't mind if my life goes in the service of this nation. If I die today, every drop of my blood will invigorate the nation."

A friend of Mrs. Gandhi, Mrs. Pupal Jayakar was interviewed on Indian television the day after the cremation[10]:

> There was in her a feel of lightness and a quickening. With all the sorrows she has faced in her life, and she has faced many sorrows, there was still in her this lightness of walk and spirit; and a capacity of great curiosity and interest in everything whether it was an Irish poet who was to come from Ireland, or the handicraft worker in Kashmir or the handloom weaver and what could be done to see that their problems were solved. . . . It seemed that the whole globe was part of her concern and I think this is what gave her stature. She was rooted in India, strong in tradition, with a modern mind she could look beyond frontiers. . . .
>
> She said, "You know my first memories of being a child was my feeling, 'Why am I here on the Earth.' In the last few months I have had a feeling that I have been long enough here on the Earth."
>
> There is a sentence in one of the Upanishads which I thought of yesterday at the funeral. "When the eye is free of death it becomes the Sun."

Mrs. Gandhi's son and successor, Rajiv, made quite an impression on astronomers at the triennial Assembly of the International Astronomical Union in New Delhi on the 19th of November 1985. It was a hectic day for him, as it was the commemoration of his mother's birthday, and he had to make several appearances before a variety of audiences. But he sat through our hour-long opening ceremony seemingly relaxed and taking an interest in the proceedings, occasionally scribbling on a cue card. And from that cue card he spoke. So well, in fact, that I inquired later of the Indian astronomers and organizers if they had actually written that speech for him, as is usually done on such occasions. But, no, they had not. And so he spoke in a quiet manner, unlike the oratory of a politician and almost like an old friend or colleague addressing us personally—in a hall with 2000 people!—carefully reviewing the background and future of astronomy in India. He explained the importance of basic science to raise the level of the nation and the livelihood of the poor. A fine moment came when he spoke of the people of India. He hesitated a moment as if to recall the scope of his responsibility, and then he said, "... that is 750 million people!"

21. Impressions of India

"East is East and West is West and never the twain shall meet." Thus we were instructed in 1945; the British generally believed it, as did the Dutch in Indonesia. Many Americans may still think of India in the same way as the immigration officer who said to one of our Indian students, Rajesh Vaidyanath, upon his entry into the United States: "Isn't India the land where the snakes and cows are worshipped? You people should start praying to Jesus. He is a beautiful God." But all nations and their people are coming closer together and know more about each other through rapidly increasing communications. Putting it strongly perhaps, but my experience has been that India is just another country. There are great similarities as well as differences between our nations and backgrounds.

In India, as in Europe, darkness and aggression entered repeatedly through wars and invasions. But the colonial suppression by the West was finally followed by Independence, and the largest democracy in the world is awakening to find its own might of the millions.

Fifteen thousand years ago, similar pictures were drawn on the walls of caves at widely separated Lascaux and Bhimbetka (Chapter 18). The first writings of the more advanced societies are found some 6000 years ago in Egypt and somewhat later in the northwest of India, near the Indus River.

Our modern religious heritage in the West is based on writings of semitic tribes in the near east, some ancient, others dating from the early Roman Empire, from which a few were chosen to make up the Bible. A rich variety of legends of the Greeks and Romans and of the various tribes of northern Europe has also influenced Western culture. The reformations that followed the Middle Ages came during a period of enlightenment, but still left rather strict religions.

In India the old scriptures of the Vedas, the Mahabharata, and the Ramayana molded the civilization. From the Vedas a small part called the Upanishads and from the Mahabharata the Bhagavad-Gita provide the core of Hindu culture. The Mahabharata is so extensive, some 100,000 couplets, the longest of all poems, that I found it difficult to read and gain an overview, until I found a poem by Mrinalini Sarabhai, which she kindly permitted us to reprint in this book. At first the Mahabharata had seemed an endless story of killings, so uncharacteristic for the present Indian people, but the country has been invaded many times in history; the poem itself is largely about a great battle in the area of Hastinapura, near the present Delhi. In any case, those scriptures are used to study human nature, life and passion, and the eternal conflict of evil and divine. The children of India hear and read the stories and all over the country the people were watching the adventures of Rama and his Sita from the Ramayana on national television in 1987.

Hinduism does not seem to be so restrictive, unlike the religions of the West and Middle East with their fear of God and punishment. India has restriction and regulation in the social layering of the caste system, much of it still seen today, but that layering seems to yield also a security through the acceptance of and compliance with one's status. This in turn is balanced by a tradition that stimulates action and aggressiveness to a greater extent than, say, Buddhism.

The Lord Krishna teaches some basic tenets in the Gita, among them that one must accept that good and bad are intertwined. And when a Hindu comes to a worthwhile stand, he must defend it, at the price of blood if necessary. One should understand this of India in world politics today. Mahatma Gandhi is known for his nonviolence, but he has also said[1]

Where there is only a choice between
cowardice and violence,
I advise violence.

I am not pleading for India
to practice non-violence because she is weak.
I want her to practice non-violence
being conscious of her strength and power.

In astronomy also, the Indian pattern is one of vigorous action. It seems that these three aspects of Hinduism—security, acceptance, and activity—made India special, a country where all and sometimes contrary grades and shades of society live in relative harmony: rich and poor, gentle and harsh, beauty and filth. In its infinite diversity there is a unity as I have seen nowhere else.

While the philosophical background in the West became strongly based on reason, under the influence of Descartes and other philosophers, a gentler, more spiritual society has developed in India. This was most likely due to the warmer climates, so that its philosophers were

India

exposed to people and nature, whereas Western Europeans lived in small rooms by the stove and so tended to closet themselves to think undisturbed and pursue a clear stream of reason. The opposite of the softness of India may be seen in Russia, where the climate isolates the people within their shelters and tempers them, especially during the Russian winters with their extreme temperature variations, like steel that becomes strong through heating and cooling. There is, however, a toughening aspect in the Indian climates also, with extreme heat and sometimes drought especially during the months of April through June. It seems remarkable that in India, where it can be so hot, the daily routine does not include the siesta so common elsewhere in hot climates, and people keep moving throughout the day. The climate may then also explain the difference between the more aggressive Hinduism that developed in tougher India and the gentler Buddhism that developed in the softer climate of lush Sri Lanka.

My hope and ultimate motive for working in India is that new science will come from the merging of Eastern philosophical depth with Western rational analysis. A combination of deeper metaphysical background with expertise in solving technical problems could result in a new scientific approach. How can this come about, since the mathematical expressions and physical laws are the same for scientists trained in the East or the West? The answer is that an essential component in the selection of problems and approaches is our intuition, which has been determined and developed by our background and environment. Intuition brings insight and makes science creative. Intuition, the integration of a lifetime of molding and experience, is the subjective complement to Descartes' reason. It is a tool that can be trained to "feel" a sentence in a manuscript, to deal with human relations and sensitivities, and to point toward new directions and approaches in research.

Will a new scientific approach actually emerge? Or will the Indian astronomers merely copy ideas from more advanced observatories in the West, as P. R. Pisharoty fears? I do not know the answer yet. In much of India there is indeed a mind-set to copy Western science and standards. One sees this symbolized even in clothing. For the IAU meeting in Delhi in 1985, I had promised my American friends that they would see the most beautiful and practical Indian clothes: colorful saris of the women and finely cut Indian suits in great variety on the men, particularly during the social functions in the evenings. They wear these clothes at home and that is where I had seen them. But I was wrong. Although the women did come in beautiful saris, the men, other than Prime Minister Gandhi, nearly all came in Western coat and tie. Imagine that the worst item of Western clothing, put on us men as a joke by our women—the choking tie, the prime symbol of colonialism—is being copied by the people of this new nation! Also, the management structure I see in their scientific institutions is still the British colonial one, with a Senior Professor Director, to whom Professors report, to whom junior scientists report, to whom the assistants and students report. In the United States the organizational table is not held to so strictly, and the young researchers and teachers have their initiatives promoted and are given the opportunity to branch out into new directions.

Traditions are prevalent in India. Astrology is still strong, even among the educated, and more so than in the US. One hears so many tales of when the swami's or astrologer's prediction was correct. Cases of failure or vagueness seem forgotten. I give public lectures occasionally on comets and asteroids, but in the question period astrology always comes up, and before March 1982, there were questions about the "Jupiter Effect": on March 10, 1982, there was going to be a fateful lineup of the planets to cause disasters and earthquakes. My protests that the idea had been debunked, and that the gravitational pulls of the distant planets are negligible compared to those of Sun and Moon, seemed to make no impression. If on that day *anything* had happened, it would have been an ever resounding proof of astrology. And there was, of course, a statistical chance of an earthquake somewhere. As it was, our globe was dead still for months around that fateful date.

My contacts in India are mostly with educated people, rather than with villagers. I can therefore not say that I know India, for the roots and the heart and by far the larger part of India lie in the villages. There is little I can do about my ignorance other than learn a local language and live in a village. Lacking that, one can only observe the people, for which there is plenty of opportunity as they are everywhere. And many are poor. Even a relatively prosperous city as Ahmedabad has enclaves of crowded hovels. The cheerfulness of the poor is always striking. It is probably because of their caste system and the well-defined social layering with its inherent beliefs in predestination and dutiful compliance. They seem to have accepted their position and circumstance in life, instead of striving for improvement, as they might in Western societies where there is greater stigma on being poor.

Around the Bombay airport we can observe the worst poverty while waiting for the next flight in our extravagance that we take for granted. The people live right by the runways in crowded hovels, but most of them seem well fed, with clean clothes. Except in a few cases, mostly near traffic intersections when tourists are seen, where there was aggressive begging, Jo-Ann and I have walked through the shantytowns rather unnoticed and without being disturbed.[2] The government has provided wells and water faucets, but the "untouchables" were not allowed to use the well in the villages, and this probably explains that one sees children collecting water from the sewers of hotels; such discrimination is now outlawed, but traditions seem slow to die out. The children, however, do not look desperately poor or hungry, and we have seen them collecting that sewer water in brass vessels that may have been given to them, but which we would consider valuable heirlooms. The contrasts are enormous between them and us, their hovels and our airplanes, their cheer and our harriedness. A marked improvement has occurred in the forty-two years since I first saw India. A large middle level has arisen that is thriving with their own businesses and trucks and scooters. The exceedingly destitute starving in the streets of 1945 are now nearly absent, with exceptions at some railroad stations.

Marriages in India are generally arranged by the parents, but the degree of choice by the couple depends on their caste. Modern young people can find each other, fall in love, and decide to get married, after which the families get together and arrange the wedding. But other arrangements are more common. My yoga instructor Vikram Panchal, who studies chemistry at the University of Gujarat and is from a caste of businessmen, first looked himself for a suitable girl. When he was not successful and getting toward the age of 25, he was told by his mother when and whom to marry, after which he met with his bride-to-be a few times, they liked each other, the family preparations proceeded, and a guru advised, using astrology, regarding circumstances and the appropriate date. Until this time, the arrangements could have been cancelled, but that would have had to be done carefully, as the families might have become reluctant to make arrangements again in the future.

Vikram's marriage ceremony appeared to me much more explicitly oriented toward reproduction than those in the West. One of the last parts of the days of festival, witnessed by family and friends, was to untie a string around each other's wrist which meant that they could now be bedded together, after which they left to do just that. Vikram's first son was born on schedule and named Om, a name that is frequently used. The naming after family members is rather rare in India. Vikram did have a problem persuading his youngest sister to approve the name Om, as she had another one in mind and it is her right to name her brother's children.

A more disturbing marriage arrangement, by our standards, affected a graduate student at Arizona. Her mother became alarmed over the living conditions she described in her letters, with three girls in a rented house, and mother therefore arranged for a spouse. It happened on short notice: an unknown man from India but now living in Chicago came with his father and a sister for a day to Tucson to meet the prospective bride, and he sent flowers a few days later telling her the wedding would be held the following week: "Sorry for the rush, but father and sister have to return to India." The girl was highly upset, for she had to give up her studies at the University of Arizona and possibly her career, but she felt obligated to her family and tradition to go through with it. So, off she went to Chicago. But she did start another new study program at the University of Illinois and seems to be happy there.

In the lower castes, marriages are arranged entirely by the family, and the couple may not get to see each other until after they are married. I have been told that there are numerous types and variations of wedding arrangements and ceremonies. We watched one of them in the courtyard of a Delhi hotel, during the IAU meeting of November 1985, from a balcony, so we could see what happened on the side of the bride as well as that of the groom. During the preparations she did not get to look at him nor he at her. She did not smile and said not one word, nor did he. It was Serious Business. He had more to smile about than she. While she was being dressed and fussed over by the women of her large family, he was put on a horse. A horse! Probably the first and last time in his life, for he was a city businessman. It was evening, and there was a nice red sunset; the groom was decked out in colorful garb and so was the horse. A band arrived and we went out into the street with all the hotel guests who were willing to join in the dancing, which was a bit risky because trucks and buses kept roaring by on that crowded street. Even this was a serious affair, however, and no one smiled except the dancers; the musicians and torchbearers seemed concerned mainly with how many 5-rupee bills the father of the groom had brought. The racket continued for an hour or so and finally the groom's procession stopped in front of our hotel. The groom came off the horse, his mother put something in his mouth, he had to caress a fruit, and a priest put a dot with holy ash on his forehead. It all had a deeper meaning, of course, but the crowdedness and rowdiness made it appear rather coarse. The men proceeded to the tent in the garden, more solemn now, with the priest chanting, while rupee bills, bigger ones now, were being passed; the priest got some and the groom seemed to receive a lot of them from the father of the bride; and they did some eating and drinking as well. Not so for the women and the bride, as they were outside of the men's enclosure and could not see what was going on. Finally the bride was guided in, and tied to the husband-to-be with shawls. No glancing sideways. The blessings were said and the bride was removed; we did not see her again until just before sunrise. But the fellows had a wonderful time, yelling and laughing all night—falling asleep might bring bad luck. Until 5 in the morning. Then the bride was fetched and all went laughing and shouting to the husband's family home where she was delivered to his room just before sunrise. The weary guests returned to catch some sleep. But that was not easy to do, as the hotel was close to a mosque, and dawn is a time for muslims to pray. The call to prayers comes from mosques all over the city, from powerful loudspeakers: "aay Allaha. . . aay Allaha. . . aay aay aay Alla. . . Allahaahaahaa. . . aay aay Allaha Allaha!"

Ahmedabad is called "Ahmadabad" in most atlases, which shows how much mapmakers copy each other, because nowhere in town does one see that spelling.[3] It lies in Gujarat, the state of Mahatma Gandhi. His Ashram is a museum, and his influence is still felt in many ways, inward and outward, as in the pervasive vegetarianism and prohibition of alcohol. Some thirty million people speak Gujarati; it is one of the Indo-European languages, and to me it sounds familiar enough that I can usually catch the topic and trend of conversation. In Ahmedabad one sees everywhere that traditions and livelihoods are shaped by the environment. One January day, I went into the center of the old part of town to take part in the yearly festival of flying kites. Old and young, rich and poor, they were all on the roofs or in the streets that day, and on into the evening when candles were carried aloft by an assortment of kites. Each kite flier tried to cut off the ones of the neighbors, and the cheers and laughter sounded all over town. Once I was walking by myself on a narrow street and a family invited me to fly kites with them from their roof. I went in and we had a great time, and I was also shown their home. It was rather narrow but five stories high, with different floors for uncles and aunts and nieces and nephews. The most intriguing was the basement, where there were a dozen or so large brass storage pots filled with an assortment of grains; there was also a brick cistern for pure drinking water underneath the floor. Such storage must have proved its worth during the drought of the late 1980s; even though the Government now provides major relief in case of famine or drought, the city of Ahmedabad had a severe shortage of water, providing it at the taps sometimes not more than one hour a day.

The traffic in the city has pedestrians, bicycles, donkeys, cows, and camels, honking cars and buses and trucks with a lot of noise of bells and claxons, but hardly a yell or oath. There are people everywhere, in a wide variety of clothing that shows their status in life. The farmers, with a proud bearing, wear white clothes and a turban around the head, leaving only the hands, face, and lower parts of the legs showing. The women wear saris in various colors, and one can quickly learn the differences in their status and in the quality and expense of the saris; the most expensive are in the $2000 range, but these are not walking in the street. The vehicular traffic also has a caste-type separation, the priority being trucks over buses over cars over motorbikes over scooters over bicycles over pedestrians. The animals have ultimate priority. They may travel in groups, or they may be single, such as a cow stepping serenely through all of the traffic. Sets of donkeys carry sand to construction sites, some large monkeys may be galloping across the street, a single camel pulls a wagon, or a family of camels strides majestically to another destination with the most senior male in the lead. The cows find things to eat along the streets, including paper, and their droppings are collected by the farmers' women to be made, by hand, into patties that are set out to dry and will be burned for cooking and heating.

The flow of the traffic is fabulous. There is a system to it. In Rome or Paris one looks the other driver in the eyes to see if he means it, and if so you let him go first. The Indians, on the other hand, seem to have an inbuilt sense for the speed and direction of everyone and everything whereby they know where it all will be the next moment, and then they maneuver around these predicted positions. When we were young we learned that too, in the navigation of our sailing ships.

The Indian drivers are kinder than the ones in Western Europe. I have a pet theory that this is because in their lives at home they are being hassled less by their women, and that they have fewer frustrations to react out—a result of the acceptance of life and their arranged marriages.

When I am in India, I get around in the cities with three-wheeled, front-wheel-driven rickshaws piloted exquisitely but hair-raisingly by young fellows. They speak Gujarati, and to communicate with them was a challenge. First I used to be polite and would say, "Good morning, please take me to the Physical Research Laboratory in Navrangpura"—and that would get me nowhere. The man would pull up to someone who might speak some English as well as Gujarati, and a confusing debate would be the result. Next I learned to say in Gujarati to go left (dabee), right (djemonee), or straight ahead (seeda)—but that was not quite expected of me either, or maybe the accent was not right. Next was a method of imperialistically handwaving to indicate left, right, or straight ahead and it worked if I knew where I was going. Finally, since the man is watching his passenger in a little mirror, I can simply point my beard with a little jerk in the desired direction.

At the sides of the streets one sees many stalls that have an abundance of goods and fruits on display, with the owner sitting quietly waiting for his customers. There will be no haggling. There are also children playing cricket, their national game, and one sees abundant flowers that add to the colors and fragrances. More difficult to describe are the sounds, which have tremendous variety but are not offensive, except for some horns of trucks and expensive cars. The charm of it all is the flow, the acceptance, and being yourself in whatever caste or place in life you were born. I have walked in the middle of the night through the poorest parts of Delhi without anyone paying attention to me. But be sure to plug the keyhole of your hotel room door for privacy! Once I waved from the inside at one of the bellboys peaking in through the keyhole, which gave merriment on both sides, but when I told the story later to a young American couple, they did not find it so funny.

In the big cities there may be parts where the local Indians are corrupted, near the expensive tourist hotels for instance. Everywhere in the world I try to avoid hotels anyway, and I have with great success written to local hosts, or even to the mayor of a city, asking to be put up somewhere as a paying guest. Short of that, a "tourist home" is the place to stay, as these are not really for foreign tourists, but are rather family hotels where the Indian people and businessmen stay overnight.

Keeping in good health is a challenge in India; it is something to learn and to make into second nature. One must get into the habit of not touching hands to mouth and being careful with drinking water unless one can be sure it has been boiled. The penalty for carelessness will come quickly. One eats mostly without utensils with the right hand, as the left hand is reserved for soilier chores. Other than that, the visitor can relax and leave the rest to India's charm. It is a country for learning acceptance and an integration of body and mind. One sees spiritual reminders everywhere along streets and highways—temples and small memorials to many gods representing different aspects of the ultimate unity of life. In India it is easy to come to the realization that this is what I am, a unity of diverse aspects. *Tat twam asi*! That thou art! This is what you are too.

I have lectured at various institutions in India, such as the Indian Institute of Astrophysics in Bangalore; a major involvement over the years has been with the Physical Research Laboratory. Until recently PRL had about forty-five scientists, but a new institute was split off from it for the study of atomic fusion for peaceful energy generation. Such splittings are a tradition in India, whenever the institutes and industries come to maturity. PRL itself is located in a large building near the outskirts of Ahmedabad among trees and soaring birds. I have an office on the eighth floor, overlooking the city which lies brightly in the sunshine of a desert climate as dry as that of Arizona—only in August may tropical torrents of rain come down. But we do the same science, observe the same stars, and speak approximately the same language as in Arizona. Near the PRL building one sees a newer part of Ahmedabad, a modern city that looks like Tucson, but with a mosque, a Hindu temple, and monkeys and donkeys and camels and rickshaws and honking traffic and nomad families actually living along the streets. Everyone is minding his or her own needs and doings with surprising dignity, in circumstances that appear to us as groveling in the dust. This always reminds me to go do my own thing on comets or asteroids or reading or studying or lecturing or writing a paper.

At PRL I have been asked to teach "Chandrasekhar style," which is to put the students under a stricter discipline than they are used to. They know Chandra because he also came as a Sarabhai Professor. For an ordinary lecture in India it is not uncommon to have people come in five minutes late, ten minutes late, twenty minutes late. They give a friendly smile as they enter, but they have missed the introductory parts and are therefore less able to follow the lecture. This only adds to the problems of language misunderstandings. We both speak English, but the intonations are totally different. To me their English sounds like the rippling of a brook without stress on any syllables. I try to speak slowly and clearly, but I may not be understood by students for whom English may be the third or fourth language. They grow up with several languages—a local tongue in their towns, sometimes also their mothers' language, some Hindi, perhaps English, and for the better educated Sanskrit and possibly French—but in Ahmedabad they are mostly educated in Gujarati, and most speak that at home.

I am also supposed to insist that assignments are handed in. In the beginning I would assign a problem to be worked out before the next lecture, but that again would bring a friendly smile, without homework done. The goal is to get through to them that Chandrasekhar got his Nobel prize not just because he is bright, but also because he is rigorous and works some eighty hours per week.

Astronomy is well established in India. The climate is excellent, with clear skies in the middle and northern parts of the country ten months of the year; only the August monsoon brings in clouds and rain. Four observatories have 40- or 48-inch telescopes, and there is a 93-inch in Southern India; there is a variety of radiotelescopes and there are plans for larger ones yet. In Ahmedabad I have given advice for the construction of an astronomical polarimeter, while at the Indian Institute of Astrophysics we are developing a joint program to search for comets and asteroids. My financial support comes from the Smithsonian Institution; Francine Berkowitz is the monitor in Washington.

As a Sarabhai Professor I became friends with Vikram's immediate family. Kartikeya is the son of Vikram and Mrinalini; with his wife Raju and their young boys Mohal and Samvit they came to Arizona for a lively visit, quickly becoming familiar with the birds and the cacti better than I ever managed. On the way to the Grand Canyon and Meteor Crater we had never-ending debates about arts and sciences and philosophy and vegetarian foods. Whenever the discussion became intense and apparently interesting, Samvit's little voice would come from the back of the car: "Shoo" (it sounds like zhoo, with a little halt at the end), Gujarati for "What is it?" and papa Kartikeya would patiently explain it to the little one. My favorite name for him will therefore always be Shoo. In Ahmedabad I usually stay with them and participate in the lively discussions around their dinner table, where Mrinalini presides. She provides a standard for many in India as a leader in the arts and cultural enterprises.

Kartikeya and Raju are involved in an animal park near Ahmedabad and in environmental affairs, building a large Environment Research Centre, as well as textiles and electronics. The British suppressed crafts and home industries to favor their own exports from Manchester's mills, but the villagers still know how to make useful and beautiful cloths and garments, and these are now again promoted by the Sarabhais—a movement begun by beloved Gandhiji in his Ashram not far from the Sarabhai house.

Kartikeya's sister Mallika is one of the finest classical dancers in India, as well as a star and producer of films; she is also involved in the book-publishing business with her husband Bipin. Their young son Revanta will probably be a dancer too. A great challenge came for Mallika in 1985, when she was asked by the famous director Peter Brook to play Draupadi and other parts of *The Mahabharata* in France for which she had to learn the long roles of the nine-hour play in fluent French; it was done in English in 1987–88 and toured the United States, Australia, Japan, and other countries.

The children of India are fun to make contact with because they are lively without becoming begging nuisances. One day in Ahmedabad I was sitting outside reading a book, and a little boy came by who hardly spoke English, but when he saw a photograph of Vikram in my book he said "Mallika's papa." He may have seen Mallika as a movie star, but her father had passed away before the chap was even born!

In addition to Mrinalini's immediate family she has many visitors and house guests under her queenly control. There I met Peter Frye, who has an interesting history. He fought with the International Brigade in the Spanish Civil War, against the fascists and side by side with the communists, as the US was to do in World War II. In his case, however, he became a victim of Senator Joseph McCarthy's witch hunt for communists, and he lost his job with CBS Television in 1951. Later he worked as a professor of drama and theater in Israel, and now had come to Ahmedabad with his wife, the well-known British actress Thelma Ruby, to write and direct a biographical play on Mrinalini Sarabhai. At the dinner table they would sing, those two, for instance from *Fiddler on the Roof*, and Peter would tell stories of Arabs and Israel. He had become acquainted with a wealthy Bedouin. The sheik affirmed that he had the prescribed number of wives, four and no more. He could divorce by walking around a wife three times saying "I divorce you," and he had done that. "When I look at a girl, and she looks at me, I get the green heart, and what can I do?" Thus he was the father of many children. How many? When Peter asked him if he knew the children by name, he indeed managed to recall quite a few of the sons, but none of the daughters.

People's first names in India have meanings in tradition or mythology. Vikram was a victorious king. Mrinalini is a lotus flower. Kartikeya, the son of Shiva and Parvati, was the god of wars to bring peace. Raju, an abbreviation for Rajshree, is a queen among women. Mallika is a jasmine flower. Mohal is the loved one, Samvit has the highest consciousness, and Revanta is the god of nature, the son of the Sun and the shade. Mrinalini believes that Divine Guidance made her decide to become a dancer at the age of four. It was indeed a religious choice, for in South India the dancing is in the Temple. By the nineteenth century, however, the Temple dancers belonged more to priests and pimps than to Lord Nataraja so that the British regime ruled temple dancing illegal in 1884.

When Mrinalini was young the classical Indian dance was still in a low state of grace. In her life she was to do more than anyone else to restore this ancient art form to its spiritual glory, but it did not come easy. The life of a dancer is for the gifted and dedicated; even the ones

Mrinalini Sarabhai: A cry from the heart for understanding, from a dance called "Memory is a Ragged Fragment of Eternity." (Reproduced from M. Sarabhai, *Creations*, Mapin, Ahmedabad, 1986, with permission)

who have the grace and coordination must practice during long days for years. Each position has to be perfected—legs and feet, back and head, arms and fingers, and face and eyes—over and over again. Mrinalini's mother had shown an apathy and lack of appreciation toward the youthful obsession. How could the child be so obsessed with something that socially was hardly acceptable? Mothers are important in South Indian families. It is a matriarchal society, the senior woman being the "Amma" who rules the daily life and the marriages of sons and daughters-in-law. (In India, however, I heard for the first time the title "Daughter-in-Love," which would be a fine term to adopt everywhere.) Mrinalini's mother, Ammu Swaminadhan, was involved in politics and eventually became a Congress worker in New Delhi. Young Mrinalini was closer to her father, a lawyer in Madras where they lived, but he died early, and that left a bitter void. It happened again, years later, when husband Vikram suddenly died.

Mrinalini and her two brothers and sister[4] had received a diversified education, some of it abroad. She studied the dances with eminent gurus in South India. She also attended a girls' school in Switzerland, which was hard, for these were the times of colonial arrogance and racial prejudice. A stay at the school of the great poet and painter Rabindranath Tagore, at Shantiniketan near Calcutta, was a stimulating experience, as were the studies of dancing in Java and acting in New York. There was always extensive travel, first with her mother and later her husband, or later yet alone or with her students, continuing professionally as a famous dancer and choreographer. Vikram had helped her to establish a school for South Indian dances in Ahmedabad. It is called Darpana, the Mirror: "You are It. Look at the Godliness within Yourself."

The Sarabhais live next to Darpana in the residence called Chidambaram after one of the great temples in southern India, 245 kilometers south of Madras, where Shiva as Nataraja is the Supreme Dancer. The dance of Shiva represents the rhythm and movement of the universe and of every being. Shiva shapes the universe in his dance, which shows creation, preservation and duration, and also taking back in destruction, hiding, and veiling, but with an ultimate bestowing of grace. Not only classical South-Indian dances of Bharata Natyam, Kathakali and Kuchipudi are done at Darpana, but also folk dancing, acting, and puppetry. The dances are accompanied by about four musicians, using voice, drum and other percussion, a string instrument, and usually a flute and small harmonium. The drums sound dominant while the singer is in close communication with the dancers and with the audience as he describes what is happening. It is not difficult for a newcomer to follow the meaning of the classical dance, even though the languages used are those of southern India. The art form seems to reach beyond opera in intricacy and religious involvement. It is quite explicit, usually telling the story of Krishna as a child, growing into deity and manhood, with the dancer becoming more entranced. The following description is taken from Mrinalini Sarabhai's "The Eight Nanikas"[5]:

> It is the night of the autumnal moon. The forest is dark and hushed. The birds have long since ceased their twittering, and folded their wings. Brindavan (the forest that symbolizes the human heart where God resides) is deserted.
>
> Suddenly, as the rays of the red moon light up the shadows, a thrilling sound is heard. It is the sound of a flute. It is the flute of Krishna.

. . . all the forest trembles.
Mystery walks the woods once more.
We hear a flute.
It moves on earth, it is the god who plays
with the flute to his lips and music in his breath:
The god is Krishna in his lovely youth.

> Krishna, the child whose mischievous deeds excite the wrath of his mother, Yasoda; Krishna, the cowherd, who plays with his friends in the woods and by the river Jamuna; Krishna, the eternal lover, beloved of the *Gopis*, the milkmaids of Brindavan; Krishna, who is no other than Vishnu, the Supreme One, Lord of the Universe, born amongst man as a promise to Mother Earth, born to redeem, to bestow, to bless.

At Darpana since 1949 more than a thousand young people have been educated in the arts and imbued with lifelong grace. The best students and the professional staff perform all over India and the world. In 1987 they danced in the large Auditorium of the University of Arizona in Tucson and received standing ovations. As a choreographer, Mrinalini designs new dances and plays. Darpana can therefore perform for you the life of Jesus Christ, the need for emancipation of women, Peter Frye's play after Mrinalini's autobiographical novel *This Alone is True*,[6] an outcry against nuclear proliferation that can bring the audience to tears, and an appeal to fight pollution in the Ganga River while pointing toward the need for the purification of the human soul. Walking in the streets one morning I was blessed with the friendliest smile from Jesus—yes, Jesus Himself—on a bicycle: One of the dancers in the depiction of Christ did not merely play well... he *was* Jesus! A young seminary student, he had no trouble with his role, not with Mary Magdalene, not at Gethsemane, not on the cross. And when he smiled at me that morning from his rattly old bike among the trucks, cars, donkeys, and people... it was Jesus Himself blessing my day.

The Mahabharata

Retold by Mrinalini Sarabhai

A few thousand years ago
at the gates of Hastinapura
a woman stood alone, a widow,
with her five sons, born of Gods:
Yudhishthira, the son of Dharma;
Bhima, the Windgod's child;
Arjuna, the son of Indira;
the handsome Ashwins gave
the twins Nakula and Sahadeva
to Madri, now dead. The boys
stood behind courageous Kunti,
resplendent together,
waiting to share the kingdom of
their royal father Pandu
with Dhritarashtra and his hundred sons.

Duryodhana the eldest, evil incarnate, sought
to destroy his cousins in myriad ways.
"Erect a house of lac," he told Purochana,
"My gift to them," he laughed,
"Then burn them all with torches."
But learned Vidura warned Yudhishthira
in secret tongue. The brothers dug a deep tunnel
  to the woods and on a certain night
when fire lit the palace, escaped,
disguised as men of wisdom, dressed as brahmins.

These Pandavas were noble, unafraid and true.
Their mother Kunti enveloped them in love.
They fought to prove their martial worth
protecting everywhere the poor and meek:
as when the ravenous Vaka, Bhima slew.
The Princes, best of men, these five
were still disguised when one auspicious day
they heard that Princess Draupadi, most beautiful of women,
Shakti herself, destined to sway the earth,
born from the sacred fire, a flame of light,
was to select her husband in an open court
from those assembled, all of highest birth,
each one to contest with his shooting skill.

Vyasa, the wise, came to their home at dawn:
"Go to her svayamvara but secretly," he said,
"Let no man know or gauge your strength
till you have won the beauteous Draupadi."

The air was luminous and gay,
great princes stood in royal lines.
Attired in gold ablaze with jewels,
while the five heroes sat upon the earth
as humble priests, waiting for alms.

The mighty bow was brought. A circling wheel
strung in the air, swayed in the breeze.
"Come heroes bend the bow, aim at the turning ring"
Drupada spoke to all these royal princes.
Boldly they came, each eager to compete,
faltered and fell, defeated, to the dust.
So strong that mighty bow, so stiff!
Sisupala, the vain, pulled till his muscles burst
but no one smiled. Only a murmur rose!
"No king, no prince, no nobleman left?"
Drupada asked but answer there was none!
A haunting silence overcast the leaden sky;
thought was a turmoil in each heart.

Then Karna son of Kunti by the Sun
whom she at birth abandoned, in her youth,
came forward and resolved to shoot the mark,
but Draupadi rejected him and cried
"I do not want a Suta for my lord."
So Karna threw aside the bow,
glanced at his father Surya and
knew she was not born for him.

Nor did he own his lofty parentage
but left the gathering, sad and angry.

And silence fell upon the crowd till
one young brahmin standing up, gracefully
strolled towards the revolving ring,
lifted the bow with ease and in a moment's
breath, strung five arrows and let fly
into the chakra, smiling all the while!
A roar went up of anguish and of joy!
The princess placed her garland round the man.
Someone came swiftly and gave Arjuna a robe
which he in turn, placed round his bride.
Both vanished in the crowd, leaving the kings aghast.
Arjuna led the lovely girl into the forest.
Nearing their hut he called to Kunti joyfully
"Rare is the fruit we have in alms today!"
And Kunti from within spoke lovingly
"Enjoy together all that you received."
But seeing the gentle girl she hesitated,
then welcomed her and spoke. . .
"You are my children's bride as I am mother
and we will share our lives."
Draupadi bowed and said, "I understand."

There came a knock, the door ajar;
a radiant being stood there like the moon
illumining the dusk of cowdust time,
he spoke in tones of honey dew:
"You are the Pandavas, cradled in my love
even before the day that you were born!
I am KRISHNA of the Yadavas
and he beside me, my brother Balarama."
Arjuna gasped and Krishna laughed!
In the far distance, a temple bell rang out
"I shall be near you friends, fear not!"

Dhritarashtra, now wishing to make peace,
sent trusted Vidura to appease the Pandavas.
Vidura said, "The king donates the land to you
upon the river Yamuna; there build your capital
and be a widespread banyan tree
that shelters all beneath its shade!"
Thus was born Indraprastha on arid desert soil,
city of gold and silver; people lived in peace, contented
that Yudhishthira the righteous should be king.

The years went by. Arjuna
took Subhadra as his queen and
built a mighty palace of his own where Maya
mingled real and unreal in glorious ways.
The Pandavas, foremost of virtuous men
ruled for many years, with honour,
till fate unheeding sought a time
compelling these young heroes to invoke
the wrath of those who bore them ill.
Minute the cause, fierce the destruction.
Thus it is with all existence
we think we know, but yet alas we know not.

Duryodhana, the wicked, pursued his cousins
hating them and jealous of their love,
he sought all ways to dishonour these
illustrious Pandavas who reigned with love.
The weak king Dhritarashtra,
persuaded by Duryodhana, sent messengers
inviting all the Pandavas to a feast,
to play a game of dice. Thus by
destiny impelled and obedient to the king
the prince and all his kin set forth.

Yudhishthira, the Kshatriya, could not refuse
the game of dice Duryodhana had planned.
Now welcoming the son of Pritha he remarked
"My uncle Sakuni (well trained in foulest play)
will stake my wealth and all I own."
Arjuna spoke a word of caution but adamant Destiny
was stronger than his words,
sweeping the Pandavas into a whirlpool of disaster.
Yudhishthira, as though in trance, lost all.
The only sound was Sakuni's "I have won!"
Brothers he lost and then himself,
now taunted by Sakuni and desperate to win
spoke "I stake that fairest woman Draupadi."
A hush, a breathless silence, till Duryodhana roared
"Fetch Draupadi, for he has lost!"

A thunder broke the stillness, lightning flashed,
voices were heard and portents lit the sky.
Vidura tried in vain to stop this madness;
can ever man arrest the hand of Fate?
The cosmos then spun round
raining catastrophe upon the earth.

When Draupadi, the fearless wide-eyed girl,
confronted by her husband's foolishness
asked, "Whom lost you first O gambler, yourself or me?"

Monstrous Dusshasana seized her unbound hair;
to outrage her tender body
he pulled away her robe. She prayed to Him
the Omnipotent One, creator of the Universe
Who saves all those who know his strength.
A pile of robes grew high upon the ground;
Dusshasana appalled sank breathless to the floor.
Her loosened hair contaminated by his touch
Draupadi's blazing eyes swept round the room,
"This hair untied, Dusshasana's blood will knot
I swear to this!" "And I will drink his blood"
the irate Bhima swore.
Duryodhana laughed, and in a vicious mood
bared his left thigh as though enticing Draupadi.
Vrikodara's anger knew no bounds;
burning with wrath he shouted
"Duryodhana's thigh I'll break when
we shall war with him."

At last the spell was broken for the old king
crushing the dice in blinded fumbling hands
called to the young woman pleading with gentle words
"Seek what you wish of me, I am the king!"
Draupadi freed her lovers one by one
overpowered by anger, she cursed the evil clan
"Bhima will drink the blood of cruel Dusshasana
your ally Karna perish, killed by Arjuna,
cheating Sakuni destroyed and slain by Sahadeva
this house will fall and never rise again."
Vidura listened sadly and as she ended
"Dragging the darkeyed Draupadi into the court
will bring the dissolution of our entire race" he said.

They left the palace, Draupadi leading them.
The queen Gandhari watched with inner eye
"From evil cause, evil results, Duryodhana,
What you've begun today," she said, "can never end.
Avert destruction, let them play again"
Hoping to make the peace Dhritarashtra sent word;
as though compelled, Yudhishthira turned back;
calamitous times do cloud men's minds.

All was lost. The Pandavas willingly consented
to live in exile rather than surrender
for twelve long years and one more added,
the thirteenth year, unknown, unrecognized.

Their mother Kunti went with kind Vidura;
parting from her sons she wept and said
"Look after all, Yudhishthira," and blessed
the sorrowing Draupadi. The sun in total darkness plunged, while
lightning sparked the latticed sky;
the old king Dhritarashtra, bowed his despairing head.
His friend, the charioteer Sanjaya, spoke
"Duryodhana knows the good but follows evil
so men destroy themselves by wicked deeds
as if in joy they welcome Yama, Lord of Death.
Who can avert disaster?"
That same moment far away in the forest
Krishna consoled the grieving Draupadi: "This is the wheel of life,
from birth to birth,
you must accept the good and bad entwined."

Thus twelve years passed, of exile in the forest.
Only the warrior Arjuna wandered, seeking
adventure and winning arms of magic skill,
bringing these to his brothers someday to use
against the evil Kurus, when the time
of exile passed. The thirteenth year began:
the brothers and their courageous bride
decided on the roles they were to play.
She, as a serving maid named Sairandhri
to queen Sudeshna, who marvelled at the beauty
of graceful dark-hued Draupadi. She stood like a goddess
voice soft as love, her lotus eyes downcast.
"Surely" Sudeshna said "you have no need to serve:
no man is there who will not covet you
and willingly agree to be your slave."
"Parted I am," Sairandhri said "from five husbands,
handsome Gandharvas who wait for me in heaven,
but I must wander on earth a little longer;
cursed by the gods, alone I wander here."
So the gentle princess became a serving girl.
Her husbands too around her worked
in the court of Virata, king of Matsya.
Yudhishthira, ironic fact, was teacher of diceplay,
Bhima, a cook, Sahadeva, a cowherd,
Nakula a horseman of marvelous skill
and Arjuna, greatest warrior of them all,

taught dancing to the princess Uttara. It happened thus,
while sojourning in heaven in his travels,
the handsome prince was seen and much desired
by Urvashi, fairest of celestial apsaras.
Ageless she was, yet when she went to Arjuna,
he smiled and said, "Most wonderful of dancers,
Urvashi watching you perform
makes me forget our troubles and our grief:
yet take you to my bed, I cannot, for you
and mother Kunti I revere as elders."
Furious and enraged Urvashi cursed the prince:
"You too will dance if that is what you love
not as a man, nor woman but as hated eunuch
for one full year, scorned by both men and women!"
Arjuna pondered now upon her curse,
the blessing in disguise that hid his manliness,
dressed in a woman's garments, his hair untied,
as Brihannala skilled in song and dance
he charmed the old Virata, who engaged him then
to teach his daughter Uttara the divine arts.

A terrible incident then occurred
to Draupadi, whose beauty stirred all men.
The Queen's own brother Kichaka a lecher
called her to him in spite of her protests,
the Queen in innocence urged her on,
but Sairandhri fled to her husband Bhima
calling for help, and he prepared a ruse
luring the wicked Kichaka into a room
where wrapped in clothes, upon a scented bed
he stretched himself. Kichaka, expectant, entered
determined to seduce the timid doe,
he thought no woman could resist his will
nor any man the strength of his desires.
The kingdom's power lay with him. In this way
men boast, forgetting highest Dharma.

Vain Kichaka set the lamp upon the table;
rubbing his hands he said "no one my dear
will teach you all delights of love as I,
handsomest of princes, and with me I'm sure
you will forget those Gandharvas who you say
are husbands to protect you ever more."
He stretched his hand envisaging surrender
when with a sudden roar, Bhima arose:
"Kichaka you spoke well, there will be soon
no one like you in this entire world."

With that he struck in anger and Kichaka
drew his sword. Bhima the strong
crushed the blustering fool, rolling him into a mass
of flesh so that those who found him dead
wondered saying "no human but a God has killed."

Duryodhana's spies in far Hastinapura,
hearing that powerful Kichaka was killed
gathered an army to attack the frail Virata.
Seizing his kingdom would be profligate gain.
Thus evil Karma leads men to their doom.
The Kurus' allies stole the cattle of the Matsyas;
Virata's forces followed them in close pursuit.
The city emptied; the Kurus stormed the kingdom
knowing that all the warriors gone, no one was left
only the young boy prince Uttara, whose sister
bravely summoned Brihannala as his charioteer
for he had told her once he drove Arjuna's horses.
Seeing those dread enemies, Uttara fearful
escaped in fright, but Arjuna caught the child
revealing his true self, and from the Sami tree
quickly snatched his hidden weapons,
his strength and manliness returned to him.

The Gods rejoiced, the crucial year was over!
From heaven flowers flew, the stars spilled out
their light, the fragrant earth rejoiced.
And Arjuna shot one long arrow through the sky
spreading the mist of sleep upon the earth.
Drona and Kripa, Karna, Duryodhana slept soundly
their weapons scattered in disordered heaps.
Uttara drove back, the mighty warrier Arjuna
stood in the rear, no longer Brihannala.
By evening all the heroes came together,
Virata's joy so great he knew not what to say
"Let me just look at you" he stammered
"How fortunate that you chose a humble man like me
while hiding from those devils."
Krishna was first to greet them. The splendid One
saw from the play of time, destruction lay ahead.
All men want peace but only on their terms.
Yet he the Lord strove to reconcile the enemy,
as messenger of love he sought to soothe
Duryodhana's flaming anger and his greed for power.
"Mere messenger I am oh king," so spoke the Lord
"Yudhishthira asks only for their rightful share;
desiring to live in friendship, they wish for land,

to lay down arms, to live again as men of worth."
Again, again did Krishna plead, yet knowing
all the while, the Kurus would not yield.
They did not even know that this was Krishna,
Lord of the Universe, who moves each straying star!
Stubborn Duryodhana refused, forcing the Pandavas to battle.
Thus wars are waged and hatred never cease;
lulled by false hope man scoffs at dogged Destiny
then calls on God to strengthen and to bless,
to mend and patch their guilty frightened hearts.

The tyrant Kurus, declaring the Great War,
chose Krishna's army, while Arjuna the friend,
asked the Lord to guide and teach,
to be his charioteer in his distress
to drive the chariot into the scene of stress.
The war began, the armies stretched afar,
friends, cousins, brothers on the field of Kurukshetra;
yet Arjuna, finest fighter of them all,
did hesitate to let his arrows fly
into the hearts, however vile, of all his kinsmen.
Turning to Krishna, he bent his head and said
"The cause of all this teach me, Lord.
I am bewildered; so every man must be
born to the earth, are these to be our acts?
Killing the ones we love, causing more blood to flow?"

The charioteer brought the chariot to a stop.
Krishna, God descended to the earth
for this one timeless moment—
each spoken word would wake humanity—
taught Arjuna and through him all the world
the secrets of a wisdom old and tried:
"There never was a time my child" he said
"When I was not, nor you, nor they,
yet mankind realizes not immortal soul
and grieves for loss which is not there.
Act wisely, work, but let result be Mine,
surrender every act to Me, yet do not be
idle or sloth; intent on wisdom, commune with me,
selfless serene, the Self is seen reflected in each one
for you are That, Arjuna, and That is you.
Fight on, your dharma leads you to this day
but keep all righteous thoughts within your heart.
Let arrows fly but free of passion; your arrows
come to Me, as do the arrows of your foes:
for I AM, Arjuna, the soul of every being
through birth and life, eternal and unchanged."

The armour shone, the conches sounded sad,
the flags of war betrayed men's foolishness.
The smell of sandal rose from elephants prepared for war,
the shuddering earth awaited gifts of blood.
Still hesitant, noble Yudhishthira, unarmed and unafraid
walked through the army to his grand-uncle's tent;
"I ask permission, Pitamaha, your gracious blessing
to fight your army and to strive for victory"
thus bending low, he touched great Bhishma's feet.
"May victory be with you," the wise one said.
Likewise did Drona, guru of them all.
The brothers Pandu ever righteous in their acts
bowed to their uncles and their evil kinsmen
yet there was still no talk of peace;
only the cries of war rent through the air;
such is the folly of mankind, the choice of darkness,
not of light. . . . Souls of God
take human form from age to age but yet
their message goes unheeded and unheard.
Dharma conquered by seductive Maya,
remains an echo to all human ears!

Brother fought brother. Arjuna, Gandiva bow in hand,
stood on the crest of the seething human sea
surging in joy and pain, voyaging through to death.
Empty the victory, immense the ravaged heaps
of corpses piling in limitless desolation.
Young Abhimanyu, ancient Bhishma, ravaged Drona,
fell like green saplings before the murky storm.
Bhima in dreadful wrath, faced Dusshasana.
Over earth and sky a shadow fell;
as though remembering the tears of Draupadi
nature was still, as Bhima rushed
towards Dusshasana, plucked out that evil arm
that dragged the lovely girl into the hall
where husbands sat like senseless phantoms
now revenged. He tore apart
the evil Kuru's heart, immersed his hands
in blood, and knotted up the lovely hair of Draupadi.

Fleeing Duryodhana ran to the water's edge
but Bhima warned him of impending doom,
and with a furious stroke severed that thigh
which Draupadi had cursed so long ago,
saying "evil Duryodhana the thigh you offer me
shall cause your death one day
these words I swear."

The time had come, the hero smashed his foe.
Thus died Duryodhana, aggressive enemy of righteousness,
yet war is war and never peace can bring
only despair and vengeance and deceit.
Ashwatthama, Drona's son, destroyed the sons of Draupadi,
the sleeping innocent ones, at dead of night.
The Kurus perished. The Pandavas had no heirs.

Draupadi, noblest of women,
shattered at the loss of all her sons,
shone brighter than the sun
when she forgave the erring Ashwatthama
saying "Let not your mother grieve or weep as I have done."
Yudhishthira, entering the kingdom, found no joy.
His grandsire Bhishma lay dying on a bed of arrows;
death waiting for his hour, as
calling Yudhishthira to him, he spoke with love:
"Now you are a ruler let your people dwell
in fearlessness and in prosperity. From avarice, sin proceeds.
Comfort the weak, teach self-control, above all, truthfulness;
harm not a single being in act or mind or speech;
equanimity comes through selfless deeds for others;
falsehood is darkness, truth eternal light;
power and wealth are shadows in a dream,
righteousness alone can save the world my child."

With gentle smile Bhishma turned then to the Lord
"Krishna" he whispered "I know you well, Narayana!
Protect them all and let me now depart,
blessed by you, O lotus-eyed, I shall attain my joy."
The Compassionate One looked at the ancient warrior
whose feebly uttered words pierced every heart.
Truth is the greatest highest power. Truth! Truth!
Mountains and valleys re-echoed with the sound
as Bhishma, freed of earth, in a shaft of light
became one with the glowing firmament.
So ended the great great war,
a fire of hate devouring all,
no life untouched, sad was the victory.

Time ran its course, the winds blew cold;
Dhritarashtra with his queen and Kunti
were first to go into the forest, then beyond.
Yudhishthira the honest ruled the land.
Krishna the man knew it was time to leave.
Returning to the woods, silent, he sat alone
where are all the values gone, who speaks of Dharma?

Yet, heavenly power must come and man must understand
so Krishna pondered.
God knows it all
His words, His deeds, His truths will live
as onwards in the stream of time we whirl
for we are simple players acting out a dream.
Yet awakening know that all is One.
Tat Twam asi! Tat Twam asi!
Om namo Narayanaya

Experiments in Health

22. A Communion with Atoms and Evolution

The previous chapters have described what has already happened; the remainder of the book develops some thoughts for the future, bringing together East and West, experiences in India and in America that may lead to new vistas and stimulation.

In the late 1970s, problems of aging had started to appear (I was born in 1925). Eventual death is unavoidable, of course, but it is fun to try to postpone that a little. In fact, how one reacts to trouble is often more interesting than the problem itself, designing new experiments more entertaining than thinking about death. Astronomers Serkowski and Bok taught us how to approach death; the latter was once quoted in a local newspaper[1]:

> On the day I pop off, do not say how sad for the world that Bart Bok popped off. Say it's about time. He had a hell of a good time, and lots of fun with the Milky Way. He has contributed here and there. He has not been one of the great men, but he has been a good worker in the field, and he has been a damn good propagandist for the Milky Way.

So let us make some experiments, first with our perception of ourselves and the atoms within us, next with eyes and heart, and finally with exuberance and mental alertness. The idea for the title of this part, "Experiments in Health," is taken from *Experiments in Truth*[2] by Mahatma Gandhi who increased truth in human relations.

In December 1978, the South Indian Association of Ahmedabad celebrated its yearly Cultural Festival. For that occasion, Mrinalini Sarabhai wrote an article on choreography in India, and one of her themes was our communion with creation:

> Primitive man had a communion with cosmic creation that is completely lacking today. He rejoiced in the rising of the Sun, the enchanting flight of the birds in the evening, the starry magical nights. Life was religion and there was complete contact with nature, with existence itself. But ours is essentially a century of the intellect. Science, with its involvement with the problems of the universe, its emphasis on man and his reaching out into space is a tremendous challenge to the artist, whose reliance is on an inner vision and whose message is of man's soul. Knowledge at our finger tips, mechanical inventions, better living, fast traveling, all these have contributed to man's progress and also to man's confusion. It is perhaps fortunate that in India the vestiges of our cultural inheritance still remain partly intact. Our civilization though fairly rigid has a basic philosophy that is the most tolerant in the world. Though this freedom may be abused, it is still better than having to be a cog in a purely mechanical world.

Mrinalini challenges us to examine our motives in science. What do our hearts have to say about these matters, in addition to our analytical minds? What should be our goal as scientists and observers of the universe? Is it not the search for an understanding of our world? When we are at work, in technical development of new equipment, during long hours at telescope or computer, or in thought and study in our office, are we not in communion with cosmic creation? Our task, our work as scientists, is to understand the origins of ourselves, of the Earth, the solar system, and the universe. Some of us may have lost the scholarly or philosophical background of our disciplines, but the narrow specialization so common these days seems again to be yielding to a wider perspective, following a cyclic trend. The increasing travel and communication between East and West may serve to enrich us in that respect.

Just as dancers practice their communion through their dance, astronomers do it through their science. There are differences in our circumstances, of course: The astronomer studies celestial objects with telescopes, while the dancer sees people from a sacred dance floor. Neither of the two is used to thinking that all of what they see in their daily lives consists of much smaller structures, atoms, and molecules, that must be taken into consideration to understand our origins and ourselves. While working on this book in June of 1987, I suddenly realized that we can make progress in our understanding by taking into account the origin and the properties of the atoms within ourselves and in our environment.[3]

As an introductory step into the smaller sizes and as a demonstration of the progress in our understanding, let us consider an ant crawling among shoots of grass. It has a vision that to us seems severely limited, but the ant does not suffer from the limitation because it usually sees what it needs for its survival. The ant does, however, have frightful unknowns such as the human shoe that may come seemingly from nowhere to crush it. The religious among the ants might be praying to the

Shoe God! Imagine now that the ants could evolve and build an instrument like a telescope. They might then be able to understand the shoe, but perhaps not other disasters yet, such as lightning in the distance; their future generation would need to develop larger telescopes and better theories.

People have similarly over the ages improved their observations and understanding, both of larger things and smaller. We could, for instance, tell our little ant friends about bacteria that kill them, and people too. During our Middle Ages there were plagues, and the frightened people prayed to God for delivery from that scourge. Now we understand more, we have careful hygiene and proper medicine to keep bacteria under control.

For an appreciation of the atoms we must first get used to the smallness of their sizes; scaling down to the ant was only a beginning. The atoms are so small that within us we have about 1,000,000,000,000,000,000,000,000,000, or 10^{27} of them.[4] We are totally made up of atoms: they make up the molecules in our cells, our bones, and our brains. Their components, the subatomic particles, have astonishing properties that they acquired at unimaginably high pressures and temperatures—at least 10^{21} atmospheres and 10^{21} degrees Celsius—at the start of our universe.[5] We think we know from the currently observed expansion of the universe that some 15 billion years ago it was compressed and hot, and that it has been expanding and cooling ever since. At first the universe presumably consisted of nothing but highly energetic subatomic particles, and it was only a little later that the nuclei of hydrogen were formed, and some helium too. The nuclei of heavier atoms, such as carbon and oxygen, were formed later, from hydrogen and helium inside stars, and they were dispersed throughout space by supernova explosions of some of the stars. In this way, long before the origin of the solar system 4.6 billion years ago, space obtained the interstellar clouds of gas and dust.

The astonishing conclusion follows that every one of the 10^{27} atoms within us is the result of a celestial inferno of energy. People do not have such high pressures and temperatures within them, of course, and they therefore do not produce individual atoms, let alone subatomic particles. We can "make a baby," but we cannot make a single atom. We can make the baby and sustain its growth because we eat and drink and breathe, taking in atoms from our environment. Seemingly lifeless, "inorganic," matter is consumed from our foods and the air, regenerating life. When plants and animals die, they too can be used for the generation of new life. There is a continuing reincarnation as the atoms we exhale or cycle through us or leave behind after death are taken in through breathing and feeding and growing again by other living things.

There is a conclusion that immediately comes to the fore, and that is of the interconnectedness of people. Some of the atoms within us have been in other people before, and they will be in others in later generations.

The atoms I exhale are mixed into the atmosphere and can be inhaled by a plant or someone else. Thus we are related, you and I and Buddha and Jesus Christ!

The interrelationship reaches much farther yet because all the subatomic particles in our environment came from the same universal source, from that origin of the universe. We saw in Chapter 16 how the gas and dust in the space between stars produced the material from which the solar system formed. The atoms within us are therefore connected to the interstellar material, to the stars in which the heavier atoms were formed, and to the formation of the atomic nuclei and their subatomic particles at the beginning of the universe. Every one of those fiery atoms within us is billions of years old. That makes us to them quite temporary configurations! If we live 94 years, as Van Biesbroeck did, we think that is a long spell, but what is it compared to the age of the atoms of billions of years?

One way to think of an atom is as a heavy nucleus with shells of orbiting electrons; it is a rather old-fashioned model, but still useful. A hydrogen atom has one such shell in which one electron races around. Heavier atoms, such as carbon or oxygen, have several shells and a total of a dozen or more electrons in constant whirl. The nuclei and electrons are incredibly small compared to the sizes of the shells, and therefore of the atoms. The radius of a nucleus is about a ten-thousandth of the radius of an atom, and the electrons are essentially points, as far as we can tell. Imagine a hydrogen nucleus scaled to the size of a pea. If the pea were at the center of an olympic stadium, the pinhead electron would be orbiting at the outer perimeter of that stadium! An atom consists of nearly nothing but space, at least 99.99999999%. Since everything is made of atoms, it is clear that every object consists of nearly hundred percent empty space—and we also. A deflating thought!

But it is just this empty space in which the atoms and their electrons are moving and vibrating and flowing and whirling, like dancers, in never ending and unbelievably energetic gyrations. The electrons race around their nucleus some 10^{14} times per second! That is what makes objects solid, giving a hard appearance to a telescope or a dance floor, even though they consist primarily of space.

The subatomic particles, confined within the nucleus where we cannot see them but can only surmise their behavior, must reach even greater speeds than the electrons outside, nearly the velocity of light, 300,000 kilometers/second. Then to realize that this is going on all the time, whether we are asleep or awake, and not with just a few but with 10^{27} of those extraordinary engines inside us! Physicists of the early 1900s, such as Albert Einstein, Niels Bohr, and Werner Heisenberg, had fascinating times discovering the properties of this atomic world. Perhaps their lives would have been easier if they had known earlier about the origin of atoms, during the beginning of the universe with its high pressures and temperatures. They would have realized that under such conditions the particles were forced to become highly energetic and have esoteric properties. Heisenberg[6] has described how puzzled they were:

> I remember discussions with Bohr which went through many hours till very late at night and ended almost in despair; and when at the end of the discussion I went alone for a walk in the neighboring park I repeated to myself again and again the question: "Can nature possibly be so absurd as it seemed to us in these atomic experiments?"

Heisenberg concluded that at the atomic scale there must be uncertainties in what we can know. For example, at any one instant we cannot know precisely both the position and the velocity of an electron. That is Heisenberg's Uncertainty Principle. For the components of each atom there exists an indeterminacy as if there were a freedom to be different from the others, an individualism or a free will. The model of the atom as shells of orbiting electrons is therefore not entirely accurate, because an electron cannot have a well-defined orbit. In fact, the electron can be thought of behaving instead just like a set of waves, like radio waves, instead of as an orbiting particle. Light, our ordinary daylight, has that same dualism, which actually is an indivisible holism: it behaves like waves and particles, at the same time. This is not so difficult to grasp, however, for we the people have the same characteristics. We also are an inseparable combination of waves and particles, of mind and matter. Descriptions in atomic physics are made in terms of relationships, closely interwoven in space and time. It is similar to the way people are known by their relationships to each other. The electrons are thought to occur in orbital waves, and they change their energies in quantized jumps from one wave pattern to another; overlapping waves, with electrons sharing the same pattern, gives rise to the bonding between atoms. The waves are thus imagined as a flow of interconnections with other atoms to make molecules and ever larger associations in molecular strings and cells and crystals and all the objects and forms we are familiar with.

The atomic dance goes on tirelessly, seemingly forever. Although some nuclei decay into others, the hydrogen nucleus, the proton, lasts longer than 10^{31} years; that lifetime is being measured in a deep gold mine in Southern India and elsewhere, in experiments that were known to Andrei Sakharov and Indira Gandhi (Chapter 20).

Imagine the myriads of whirling atoms everywhere! Materials may seem hard to our touch, but remember we are touching and treading on mostly space with atomic waves and particles continuously in motion. It feels firm, but it is alive with vigorous movement and esoteric properties. Even inert material still is alive in atomic motion, and ready to move to other combinations. Dead leaves that fall to the ground continue as fertile compost.

The atomic drive is not poetically gentle, however, for it continues relentlessly, in all permitted mutations. It is no wonder then that evolution occurred—it had to happen. Atoms must join to make molecules whenever possible. Molecules must interconnect to exploit all feasible combinations such as the long molecular strings of DNA. And so it went in making the variety of species: some worked out well and others not, some varieties branched out richly while others vanished, as nature selected those that were fit. Climates and environment made plant, fish, and animal species adapt and develop differently under different conditions. Some apes remained apes while others developed into humans. Even among people there are variations in appearance—light skins and dark, flat noses and long. There are remarkable individuals within groups and families, but education and conditioning play a role in addition to genetic variation.

Life and matter are interrelated, as are the characteristics of each individual entity, be it an atom, plant, animal or human being. The unity of atomic structure is striking; there is a completeness, a wholeness, about the ensemble of the participating components and esoteric properties of dualism, indeterminacy, and so forth. When atoms are combined into molecules, again there is a completeness about that system, with indeterminacy still present. At the end of each stage, a wholeness is reached that includes new, esoteric properties not present at the lower stage. These are most strikingly developed in the workings of the brain. Everything in the brain is based on chemical and electrical processes, but the result of connecting ten billion nerve cells together may be greater than the sum of the parts. Consciousness, thought, personality, and free-will are no longer entirely unexplained miracles, but rather natural consequences of the workings and properties of the 10^{26} or so atoms in action in the human brain.

Thought, free will, determination, and peace of mind are as much a part of the person as physical strength and well being. They all are implicated, following the example of each of our atoms, which has its energetic wave motion interwoven with material substance, yielding the restless drive that originated us and still sustains us. A healthy person is an integrated whole with spirit and body in balance, as I will demonstrate in the following chapter.

23. Eyes, Heart, and Yoga

In speaking of health and physicians, my father had warned us to seek a second, independent opinion before agreeing to a major cure or surgery. When I was an undergraduate student in Leiden surprisingly little basic science was taught to medical doctors. The pre-med students used a physics text that had been written especially for them, with mathematics removed. Since then, I have made a hobby of observing medical people.

When Neil was in grade school in Indiana in 1959, he participated with a thousand other schoolchildren in an experiment to determine the effectiveness of fluorine in toothpaste. Half of them were given toothpaste with fluorine, and the other half an identical tube but without the chemical. The result has been advertised widely, and nearly all American toothpastes now have some fluorine compound in them. But what about long-range side effects? Fluorine is highly toxic! The experiment seems strange to a physicist, whose approach would have been to study in detail the interaction of fluorine with the chemical substances and physical structures of teeth, gums, and other parts of the whole body. The medical world argues that such an approach would be too difficult, that the human body is too complex for such detailed analysis. But the argument does not make much of an impression on many atomic physicists, or astrophysicists studying the complexities of the universe.

Back in our Indiana days our dentist had a newfangled imaging tool: x rays! They were used abundantly, needed or not, and the dentist would stand next to his patient and be exposed to the radiation. My comments that the x rays destroyed cells were countered with, "The dosage is low enough; it is safe." That there is a cumulative effect, a number of cells destroyed each time, penetrated into their literature only later, and x-ray technicians now avoid exposure to the rays. But seemingly nonessential x raying of patients is still going on.

I had another jolly encounter, this time with an optometrist who tried to sell me glasses a decade ago. If the muscles of a leg are getting old and stiff, would one put them in a cast? How about exercises? If the eyes are getting old and stiff, should one harness their performance with glasses? How about eye exercises? But now I had entered a no-man's land between common sense and a lucrative industry.

Eye exercises are said to be common in the People's Republic of China; school children and women who do fine-point needle work in textile plants have pauses for eye training and relaxation. My eyes are used heavily and suffered a special hardship in the early 1960s: For observing the Moon with the 82-inch McDonald reflector in Texas, I would set the main telescope at precise locations on the lunar surface by looking through an eyepiece. That light was too bright. Astronomers do not look through telescopes anymore; even photographic emulsions are becoming obsolete, because we have other detectors of light such as photoelectric photomultipliers and charge-coupled devices.

In the 1970s there were few optometrists in the United States with training in eye exercise, and they kept that practice quiet. Their professional organizations might have disqualified them and have their licenses taken away. In Tucson, a city with some 300,000 inhabitants at the time, I could find only one optometrist willing to coach me: Dr. David Friedel. The encounter was an eye-opener.

David explained first that exercises are not applicable to all cases of eye disorders and may in fact be detrimental or even destructive for some. When there is an eye problem one must not begin such training without consulting a doctor or optometrist. The problem might be glaucoma or cataracts, for instance, and eye exercises would be no good and might cause a disastrous delay in treatment. Furthermore, my problems were connected with aging so that I may need eyeglasses eventually anyway; indeed, I had noticed a gradual degradation in needing stronger lights for reading. And finally he explained the concept of interaction, of considering the body as a whole. Delicate organs, such as our eyes and ears and heart, are connected with functions elsewhere. The heartbeat, for instance, is controlled by the brain, the brain's condition depends on that of the entire body, and the body's condition depends on interaction with other people and the environment. No part of the body should be treated by itself, for it is an integral part of the whole. There is a holism: the whole body and its environment must be taken into account. This is, of course, the theme of this part of this book: to become a more integrated person, with mental and intuitive capacities enhanced by physical exercises.

Friedel explained further that aging problems of eyes are often caused or accelerated by nervous stress. The first step in eye training is therefore to learn to relax. I had to lie flat on his office floor, consciously and totally relaxed. Then fiercely contracting all of the muscles that one can possibly think of in the most intense concentration of effort for a minute or less. Then letting go again, exhaling and relaxing totally. And so forth, maybe five times.

Such exercises have now become part of my everyday routine, and I will tell about these experiments with health in this chapter and the next. They will be described in three stages, as I went through them, of increasing enjoyment and perfection. First, it was easy to use a little imagination and do what feels good, without a formal exercise period, anywhere—in an office, in an airplane, or lying in bed. The eyes themselves are not touched, except for maybe a light shower massage on the closed eyes. They are not exposed to bright lights; Aldous Huxley disagrees with that,[1] but I cannot see the sense of exposing our eyes to light that is many orders of magnitude brighter than that for which they evolved. Reading fine print is a simple exercise. Another is to cover one eye with the cupped hand of the same side, without touching the eyelid, and keeping both eyes open. Look at the thumb of the other hand and follow it as you move it around a big circle such that the eyeballs roll around as far as they can; after a few turns, switch eyes and cover the other side, and so forth, maybe three times on each eye, rotating both eyeballs in alternate directions. It feels good to have the eye muscles stretched, without overdoing it; and not just the ones that move the eyes exercised, but also those that open and close the iris for bright and dark. Next, stretch the thumb as far away as possible, keep looking at that thumb as you move it back to the tip of the nose and then slowly moves out again; follow that with a fast-focus exercise by looking at a distant object, at that thumb, at the tip of the nose, and on out again. Friedel recommended doing that exercise by putting some beads on a string and tying it to a point a few yards away, so that with the other end held at the tip of the nose the beads are at distances of about 5, 25, 100, and 200 centimeters away; focus quickly on each bead in succession. These exercises I also do sometimes with one eye at a time, covering the other one. By rolling the eyeballs to the extremes of their range to keep looking at the thumb as it moves up and down and left and right one can extend that range remarkably far backward in peripheral vision.

Curiously difficult is to scan along a line, such as where a wall meets the ceiling, from extreme left to right and back again, in a slow, exactly uniform motion; it is more difficult than expected because the eye is used to scanning in jerks.

With fingertips I give a firm massage to muscles and skin all over the head, neck, and jaw; soon the body reacts with a deep sigh of relaxation. I am trying to extend these massages to the muscles[1] around the ears and eustachian tubes to see if they improve hearing as well. The rubbing therefore goes also into the hollows behind the cheekbones and up around the ears and strongly so, massaging around and pulling the earlobes.

A refreshing feeling for tired eyes is achieved with an eye lotion, some neutral borate solution. There are various brands on the market, some of them equipped with a cup that fits over the eye to spread the eyelids open and away from the eyeball. In the shower or with cupped fingers one can do the same with water; I do this often in Ahmedabad, because of the air pollution caused by illegal use of soft coal without precipitators, and in Tucson, because of microspores in the air. It is some experience to look at those pollen grains in a high-resolution microscope; their hairs and hooks look spooky.

All the exercise has worked out well and I need no glasses as yet, more than ten years since I went to Dr. Friedel.

A few years ago I began another exercise regime, for the whole body, because late in 1983 I got the impression that NASA might select a space scientist to fly as passenger on the Shuttle. I had kept alive the wager I had made with Hans Mark in 1976 when he lectured in Tucson about the exciting future of the Shuttle.[2] His enthusiasm did not seem to move the audience—being planetary scientists, they feared the Shuttle would drain funds away from proper exploration—so he added for pizzazz, "I bet that our host Tom Gehrels and I will fly together on the Shuttle," at which I jumped to my feet: "Your bet is on." In 1983, Mark was Associate Administrator of NASA, so I sent him my application. The passenger astronaut had to have proven ability to tell the world of his or her experience and I therefore asked people who had heard me lecture about Pioneer, such as Arthur C. Clarke, to support my application. My case seemed strong, and I was sufficiently encouraged to start an exercise program to get into shape for the examinations that might come soon. That actually did not happen because President Reagan ordained that the Shuttle passenger would be a school teacher, but I exercised with zest anyway.

The University of Arizona has a jogging course through the campus about a mile long with some seventeen stations where one does knee lifts, pushups, pullups, et cetera. The first morning I ran with utter delight, to the end, where instructions are posted to count the heart rate. The maximum permissible for my age was 161 beats per minute; the formula (between 20 and 65) is 220 minus the years of age. My count was well over 200 and I noticed with some consternation that every fifteenth pulse or so seemed to be missing. Back in the office the defect disappeared when the pulse slowed down to a normal 60. Just to be sure, I made a phone call to a doctor, who was not available, but his nurse told me not to worry: "We all have that at times." So, the next morning I ran the course again; this time the heart rate rose to near 220, and now every fifth pulse or so was missing. This defect did not go away until nearly two years later in Ahmedabad.

The phenomenon of apparently missing heart beats is called Premature Ventricular Contraction; the beat may actually not be missing, but superposed onto another one. I could feel the effect of the PVC sometimes from the top of the head to the tip of the left big toe. Three doctors and specialists made tests, with a monitoring recorder attached for twenty-four hours and having me run on a treadmill. They all said the same, not to worry, we all have it occasionally, and nothing can be done for a cure. My problem would, however, be severe at times, especially at Kitt Peak where the Spacewatch observing requires hard physical work at the 6700-foot altitude, and is accompanied by stress caused by the planning of new programs, occasional equipment problems, and wanting to use all of the night efficiently. My particular problem is not knowing limits, not recognizing when to slow down. Sometimes there would be more than one successive heart beat missing, or the pulse would be irregular and at the same time there would be a feeling of nausea and weakness; occasionally the PVCs would wake me from sleep. The deficient heartbeat might cause a shortage of oxygen that brought chills and tiredness. A few times I seemed to pass out. I was alone in an isolated dome.

Reporting back to the physicians in Tucson brought only the usual comments, and I felt from them a funny kind of doubt whether the patient's observations were correct or perhaps imagined. The medical world in the United States seems to show some disdain for patients, especially those with a foreign accent. A fun experiment is to call a doctor's office and ask to speak to Him. If I say "This is Tom Gehrels," he is busy. If I say "This is Doctor Gehrels," the office comes to attention and he to the phone. We have to wait long times in their waiting rooms to accommodate their schedules; our time is considered less valuable than his, even though we are the customer and pay him. Patients are called by their first name. By nurses and receptionists half my age! Me, a big-shot Professor! "Antoan?" and my resentment against that Anton name comes right back!

I have no argument with the idea that stress and imagination bring about PVC and other sickness. That is in fact the central idea of holism. The heartbeat is regulated by the brain, and the brain is affected by the condition of the whole body and its environment. I finally consulted Dr. J. K. Shah, the respected physician of the Physical Research Laboratory in Ahmedabad. He asked if I would consider yoga, which is especially effective on stress-related problems, and submit to its full physical and spiritual regime. Of course I would. He then referred me to Swami Manuvarya at his Yoga Sadhan Ashram for treatment and exercises. That became a delight.

Soon after 6:00 AM, I sneak out of the Sarabhai house and find a rickshaw to the Swami's Ashram. It takes a little patience, since at that time of the day the city is still rather dark and asleep. But the trip has a special charm in its freshness of winter morning twilight. A few people jog. The poor make small fires to warm themselves by. That is all, in contrast with the crowds of the daytime. At the Ashram some are already exercising, but there is little talk, and the atmosphere is filled with gentle hues and incense. A few more men come in gradually and change to shirts and shorts; the women practice separately, but a few of them attend the starting ceremony.

Candles are lit on a flowered ledge that looks like an altar in a small room; on the walls are a few pictures of people who have reached a knowing and exalted state. Our Swami, who is a kind, elderly sage, is arranging the flowers; he is the only one on a chair. His back is to us as

The Eagle posture of Yoga.

we sit in lotus pose on the carpeted floor. Soon after 7:00 he will start the short service of morning chants with three times "Hari Om," the basis of all the mantras, for which one inhales deeply to let the ooooom (as in "home") extend as long as possible while exhaling the full breath. Rhythmic and lively songs follow as a rejoicing about life and yoga and our teachers and peace of mind; the singing is accompanied by the ringing of a bell and shaking of a tambourine and a harmonium played by one of the women. I feel perfectly welcome and at home; no one pays attention to me as I happily sit with them and chant along the unknown words in Gujarati, not yet quite familiar with the exact meaning of it all. What a contrast to the Sunday I attended a Dutch Reformed Church in South Africa, where I was ostensibly the intruder to be closely watched! Here are no weighty dogma or allegiance to Commandments—no "Thou shalt do that" nor "Thou shalt not do this"—but rather a communion with health and wholeness. It is solemn, but at the same time celebrating life and the godliness within us, a relaxed occasion lasting some ten short minutes. The end of the service comes too soon but gently, after which we pay our respects to the Swami. Some of his followers kneel down deeply to touch his feet while others merely bow, hands together in Namaste greeting. The Yogi puts his right hand on each head for a short blessing and greeting word. Then we walk by an assistant who has a tray with a candle that is used to symbolize a source of energy by moving the right hand over it in a circle; next we cup that hand in order to receive a bit of milk to be sipped and some fruit to be eaten.

A shoulder stand.

Now my young teacher Vikram Panchal joins me. We start exercising in a nicely balanced series of motions, swinging and stretching all parts of the body in turn, accompanied by controlled inhaling and exhaling. Sometimes a full breath is put under such pressure that one can feel it tingling at the back of the head, while the body stretches can be such that the effect is felt for the rest of the day. Success with the "eagle posture"—balancing on a single bent leg around which one hooks the other leg, while the arms are also intertwined and held in front of the torso—is an indication of one's inner balance and peace of mind, which may vary from day to day. There are extended shoulder stands with the body and legs straight up, and head stands for the ones who can do that; the change in circulation brings a refreshing feeling. The procedures are supervised by the Swami who watches us from a distance and occasionally calls his pupils in for another series of chants, for which we stand in a semicircle around him, with our hands together in Namaste and eyes closed in concentration, repeating after the Master a variety of mantras, with *Om* repeated over and over.

The deeper cause of my PVC probably are stress, getting too hot, overdoing the labor, and also a lack of oxygen due to breathing with the chest muscles only. I was therefore taught to breathe with the diaphragm primarily, that is with the midriff and belly muscles. Since then I have been watching people and have seen that many breathe as incompletely as I used to, with the chest muscles only. The Ashram teaches a variety of breathing exercises: inhaling and exhaling through one or the other side of the nose; six counts for taking the breath in, twenty-four for holding it, and twelve for a controlled letting out. Another is to exhale quickly in bursts from the midriff for as many as sixty times in a row. In all exercises the mind concentrates on the body part and activity; in the breathing one obtains a sensation of communicating with deep down inside the body where the oxygen and carbon dioxide exchanges take place. The result of such practice is a deeply founded feeling of peace.

Learning to relax is again an essential part. It is a challenge to sit still in meditation, for five or more minutes, by concentrating the mind first on the breathing and then ever more inward toward an area between the closed eyes. One sits in the lotus pose—I think of it as the Buddha's pose—with legs crossed. Or one lies as dead, flat on the back; in this case I begin with Friedel's tensing, then consciously relaxing parts of the body successively from toes to tip of the head, the mind thinking of nothing but the breathing, and drifting slowly off and away. While lying down we also do leg lifts, just a little off the floor, but for longer and longer times, exercising midriff and willpower simultaneously. No flabby bellies for yogis!

We quit at about eight and by that time have gone through exercising all parts of the body. By 8:15 I rickshaw toward PRL with a glorious feeling of having rightly started the day.

The sum of these experiences is a feeling of being on top of mundane affairs, floating above them, perhaps even a soaring toward an exalted state, a Hindu nirvana, but willing and prepared to act vigorously in worthwhile pursuits. After only four mornings of this wholesome treatment, the PVCs were gone.

24. Sunercise and Serendipity

The combination of mental relaxation with thorough physical exercise brought a feeling of radiance, another liberation. Whenever I came back to Tucson from a trip to India, friends would ask how it was. At first, years ago, I answered with: "It feels like I have been set free, totally free." Later it was: "It feels like the Sun is shining within me," which is quite a claim for an astronomer to make! And more recently, "It feels like there is a oneness with nature and with problems and solutions; there is a peace and yet a penetration as if I can see through it all." People have looked puzzled, but my descriptions were spontaneous, the effects were strong, and they would not go away. So I tried to check if my answers were consistent with the principles of yoga as described in the Upanishads. I found there that the use of the sacred syllable *Om* is urged as a means of meditation, a "yes" to life and the world. They refer to the *Om* as a tool, a bow and arrow to pierce the darkness within the body. One imagines in the heart a tongue of flame that forces its way out. Elsewhere the Upanishads become more technical, and one reads that by means of *Om* one can penetrate the Sun of the heart and our 72,000 veins,[1] the flame journeys upwards through the carotids and pierces through the head, and the yogi continues to exist as the giver of health to all beings. The freedom, vision, and radiance bring a submission toward a goal of journeying upwards, taking all of the world along. It was impressive to realize that millions of people use that word *Om* to celebrate their joy of life, and it seemed a privilege to join them.

So I forgot about my PVCs, never thought about them anymore, until one night six months later at the Spacewatch Telescope. I was working and running up and down its stairs, all out, in utter delight, pushing a full program of observations through a beautiful night. The PVCs returned with a vengeance. I seemed to pass out briefly and series of heartbeats were missing. Unlike what the doctors had been saying about

PVC, this was a sickening experience, with nausea and fatigue, and a warning that next time it could be the end.

What happened in the following days is more difficult to describe. My Yogi and Vikram were on the other side of the Earth. But the Sun never stopped shining within me, and I began a period of bootstrapping. A nice verb, to bootstrap, understood by horse people: how one has to wiggle and pull upwards at the inside straps of tall boots in order to ride away in style. The usage in physics comes from trying to pull oneself up by the bootstraps, which a theory can do only if it is self-sufficient. The word implies a self-reliance, and indeed my recovery involved a building up, step by step, with meditation and exercise, gradually decreasing the PVCs while increasing the work. All of this brought about a new outlook, with fresh research plans for years.

The Sun shining within began to connect with the Sun shining outside in its radiant glory around the Spacewatch dome on Kitt Peak. My astronomer's imagination felt the presence of those fiery atoms and photons that had originated in the early universe and seemed to radiate inside and around me that energy they had never lost. The practical need to exercise and get rid of PVCs led to a regime of practicing all that had been learned from David Friedel, Swami Manuvarya, and Vikram Panchal. The exercising was done outside in living sunshine: an exalted set of sunbaths, not passively lying down but massaging and meditating and exercising and relaxing and breathing and inventing postures and stretches, and all the time exposure to the Sun.[2] I call it Sun-ercise, in analogy to Jazz-ercise, in which people practice to jazz and other music, though Sunercise cannot be done with music because it involves meditation and yogic concentration on the exercising parts of the body.

What is the direct benefit of sunshine on the body? Why does it make us feel good? I have found no proper answer from the medical experts as yet. More seems to be involved than the feeling of warmth. The Sun's curing effect on colds, sore throats, and eye and ear infections is immediately noticeable. Is there a nutrient value, vitamin formation, or energy generation by the human body directly from sunlight's photons? Ants walk slowly in the shade and fast in the sunshine; Jo-Ann has done the experiment in a science project for her junior high school. Medical people agree that some sunshine is necessary, and in dark northern winters or cloudy climes they recommend compensating with sunlamps. The other extreme is also obvious and to be taken care of: in places like Arizona, overexposure to any one part of the body could come already in ten minutes and cause skin cancer.

The development of Sunercise was the second stage in my experiments; the first was the immersion in yoga, described in the previous chapter. I still felt too busy to spend much of time on the Sunercise. Even so, it was fun to do a little here and there during walks up Kitt Peak, bike rides or climbing stairs, or in the office, during showers, or in the back of a Boeing 747. The awareness of the whole body in touch with the daily

A few Yoga exercises demonstrated by Vikram Panchal.

environment lies at the core of Sunercise. One can also obtain this contact with nature by bathing in the wind, or in the rain, or in the snow—which is a part of the Finnish sauna. An upside-down shoulder-stand in the office can be as refreshing as a break for coffee or a nap; in the shower it is a special experience.

An important aspect of Sunercise is the relief of stress. I reach back to the first lesson on David Friedel's office floor: There should be no regimentation, only some stretching, relaxation, and meditation. When the exercise is a pleasure, it is done steadily, without the forcing that often turns people away from physical exercising. Jogging, for example, is too regimented and monotonous, too opposite to relaxation; it is not a part of yoga, and I have come to dislike it as much as Neil Armstrong did. Joggers rarely smile. In Sunercise one improvises as the exercising proceeds, and sometimes forgotten teachings come back to be used. Instructors or instruction books are not required, nor is a previous experience with yoga; one simply starts with a stretch and deep breath and then one improvises, trying to train all parts of the whole body. The challenge is to come up with new things to do, thereby also exercising the brains. This is, in fact, the unity of yoga and the point of my writing about exercises in this book: it is the stimulation of body and mind together, at the same time. Whereas a jogger may listen to Walkman music for spelling the boredom, the yoga student concentrates the mind on each part of the body that is being exercised and develops not only the body but also self-awareness and power of concentration.

The following are a few practices I execute rather regularly. In addition to the eye and ear exercises of the previous chapter, there is a massage of the body including the soles of the feet. There may be a short warming up in the sunshine, letting the Sun into the ear canal too, which has an invigorating effect, and even into the wide open mouth. For a general warm-up, I enjoy swinging legs and arms and bending and twisting the body. A brief trot on the spot brings the heart rate to 120–140, and I probably keep it near that for the duration of the exercise. I tend to do everything three times, and usually the second time is better than the first, and so forth. Next I have a sequence of fourteen steps, developed from a daily exercise done at the ashram in Ahmedabad:

1. Stand at attention with hands flat together in front of chest, *Om*
2. Stretch hands high and backwards
3. Fetch ankles low and pulling shoulders toward the unbent knees
4. Left leg bent, hands on the floor, right leg far back, stretch head backwards
5. Left leg also back, with body a straight line supported on toes and stretched arms
6. Three pushups with the arms, down to the floor, and stretching back up (this is not done by my yogis)
7. Body far back, folded double on legs and knees, with hands stretched out forward on the floor
8. Bring head and chest forward low to the ground to between the hands with bottom peaked up and a hollow back
9. Stretch head up like a cobra, completely hollowing the back, and lift the body off the floor, supported only by hands and toes
10. As for #5, but with the back now high up and rounded, head between arms, heels on the floor
11. Step the right leg forward, keeping the left leg back, and bring the chest up with head stretched backwards (a repeat of #4, but legs interchanged)

12–14. Repeat the third and second steps and return to the first

Repeat the sequence three times.
One can freely invent other exercises and follow them until one reaches an integration of mental and physical well being.

Off the mountain, half an hour for exercise and meditation became an essential part of my daily routine in the early morning. Sometimes I do them also in the evening, or in the afternoon by one of the swimming pools of the university. My reaction to these experiments was to want to do more. The PVCs were completely under control, and I wished to know more about eyes and massage and yoga and to do more difficult exercises.

Eventually I got to a third stage of this training. My meditations are becoming more professional; each session starts by concentrating on the breathing, and that already is an exercise in holism since breathing supplies the oxygen, the primordial, fiery, life-giving atoms that are distributed throughout the body. Sometimes I focus on a person in meditation, sometimes on the atom and its electrons and nucleus, trying to visualize them as a duality of particles and waves, and at the same time integrated entities. At its best, the meditation becomes abstract, drawn inward, toward a spot inside and slightly above the eyes. In any case, there are many avenues to explore, and the practice gradually brings clarity to see through problems, greater concentration and willpower, and a peace of mind.

The notes list some references to a few books for more advanced exercises,[3] but one should proceed with caution: at least an occasional consultation with an experienced instructor may be necessary. Headstands, for instance, have been absolutely forbidden by a medical doctor for me at my age; I still do them but I realize there is a risk—of getting the PVC back, in fact. (A headstand must be done on the strong, forward part of the cranium, not on the center where the plates join and where, in a baby, one can see the pulse.) The attractive feature of headstands is the accomplishment of physical balancing, which seems to bring with it a mental balance as well. One of the great books from which to learn more about yoga and new postures is *Light on Yoga*.[4] It has a foreword by the American-born violinist Yehudi Menuhin that reads in part:

> The practice of Yoga over the past fifteen years has convinced me that most of our fundamental attitudes to life have their physical counterparts in the body. Thus comparison and criticism must begin with the alignment of our own left and right sides to a degree at which even finer adjustments are feasible. Strength of will may cause us to start by stretching the body from the toes to the top of the head in defiance of gravity. Impetus and ambition might begin with the sense of weight and speed that comes with free-swinging limbs, in contrast with the control of prolonged balance on foot, feet, or hands, which gives poise. Tenacity is gained by stretching, in various Yoga postures for minutes at a time, while calmness comes

with quiet, consistent breathing and the expansion of the lungs. Continuity and a sense of the universal come with the knowledge of the inevitable alternation of tension and relaxation in eternal rhythms of which each inhalation and exhalation constitutes one cycle, wave, or vibration among the countless myriads which are the universe.

It is a technique ideally suited to prevent physical and mental illness and to protect the body generally, developing an inevitable sense of self-reliance and assurance. By its very nature it is inextricably associated with universal laws: for respect for life, truth, and patience are all indispensable factors in the drawing of a quiet breath, in calmness of mind and firmness of will.

In this lie the moral virtues inherent in Yoga. For these reasons it demands a complete and total effort, involving and forming the whole human being. No mechanical repetition is involved and no lip-service as in the case of good resolutions or formal prayers. By its very nature it is each time and every moment a living act.

For a scientist, an experiment reaches a completion when it suggests a new idea or a new experiment. For me, then, these experiments in health are to be used to become more alert to new ideas and discovery. A healthy observer makes keener observations. Diligent pursuit always brings results, this is assured by the principle of reciprocity. If the results are fine and unexpected, it is referred to as serendipity.

The word serendipity is often misunderstood and used to indicate luck in discovery, but the original fable shows sagacity only. I found the story in an Italian translation of an old Arabian fairy tale.[5] A small part of the lengthy and bawdy travelogue of the three princes of Serendip (Sri Lanka) concerns a lost camel. The princes report to the king that they have seen evidence of it:

The first brother said: "I guessed that the lost animal must have had only one eye, because along the way he had covered I had noticed that on one side the grass had been eaten in spite of the fact it was very bad, while it was not so on the other side of the road where the grass was very good. So I thought that the camel must have been blind in the eye facing the side of the road where the grass was good, and had not eaten it because he could not see it, while he had eaten the bad grass he had seen on the other side."

The second brother said: "I guessed, Sire, that the camel was lacking a tooth, because I had noticed on the way many cuds of chewed grass of such a size that they could have come out only from the empty space of a missing tooth."

"And I Sire," added the third, "guessed that the lost camel must have been lame because I had clearly noticed the tracks of only three camel feet, while at the same time I noticed the trace of a dragged foot."

The amount of serendipity an observer encounters during his career may be an index of his or her diligence and effectiveness. Jim Elliott's discovery of the thin rings of Uranus is an example of serendipity: They were making careful observations, long enough before and after the predicted occultation of Uranus itself, and they were rewarded with a great discovery. The Pioneer findings of the thin F-ring of Saturn does not qualify as serendipity, because we were actually looking for a ring in our careful imaging, and Elliott had already pioneered in such discoveries. Of course, any discovery can be predicted beforehand if one is bright enough, but even then, other unexpected findings usually appear. Perhaps the discovery of serendipity as a test of effectiveness for observers is a case of serendipity itself, the result of searching for an original source for the tale of sages of the East truer than that of the eighteenth-century author, Horace Walpole, who coined the term.

Integration

25. A New Faith

In India, camels are still an important mode of transportation, but the trains, trucks, buses, cars, scooters, and three-wheeler rickshaws in towns and cities have made the society more mobile. There are 16,000 rickshaws in Ahmedabad, a city of about three million people, compared with 290,000 cars in Tucson, a city of about half a million people. The drivers of the rickshaws like to display pictures and sentences on their vehicles, a little like the bumper stickers on Western cars that tell us "I love my dog," or, most profoundly, "Astronomers do it at night." The statements by the new sages of the East sometimes show an unexpected wisdom. One of the rickshaws in Ahmedabad carries the thought, "Love is God." I had to look at that twice because I had often heard, "God is Love," and had agonized over that with a young widow in our Calvinist family, who had lost her beloved husband, the father of their five children, a good man without enemies. How could one believe then that God is Love and that He took such a man away from where he was needed so desperately?

Love seems to be present in even primitive forms in nature—animals show love. I see many pigeons from my perch on the eighth floor of the Physical Research Laboratory in Ahmedabad, especially toward evening when they come to roost on ledges of the building. These pigeons fight over the resting places, every evening again, even though there is plenty of space available. Each window sill is used by only one, or rarely two, birds and over this they fight viciously, strong beaks pecking and flapping wings hitting. Love seems to be rare among them, but it is not absent. In their courting they seem to kiss repeatedly in a gentle pecking motion, preparing glandular and other functions for the mating to come, after which they may stay peacefully together. A pair of pigeons roosts regularly inside one of the practice rooms of the Darpana Dance

Academy, even in the daytime; no matter how much racket the dancers and drummers make, the two sit closely and undisturbed.

Human beings also establish pecking orders and compete for more of everything, just like those pigeons, but they, too, can show love. They can also laugh and appreciate beauty. There is a godliness within them, be they rich or poor, friend or foe—something to keep in mind when one sees a bitter enemy or a crippled destitute! The motive for some of our friendliness toward others may not be pure love, but a simple matter of self-interest: doing good in our relations with other people may be more advantageous than being bad. Who does good, meets with good. A jolly attitude generally brings a cheerful response. There is a reciprocity in human relations.

The rickshaw driver exclaimed "Love is God." I have to be more guarded in using the word "God." It brings back too much of what I saw in my youth: the primitive concepts of sin and punishment, of creation by a God who looks like us, the fear of Satan and Hell and the unknown that seems to lie at the root of religions, the conservative holding back of progress, and the fanaticism that makes people so cruel in Ulster and the Middle East. We saw in Chapter 13 how important it is to define our critical terms, to use a dictionary. In India, in fact, the word "God" is used frequently, and I do not always know who or what is meant. I can celebrate along with the rickshaw driver, but only in the sense of "God" as "a thing of supreme value," which is one of the definitions in the dictionary. So, the first task of that child who looked along the steeple in Baarn so many years ago is completed: after much restlessness he became an astronomer to look at the universe with powerful telescopes, and he did find much of value.

One could not help seeing something of supreme value in the observable universe. It is awesomely vast. Its mass is estimated at 10^{54} kilograms, while my own is less than 70 kilograms. If we compare volumes instead of mass, we are even more insignificant, especially since we are mostly empty space as we saw in Chapter 22. The space between the stars is vast, but not entirely empty, as it contains radiation and life-cycling dust and gas. The volume of our observable universe is at least 10^{78} cubic meters (the farthest objects one can detect are at least 10^{10} times the distance that light travels in a year). A person occupies less than one-tenth of a cubic meter. We therefore are only one part in 10^{79} of our universe! How could one not have an awe for such a "glassy sea almost immense" and rank its value as supreme?

Mrinalini Sarabhai uses a different approach in defining God: "what *we* are but do not recognize, due to our ignorance." There is no conflict between the two definitions—even though that factor of 10^{79} makes us look so small—because we are made from the same material and have the same origin as the universe itself. The Hindu philosopher Sankara stated, about 1200 years ago: "You are rooted in Heaven; you are in the nature of God; be satisfied with no lesser status and destiny."

I have also found a religion, but only in the dictionary's sense of a "system of beliefs held to with ardor." I have no wish to return to a mythical religion, but even the one I was raised in had a useful ethical code. In this complex world it seems increasingly necessary to define a moral standard of behavior that all—religious and nonreligious—can agree on, a "moraligion," as I like to label it.[1] I was influenced by my impressions, as we all are, and for me those of India were added to those of Europe and the United States. In India, like everywhere else, I found old myths and superstitions, but also some concepts that were new to me and that I could weave into a general fabric. To strive for unity in a system of beliefs is a basic tenet of Hinduism; "yoga" is, in fact, the Sanskrit word for union.[2] Many people have reached such insights, and there is much on which they can unite, be they from East or West, religious or not. Lin Yutang, for instance, in his *The Importance of Living*[3] provides an introduction to combining Eastern and Western thought; it is a cheerful book with suggestions for joyful living. Fritjof Capra's *The Tao of Physics*[4] proposes that certain concepts of modern physics are in agreement with those in Hindu and Buddhist religions, and he shows how to combine Descartes' reason with the holistic philosophies of the East. The word "holism" was first used to express an integration into unity by J. C. Smuts, the South African general and statesman, in *Holism and Evolution*.[5] The word holism is after the Greek noun "holon" for the whole. Smuts' theme is that nature forms whole organisms that strive for self-preservation and improvement. We, together with the environment with which we interact, are such entities. Many factors enter in: instinct, brainpower, goals, willpower, personality, and also time. Smuts' book made an impression on Western philosophy; the ideas of holism have been further developed by many authors. The result is a system of beliefs that is easy to hold to with ardor:

The wholeness of a person. There is an interdependence of the physical body and the nonphysical mind. The mind drives thought and action; it is responsible for communication, sensitivity, and consciousness. Sometimes it is called "soul" or "spirit"; it sets goals and frames ideals, it determines the personality and searches for meaning and understanding. It has a freedom that is its human right, a certain amount of indeterminacy that appears to be its inheritance from atoms and the origin of the universe. Matter and consciousness are intimately interwoven. The duality of body and soul is as intrinsic to humans as the duality of waves and particles is to atoms.

The wholeness of history. Since antiquity, people have raised such questions as, "What are thunder and lightning?" "What is our origin?" "Is there a God?" People started as poorly informed as the ants in the grass, but they have found many answers. We may trust that knowledge and understanding will continue to grow. Much has become clear in the few thousand years mankind has been able to explore, and many of the problems that puzzle us now will be solved in a few decades or sooner (and new questions will arise). This is simple and certain, even if life itself is not. There is no need for teleology. No need to appease or replace fear or uncertainty with concepts of God. No need to refer to God as something as poorly defined as a cosmic consciousness, as some of my colleagues do. No need to see more than a thing of value and interesting to study in the origin of the universe, the laws of physics, or the constants of nature. Everything follows from the great pressures and temperatures in the early universe, the energy and ethereality carried through time by atoms. People create new people and new structures; they invented gods, myths, hatred, deviousness, and the means of destroying much of the life on Earth. But people also invented the arts and sciences, and they show love, common sense, will power, and a godliness within themselves. Preservation has prevailed until now; people apparently possess more "good" than "bad."

The wholeness of nature. Integration is seen everywhere in nature, in macro- and micro-cosmos both, and the interactions of the parts are expressed in physical laws. Each law and phenomenon we observe is the result of what happened before. There seems to be a holistic dynamics at work: Everything in our universe is in restless motion, striving toward combination into whole, survivable configurations. The restlessness of atomic particles lies at the root of all this motion and evolution; the atoms seek greater bonding, without stop and with penetrating energy. We see the results in the energetic whirling of star-forming nebulae and in cosmic explosions and radiations, in erupting volcanoes and teeming tropical forests and swamps, in everything we see and touch.

The wholeness of the world. The world is not a loose set of isolated parts, but rather a whole of interacting components. Large parts of the globe are now connected politically and ecologically. People are acting to improve communication. People are acting to cause global problems too. Although many suffer under totalitarian systems and fanatic and capitalistic excesses, the free will of the people tends toward greater freedom. One may thus have a long-range optimism, even when there are short-term setbacks.

The wholeness of life. There is an ultimate holism of a person in history and nature interacting with his world. The above points are inseparable; the reason for stating them is that they encourage us to act to preserve life, to strive toward truth and dignified freedom, and to live a full life with cheer, exercising all physical, intellectual, intuitive, and spiritual facilities. Here is the philosophical root for setting goals that come from past experiences and that are driven toward the future by free will and the human spirit. Holism means involvement with problems of neighborhood, country, and the world.

This is my "system of beliefs held to with ardor," which guides me toward the future in being as much as possible involved—with family and colleagues, atoms and textbooks, asteroids and the outer solar system, human rights and environmental action, and whatever next excitement will come along—and to execute all actions with an aim at professional perfection. One sees much of the subject of this chapter in a few quotes from Mahatma Gandhi, the great teacher[6]:

> If you want something really important to be done,
> you must not merely satisfy reason,
> you must move the heart also.
>
> Faith is a function of the heart. It must be enforced
> by reason. The two are not antagonistic as some think.
> The more intense one's faith is, the more it whets one's
> reason. When faith becomes blind it dies.
>
> As regards God it is difficult to define Him;
> but the definition of truth is deposited in every human
> heart. Truth is that which you believe to be true at
> this moment, and that is your God. If a man worships
> this relative truth, he is sure to attain the Absolute
> Truth, that is, God, in course of time.
>
> I believe in the message of truth delivered by all the
> religious teachers of the world. And it is my constant
> prayer that I may never have a feeling of anger against
> my traducers, that even if I fall victim to an assassin's
> bullet, I may deliver up my soul with the remembrance
> of God upon my lips.

On January 30, 1948, three gunshots were fired at him, unexpectedly and at point-blank range, but as he died he said "O God," but a better translation from Gujarati may be "Hail God," and he said it twice

> He Rama! He Rama!

26. A Look Back and a View Ahead

When I was a student in the 1950s there was little understanding of the origin of the solar system, and there was no integrated planetary science as yet. Isolated sciences, such as meteoritics, were providing some insights. Since about 1800, meteoriticists had known that they were studying extraterrestrial material—they were receiving samples from the asteroids long before the first spaceflight. There was planetary astrometry—quite a bit actually, and carried out at a dozen observatories all over the world—to measure the positions of objects in the solar system and to determine their orbits. It was Kuiper who started in the 1940s to use astronomical telescopes to observe planets, satellites, and asteroids for physical studies and to determine spectra of planetary atmospheres, for instance. I was fortunate to become a participant at an early stage, for there were ample opportunities. Everything was new and exciting.

NASA was founded in the late 1950s, with the study of the origin of the Solar System one of its goals. There was enough money to support both the manned space program and the unmanned probing of the planets. Earth-based observations, theory, and experiment were not neglected. For example, statistical studies made it apparent that collisions in the asteroid belt are much more frequent than had been supposed; this led to a great change in our understanding of asteroids: they are not individual rocks, but reconfigured piles of rubble.

Some universities, such as the University of California at Los Angeles, combined the planetary sciences into a united discipline. At the University of Arizona this was largely done under C. P. Sonett, Kuiper's successor. The study of the origin of the solar system became precise and detailed. Names and faces come back in my memory: Kuiper, Öpik, Urey, Whipple, Cameron, Anders, Wetherill, Safronov, and so many others. Please do not be offended if I did not mention your name, even though we know each other well from meetings and our space science books, so many friends and colleagues.

The 1960s and 1970s were a golden era for science in the United States. Since that glorious time, the NASA program seems to have lost some of its thrust and popularity; this was probably a natural reaction to the pinnacle of the moon landing in 1969. The effects were not felt until later in the 1970s. In order to promote the Shuttle, NASA designated it as the only vehicle to take scientific probes into space. It was a fateful decision, not only because of the Challenger disaster and the delays and demoralization that followed, but also because it was costly and improper. For science at the universities and in space programs, the US government in the 1980s seems to have more talk than action; good missions will soon be on the way, but there is a lack of new program starts and a loss of young students.

It is easy to complain, however, and this too will pass. The future always holds a sequence of ups and downs; human dynamics will eventually bring about the opposite of what people had reacted to. Space keeps beckoning, tempting us with new puzzles to solve. The current financial deficits seem to coincide with a decline in international tension, and that could bring a reasonable reduction in military spending. Would that our politicians and military leaders see the wisdom of reducing the overkill and spending some of the money saved, from quick industrial profits, on science and space exploration while not neglecting the homeless and destitute either. The United States could quickly become a leading nation again in the eyes of the world and our young ones.

I have great trust in the future, because of the renewal of life. Much to everyone's delight, Malthus, Spengler, and Orwell were wrong in their predictions of impending doom. Pessimists have always been wrong. My father was correct when he expected an all-out World War II; many people perished, including his son, but he would have admired his daughter-in-law's resilience and the reconstruction after that war. Winston Churchill, who foresaw the Iron Curtain descending upon Europe, might agree that people are finding some openings for lessening its effects. He might grudgingly grant that the suffering of the Second World War was not in vain, and perhaps even that the breakup of colonial empires has led to a fairer world. Mahatma Gandhi would agree with that, if he could see the world now, forty years after his death. We did our best in our imperfect ways, and many of the social injustices of the 1930s have been at least lessened. In the 1940s the murderous regimes of Germany and Japan were defeated. And we have not destroyed ourselves yet, though in 1945 the common prediction was that a nuclear World War III would come upon us within twenty-five years.

That the future cannot be known is its charm and adventure. Yes, Neil and Ellen, George and Jennifer, Jo-Ann, you have interesting times ahead. Do you agree that people and science are not irredeemably wicked and that human endeavor is not futile? Can you see the interesting questions, as well as our shortcomings, that we are going to pass on to you? You will never run out of ideas for asking new questions and finding solutions or for making fun and mischief. And, before we get off these fatherly remarks, you realize of course that before long you in your turn will hand the world over to your children and grandchildren.

For the latter, all you have to do is to spoil them and help them do new and naughty things that their parents, your children, will frown on.

Trusting that more good than bad exists within people, we can extrapolate a little into the future. A major nuclear war is becoming unlikely, thanks to studies of radiation effects and nuclear winter and the demonstrations of the horrors at Hiroshima and Nagasaki. (This does not mean, of course, that we should let up on our activities to press for peace. One cannot merely wish that war away.) I wager that the US and USSR will come to cooperate, as occurred so amazingly with the US and the People's Republic of China.

Natural catastrophes also seem generally within the domain of human control. Even asteroid impacts may be possible to avoid: It would be a turning point in human history if a menacing asteroid were discovered, as it would unite the great powers to do their utmost together to prevent the disaster from happening.

I wager that a genuine discovery of a planet of another star will be made before 1995. That would stimulate us into activity, especially if radio signals were received. There are bright perspectives in the exploration of space and the Earth with sensors on spacecraft. We are increasing our awareness of pollution and overpopulation.

Participation of thinking people in global problems is growing rapidly; there seems to be an exponential increase in information. There also seems to be a rise in searching for moral principles and a commitment to resist excesses and fanaticism and to promote the causes of human rights. We may take guidance and courage from the views of Anne and Paul Ehrlich, who are biologists with a broad experience in global issues[1]:

> It is entirely within the power of humanity to close the gap between rich and poor and to reduce the human population size to a level at which all people could lead a decent life without degrading the ecosphere. A transition to living primarily on income can be made; agricultural systems can be designed that would be highly productive and would also help to support the natural ecosystems in which they are embedded and on which they depend. Societies can turn their backs on racism, sexism, gross economic inequality and, above all, *war* as a mode of problem-solving. And people can learn to value political, social and cultural diversity, and to make the maintenance of organic diversity a quasi-religious duty.
>
> There are, in short, no insuperable barriers to creating a peaceful Earth in which *Homo sapiens* leads a rich existence without overstressing the natural systems that support human life—an Earth on which both biological and cultural evolution can proceed into the indefinite future. Unless, of course, the behaviour of our species

itself turns out to be the barrier. We might be cheered by the thought that many present behaviour patterns are quite new in evolutionary terms, the product not of hundreds of millions of years of biological evolution, but of a mere few millennia of cultural evolution. Humanity has the tools in hand to accelerate cultural evolution to the point where patterns that took thousands of years to develop can be altered in decades.

The great hope for civilisation lies in that fact: that people can recognise how the human predicament evolved and what changes need to be made to resolve it. No miracles, no outside intervention, and no new inventions are required. Human beings already have the power to preserve the Earth that everyone wants—they simply have to be willing to exercise it.

There *is* a problem of lethargy. A surprisingly small percentage of people cares enough to become involved, to act for the future of their children. Humankind shows a lack of love and hope for its offspring! We have found that only about 20% of the people who agree with the need and feasibility of an action (by, for example, giving financial support or writing letters) will actually participate in it. For the rest, I would like to be able to show them a poster used in our training during World War II. It shows a naive paratrooper sitting down flat in the field, happy after his successful arrival, perhaps in a daze or a fright, but not doing anything, while over the hill mean looking bullies with ugly weapons are about to terminate him. The poster said: "DON'T JUST SIT THERE. JUMP TO IT!"

I see the lethargy diminishing as global progress becomes clearer. After the war and into the 1970s the problems were paralyzing. The superpowers seemed about ready to go at each other's throat, and what could a single citizen do about that? But gradually popular actions such as massive demonstrations, as well as studies of nuclear winter and the like forced the leaders to reexamine options and consequences. Gradually they got together to discuss arms reductions. There may be setbacks again, but an awareness is rising that global war indeed may never come. When this becomes more securely established, people will grow more confident and apply their demonstrated powers to other urgent problems. It is a matter of education, of getting used to action. I did not march with the black people in Selma, Alabama, in 1965 when they needed our support, because it simply did not occur to me to do so. The idea to demonstrate at the Nevada nuclear test site in 1987 was not mine but came from David Grinspoon, one of our students, and his father, who was one of the national organizers.

Is the future bright? We cannot be naive and innocently optimistic. We have to live with limited resources and terrifying weapons.[2] The World Watch Institute publishes a yearly report, *The State of the World*,[3] on progress toward a sustainable society; it describes severe problems, but it also provides technical information toward solutions. The growth in our global awareness may be analogous to the growth in our personal lives: as we mature we gradually learn to live with reality and the world around us. We then also produce a new generation that looks at the mess we made and takes it over with renewed energy.

We should never underrate the younger generation. Chandrasekhar made a strong argument for that on the day after he received his Nobel prize in 1983. An experienced Swedish reporter set up a television interview and invited a young Ph.D. candidate in physics to join the laureates. He then confronted the older people with the young one. Did they believe that the new generation works as well as they did when they were young? There was a rather united reaction: The present generation works much less than did theirs. They described laboratories that nowadays seem to be dark and deserted on evenings, weekends, and holidays. All but Chandra. He said nothing, and I knew from Yerkes days that he probably would go against everyone else—if for no other reason than to stimulate the discussion—and that he would do it eloquently. Sure enough, the interviewer noticed something amiss after a while and asked Chandra for his opinion, at which he held forth that the present generation, any generation, will be as good and productive as the one that went before. I could not help but write him a letter then, needling him a little by saying that his students always were the better ones—like me, you know—and that we were scared enough of him that we did not dare stay away on evenings, weekends, and holidays. He wrote the following reply.

> I am, however, very suspicious of an older generation finding the younger generation inadequate. A striking illustration is provided by Lord Kelvin's despondent remark after attending the funeral of Stokes in 1907: "Stokes is now dead; Maxwell is no longer living; and Lord Rayleigh is not at the Cavendish any more. I shall never come back to Cambridge again." But the Cambridge which Kelvin mourned was soon replaced by the generation of Rutherford and Thomson, to be replaced once again by the generation of Hardy, Littlewood, Eddington, Dirac, Chadwick, Blackett, and a host of others. Similarly, when after Rutherford died and Bragg succeeded him, the generation of young Fellows of Trinity including Freeman Dyson, deplored the passing of a great period; and they became even more critical when Bragg brought in "clowns like Crick and gadgeteers like Ryle." Might I add that in time Crick and Ryle set the pattern for the Cambridge that followed. I think it is always unsafe to be adverse with respect to the future in terms of one's past experience.

This brings me to the ultimate reason for trust in the future, namely the basic atomic source of the restlessly moving force of life, the thrust to reproduce—orgasm, if you like, fulfillment is a nicer word. It drove our evolution and it assures our renewal. The basics of life and its movement therefore have nearly an infinity as their promise. As Mrinalini Sarabhai puts it[4]:

All life is movement. The planets, the universes, the elements all whirl in space. On the tiny earth, man too is constantly moving. Only in movement does he find the eternal stillness. In loneliness, in crowds, in bondage, in freedom, repeating the ordeal of birth after birth, man searches endlessly.

Thus our concept of the universe has come from its smallest subatomic entities with their ceaseless movement to the most advanced entity of people with their thrust toward fulfillment, toward an ideal that Rabindranath Tagore describes[5]:

> Where the mind is without fear and the head is held high;
> where knowledge is free;
> where the world has not been broken up into fragments
>   by narrow domestic walls;
> where words come out from the depth of truth;
> where tireless striving stretches its arms
>   toward perfection;
> where the clear stream of reason has not lost its way
>   into the dreary desert sand of dead habit;
> where the mind is led forward by Thee
>   into ever-widening thought and action—
> into that heaven of freedom, my Father,
>   let my country awake.

I first saw the poem near the entrance of the Physical Research Laboratory in Ahmedabad, posted at the place where on our building in Arizona we list the architect and the regents of the university. Each of us reacts to and connects with great poetry in a personal manner; I see this poem as a summary of past events and of a view ahead as follows. In the beginning there were fears of the unknown, of Satan and Hell, and of death and the end of the world: "A fear of the supernatural, which means that you daren't question anything or change anything. The late antique world was full of meaningless rituals, mystery religions, that destroyed self confidence," as the art historian Kenneth Clark has said.[6]

In modern times some of those fears still linger, but they vanish where knowledge replaces ignorance. The result is a liberation that brings a jubilant pride and a mature security. There must also be no fear to defend a worthwhile stand, at the price of blood if necessary. Inner peace brings up the heads so as not to miss the sunlit sky and the twinkle in each other's eyes.

I learned to read the next line of Tagore's poem as a reminder that knowledge must be free from control by religions just as it must be free from control by any regime or financial restriction. Our reason and the good within us will struggle with bigotry, religious fanaticism, excessive nationalism, totalitarian regimes, and powerful industrial lobbies. The world will never be perfect but at least the striving toward these goals elevates the best in humanity. The human spirit tends toward liberty.

"Is it the Truth?" should be the starting question before speaking, writing, dealing, or judging. I have seen that question hanging in offices of Hindu businessmen, probably because Mahatma Gandhi said truth is a godliness. It is all we can do, to aim at our honest view and recollection, even if it sometimes is embarrassing, and have others check and interact with them, for there is no absolute impersonal truth. Different people seem to remember different parts or aspects of an event. Perhaps the differences occur because there is an indeterminacy of individuals that makes them personalities. Everything we believe or remember is at least somewhat colored, distorted, and embellished by what happened afterwards, and by our bias, education, and environment to begin with.

All these ideas are useful only if turned into practical action. Without deeds there is nothing and we are nothing. We are what we do. The best within us can only be brought out in sustained practice; arts, sports, and sciences require long hours of work and concentration to attain a new goal. But we must first ensure our survival before we can aim at perfection. Complacency will not control overpopulation, nuclear proliferation, or degradation of the environment. Too many people avoid involvement out of selfishness and a lack of love for the children. The children in turn have to be taught to be involved and how to be involved, at home and in the schools as an integral part of the curriculum. Teaching them the origin and evolution of the atoms within us may help to show them the interrelatedness of people and their environment.

Everywhere we see an increase in rules and regulations, dead habits and useless laws. Not even the Soviet Union's Glorious Revolution or India's Independence seem to have brought fresh administrative approaches with less paperwork; there is a global tendency toward conformity and common denominators. There also are, however, human dynamics to recognize the excesses and to do something about them. If all thinking and concerned people would get into the habit of spending even a small fraction of their time on global and community issues, waves of renewal would ripple across the world. Bart Bok used to state that six days a week he spent in astronomy, and the seventh on our issues and environment. What we seem to need most is to seek a stronger moral background and clearly expressed ethical standards, and there are signs that the world is doing just that, together with a growing ecological concern. Ever widening thought and action will increase material and spiritual freedom, which in turn has, since the Renaissance and Golden Eras, set human minds free to pursue loftier goals.

Finally I see a heaven of truth in a universe so great, with such wonder in its energetic atoms and their diverse configurations, that it seems appropriate to focus our respect by calling that Truth our Father and the world our country.

27. Fulfillment on the Mountain

At first there is waiting, as if for a battle. The sky looks clear, the equipment is ready, the observer feels rested. Evening descends, but there is still time to look out and around from the mountaintop. The view goes far, to other mountains where other astronomers are also waiting in readiness. Valleys lie in between, with unseen people in pursuit of mischief or happiness. The eyes roam over the valleys and mountains, and as the shadows get longer and colors deepen it becomes easier to develop a sense of peace and an understanding of our place in the universe. From that mountaintop one sees a cycle from sunrise to sunset, from birth to death with terror and peace in between. The human condition is not all that different from that of the animals.

Slowly the eyes follow the path up the mountain, a trail made by early explorers, challenged and impelled to climb to its peak. They knew that on the summit they would find a fulfillment. They may have realized that the mountain is a symbol of human longing and conquest, and that on its peak they would rest and sense a fusion of space and time.

Astronomers knew that this mountain could help them to look back into the past. Here they would observe objects lightyears away to learn how the universe originated. They would build instruments to answer their questions: How do galaxies, stars, and planets originate, and what are their numbers, their conditions, and compositions? What molecules are there in space? So they built their observatory, those doers and dreamers, advisers and administrators, Indians and immigrants, too; and their engineers and technicians are still working there to maintain it all.

But the reverie must end. For an astronomer the oncoming darkness is a call to action. The telescope must be readied, like a ship for sailing or an altar for communion. Radio and telephone links to the computers on a distant campus must be established. Power must be turned on for all the equipment to translate into action the question asked of the heavens: "What are you?"

The astronomer must fit the observing to the conditions of the night, be it stormy or calm, cloudy or clear, turbulent or steady. His training lets him adjust to changes in conditions, even within himself, for sometimes he is not so rested nor at peace. But this is not the time to be disturbed, to let tiredness interfere. The turning on of drives and optics, computers and electronics, instructions and sequences, has an overriding effect on the mood and the feelings of the observer alone in his dome. Alone but not lonely, for now enter the work, the nightly chores and the data and plans for new research.

In free moments between instructions to the telescope and interactions with the computers there comes an urge to sing, quietly humming at first, but once it has started it doesn't want to stop and it goes on from one tune to another in a surprising array of youthful ditties, old soldier's ballads, long forgotten psalms and national anthems, and sometimes mere melodies that may be out of tune, the words forgotten, coming from the opera or who knows from where. This exuberant racket may continue all night, till exhaustion and with the neighbors believing that the bobcat again is in heat.

There is an urge to take a starbath, to shout from the rooftop a celebration of life, with its understanding or sometimes not knowing at all. The roof of the dome is a good place to look around, to check on the sky, to count the stars of the Pleiades or meteors dashing in, and it is superb to feel on top of the world, to dream of flying off and willing oneself, in a Superman soar, over valleys and mountains, and it is above all the finest site to meditate and listen for what one can learn from observing the universe.

Irresistibly one is forced to rise up
with the arms stretched widely out,
as if to offer back to the sky
all that was done and observed.

In the end, a jubilant cry comes out
to fuse the sunlight shining within
with the glow from the heavens above,
in extreme fulfillment and ecstasy.

This merges with all-time celebrations
of honest Christians and other religious,
who over the ages cried out to their Gods
and sometimes received a reply of their own.

Tuned in everywhere are myriad atoms
that sustain nature and people with strength
through their properties, always restless,
having evolved the trees of life and faith.

All cries seem to return as if reflected
by the asteroids and planets of stars,
like a carrier wave scrambled in time,
to bring an ultimate answer and reward:

"You are a Creator. You challenged the Truth.
So much You discovered already;
trust the future to carry it through.
Radiate Your freedom's security."

The answer brings a stream of strength
to climb back in and back to work
to continue to observe and enhance
the facts and the fate of this world.

Notes

Preface

1. *The Complete Poetical Works of John Milton*, edited by H. F. Fletcher (The Riverside Press, Cambridge, MA); the cited passage is on p. 289 in the 7th book of *Paradise Lost.*
2. F. J. Dyson, *Disturbing the Universe* (Harper and Row, New York, 1979).
3. People mentioned in the text have usually seen it and helped to improve it. The following helped in addition: D. P. Agrawal, Helmut Abt, Greet Ballintijn, Judith Bausch, Jagadish and Indira Bhattacharyya, Barend Biesheuvel, Wim Bijleveld, Helen Blenman, Ted Bowell, J. Bussels, Dale Cruikshank, Jim Cook, Steve Cox, Ellen d'Acquisto, Karen Denomy, Richard Dubinsky, David Evans, Aad Fokker, Ada Gehrels, Annemarie Gehrels, Jaap Gehrels, Jack Gehrels, J. F. Gehrels, Sr., Marijke Gehrels, Roland Gehrels, Bob Goff, F. J. Hermsen, Wil Hoogendoorn-Gehrels, Harrie Holla, John Jefferies, Janie Kamper, Inez Kleinhuis-Gehrels, G. R. La Rocque, Henk Latour, Bill Livingston, John Mathis, Gloria McMillan, Marjorie Meinel, Ron Milo, Jan Monster, Jayant Narlikar, Jyoti Nevatia, G. G. Norton, Bram Oort, Steve Ostro, Peter Pesch, Charles Peyton, Rita Qubti, R. Rajamohan, J. Rawal, Dirk Rustema, Terry Schemenauer, Jeff Schmidt, Willis Shapley, Harlan Smith, Ron Stottlemyer, Krisha Swami, Ien Swen, Barbara Timmermann, Mr. and Mrs. F. J. van Elburg, Karel van Houten, Arie D. Verhagen, Frank Verheest, E. Vernède, Mickey Wilson-van Biesbroeck, Wieslaw Wisniewski.

   Jack Frecker inspected the manuscript, and he made many suggestions for improvement. I thank the American Institute of Physics for the execution of this book. All my wishes were granted, which is rare in the book publishing business. The editor, Tom von Foerster, challenged every word and thought, often putting them in better order for a smoother flow, and yet he managed to leave in tact what I wanted to say in this book.

Chapter 1

1. However, we did not take seriously a more precise statement attributed to Dr. John Lightfoot, Vice Chancellor of Cambridge University in 1642: "Man was created by the Trinity on October 23, 4004 B.C. at nine o'clock in the morning." [Quoted in G. Daniel, *The Idea of Pre-History* (World, Cleveland, 1963), p. 25.]

Chapter 2

1. The three rivers were formidable obstacles to any north–south travel until the twentieth century, when they were bridged. They remained obstacles to troop movement still toward the end of World War II. As late as 1953, the combined effects of sea and wind caused floods that killed 1835 people, and alarmed the country into building additional waterworks of dams, dikes, and barriers. In the cities, parts of walls, gates, and ring canals can still be seen today; until the 1940s the canals served as sewers as well.
2. Eventually the Netherlands were to have thirteen provinces of which the western two—with the cities Amsterdam, Haarlem, Leiden, the Hague, and Rotterdam—are called North Holland and South Holland. The whole country is often called "Holland," but the proper name is "the Netherlands," and only the western part ought to be referred to as Holland.

Chapter 3

1. We have kept up the custom of writing poems with Christmas gifts. For instance, one of our Santas wrote the following with a gift of a sleeping bag for George working in Alaska, as is described in Chapter 15:

   Not only in the plain of Spain
   is there a lot of rain.
   Alaska has its share,
   especially when George is there.
   It is no joke,
   everything gets soak,
   hunched under a plastic coat,
   with his tent nearly afloat.

   The ink is smearing,
   the paper tearing,
   his hands are shaking,
   the clouds keep breaking.

   How to live on the strands
   of these beautiful lands?
   How about his sleeping bag?
   It looks like a rag!

   Well, count till three,
   look under the tree,
   and there you'll see,
   a gift just right for thee!

2. The strong connection between English and Frisian is explained in R. McCrum, W. Cran, and R. MacNeil, *The Story of English* (Viking, New York, 1986).

Chapter 4

1. J. Nehru, *Glimpses of World History* (John Day, London, 1942), p. 625.
2. L. Rose, *The Tulips are Red* (A. S. Barnes, Cranbury, NJ, 1978).
3. C. V. J. Murphy, "The Unknown Battle," *Life*, **17**, No. 16, p. 97 (October 16, 1944), shows a map for February 22, 1944, with Nijmegen, Arnhem, Deventer, and Enschede as industrial targets of opportunity. In *De Fatale Aanval* (de Gooise Uitgeverij, Weesp, the Netherlands, 1984), A. E. Brinkhuis describes the confusion of that day (which he learned from correspondence with crew members), the concern of Queen Wilhelmina, and the reprimands that were given the commanding officers later. The crews did not know the difference between the Netherlands and Germany. No industrial targets were hit, or aimed at, only trusting civilians as Dutch people would not seek shelter for Allied planes, and 880 of them died.
4. Soviet transmitters are similarly jamming Radio Free Europe.

Chapter 5

1. The laboratory where Cor worked at Philips is described by H. B. G. Casimir, *Haphazard Reality, Half a Century of Science* (Harper and Row, New York, 1983).
2. It could have turned out differently. If the Allies had not stopped in Belgium, they might have gone on to liberate the camp at Vught. Even if not, on the evening the train evacuating the Vught camp was to come through, a Resistance team was lying along the tracks to derail that train, but no train came. The Germans had routed it on an unexpected track toward Utrecht.
3. Others have reported differently, but they may have said so because they were reporting to our religious family.
4. V. E. Frankl, *From Death-Camp to Existentialism* (Beacon, Boston, 1959).

Chapter 6

1. Our society seems to be hard on small offenders while letting big crooks go free when they are smart enough to stay within the law. An example concerns the copper smelters of Arizona. We can see their thick smoke clearly from our mountains, while the people below, inside the smog, see it much less clearly. The smoke is difficult to regulate, but it is technically easy and economically not unreasonable to eliminate excessive pollution by sulfur dioxide. Mixed with the water in the lungs, SO_2 makes sulfuric acid, which destroys the minute cells that are essential for our intake of oxygen. The smelter smoke is therefore shortening the lives of millions of people and animals. With support from the World Bank, new mines and smelters have been opened near Nacozari and Cananea, some forty miles south of the border in Mexico which has even fewer environmental controls. Their output of SO_2 has been estimated as 1.1 million tons in 1985. The owners, managers, and stockholders of copper smelters that unnecessarily put out large quantities of SO_2 are therefore murderers by ethical standards, while being law-abiding citizens—and devout Christians, some of them—at the same time.
2. C. Ryan, *A Bridge Too Far* (Simon and Schuster, New York, 1974).
3. I was unbelievably lucky crossing the line near that bridge during a quiet spell. B. L. Montgomery, in *Normandy to the Baltic* (Riverside, Cambridge, MA, 1948), mentions the capture of the bridge the evening before and that soon afterwards there was again

fighting in the area. Winston Churchill mentions the bridge in *The Second World War. Triumph and Tragedy* (Houghton Mifflin, Boston, 1946), p. 193. It is "Joe's Bridge" in *A Bridge Too Far*, probably in honor of Lt. Col. Joe Vandeleur of the Armoured Guards.

4. F. Russell, *The Secret War* (Time-Life, Chicago, 1981), p. 43.

Chapter 7

1. The parachutists' song has various versions; my favorite is the one in the text; it was adapted for Dakota D.C. aircraft. I obtained another version in 1987 from T. H. Fitch, Curator of the Airborne Forces Museum at Aldershot, Hampshire; it is probably more original:

Oh come sit down by my side if you love me,
do not hasten to bid me adieu.
Just remember the brave parachutist
and the job he is trying to do.

When the red light goes on, we are ready
for the sergeant, to shout 'Number one.'
Though we sit in the plane close together,
we all tumble out one by one.

When we're coming in for a landing,
just remember the Sergeant's advice:
'Keep your feet and your knees close together
and you'll find mother earth very nice'.

When we land in a certain Country,
there's a job we will do very well:
We will fire old Goering and Adolf,
and those blighters can then go to hell.

So stand by your glass and be ready
and remember the men of the sky.
Here's a toast to the men dead already,
and a toast to the next man to die.

2. My last jump, years later at Ryan Field in Arizona, ended with broken toes.

3. E. Hazelhoff Roelfzema, *Soldaat van Oranje '40–'45* (in Dutch, Forum Boekerij, The Hague, no year). English edition: *Soldier of Orange 1940–1945* (Ballantine, New York, 1980).

4. After the war I learned that her real name was Wil van der Pol.

5. E. de Roever, in *Zij Sprongen Bij Maanlicht* (Hollandia, Baarn, the Netherlands, 1985), describes the missions of BBO in 1944 and 1945; this book is to appear in English translation in 1988. F. Visser, in *De Bezetter Bespied* (Thieme, Zutphen, the Netherlands, 1983), describes the Dutch Bureau for Information. Another description of the Resistance, of the fighting near Apeldoorn, and of the poor conditions in Dutch homes that last winter, is in P. Voute, *Only a Free Man. War Memories of Two Dutch Doctors (1940–1945)* (The Lightning Tree, Santa Fe, NM, 1982).

6. H. Giskes, *ABWEHR III F* (Bezige Bij, Amsterdam, 1949). It was translated into English with the title *London Calling North Pole* (Wm. Kimber, London, 1953).

7. The three principal culprits have been in court, but no convictions or punishment resulted. I have since learned that the Parachutist Wall was so named by the SS after they forced a large group of Dutch jews off that wall.

8. There were notorious traitors in the Netherlands, people who infiltrated the Resistance with generally disastrous results.

9. One cannot say that the England Spiel and other resistance and deceit swung the balance of the whole Russian war. Even with a million troops in addition to the three million they had in Russia, the Germans would eventually have lost. [The number comes from M. Vassilitchikov, *Berlin Diaries, 1940–1945* (Knopf, New York, 1987).] Walter Kerr, in *The Secret of Stalingrad* (Doubleday, New York, 1983), has documented that Stalin commanded over far greater reserves than the Germans, or the Allies, were aware of, and that the Soviet Union was never close to collapse. Ironically, what Stalin wanted so much—the Allied invasion of Western Europe before 1944 —he did not get because of his secrecy toward the Allies. A lack of information can affect world history, as Kerr showed through his remarkable detective work; this supports Arthur Clarke's thesis in Chapter 14.

10. Cieremans' report was given to me by Eddy de Roever.

11. Two were overlooked and left in their cells.

Chapter 8

1. We have not been able to resolve the date of my going to the Far East. I and some of my family remember that it was in June, to fight the Japanese. Colleagues remember that they were recruited later in August to jump to concentration camps that were still held by the Japanese.

2. In fairness to our Commander, I should say that I now know that the British did not allow Dutch military to leave Ceylon for Java, for reasons of international politics.

3. In 1973 I made another visit to Malaysia and had an interview with Prime Minister Tungku Abdul Rahman in which he lectured me on two topics of obvious pride and pleasure: He had postponed Independence for Malaysia until the transition could occur smoothly, without fighting the British (but they did have to fight communist guerrillas in mountainous jungles); and he had succeeded in having Malaysians take over many of the functions from the British at the same pay scales, thereby creating a middle class as well as the least corrupt, according to him, administration of the new countries in Southeast Asia.

4. Amnesty International has established that the Indonesian government massacred large numbers of Chinese and communists in Java in 1965 and Portuguese and nationalists on the island of Timor in the late 1970s. They estimate that some 100,000 Papuas were killed in aerial bombardments and mass executions. The Indonesian army used to receive some $50 million per year in military supplies from the United States Government, and the World Bank has supported the immigration of Javanese into Irian Jaya.

Chapter 9

1. I was tested at the Nederlandsche Stichting voor Psychotechniek, Wittevrouwenkade 6, Utrecht.
2. An updated version is: J. H. Oort, "Empirical Data on the Origin of Comets" in *The Moon, Meteorites, and Comets*, edited by B. M. Middlehurst and G. P. Kuiper (University of Chicago Press, Chicago, 1963): *The Solar System*, Vol. IV.
3. H. B. G. Casimir, one of the professors in Leiden, has given a description in *Haphazard Reality, Half a Century of Science* (Harper and Row, New York, 1983).
4. In our history classes we were taught that Leiden had been offered a choice between a university and freedom from taxation for a number of years, but this is merely a legend according to Rector Magnificus J. N. Bakhuizen van der Brink, Algemeen Handelsblad **128**, June 3, 1955.

Chapter 10

1. E. Hazelhoff Roelfzema, *Rendezvous in San Francisco* (Atlas Reeks, Utrecht, 1939), is in Dutch and out of print. It would be nice to have it in English translation as it contains interesting views on country and people, and Hazelhoff is a born writer.
2. The Russians now have a 6-meter telescope, and the Palomar astronomers are building a 10-meter reflector on Mauna Kea, Hawaii.
3. H. Wright, *Explorer of the Universe, A Biography of George Ellery Hale* (Dutton, New York, 1966), describes a more reliable history of the origin of the Yerkes Observatory. I have the story from Van Biesbroeck, a known raconteur; it may yet have the essence of Hale's scheming.
4. I do not remember the source of this anecdote.

Chapter 11

1. During the checking I did for this chapter, colleagues have remarked that there may have been too much isolation and that there seemed to be too much concern for oneself and one's credit and scientific reputation. Otto Struve in 1950 mentioned the "boastfulness and self-admiration. This, as you know, tended to undermine the success of the Yerkes organization, and since I was not able to cope with it, I preferred to leave." (See *Sky and Telescope*, April 1988, p. 30.) To us students, however, such opinions could only come later.
2. Perhaps one should include a third type: the instrumentalist who builds new observatories and their instruments.
3. D. S. Evans and J. D. Mulholland, *Big and Bright. A History of the McDonald Observatory* (University of Texas Press, Austin, 1986), 186 pp.
4. My information on Kuiper's childhood in Haringcarspel is based on a visit there in 1985. I interviewed Maarten Burger, who had been a classmate.
5. Kuiper had read Descartes when he was quite young, according to the biographical memoir by D. L. Cruikshank in the Proceedings of the National Academy of Sciences (1988).
6. The Yerkes 24-inch reflector is now in the American History building of the Smithsonian Museum in Washington, D.C.

7. There is a report of an astronomer who died at a telescope of the Bonn Observatory in Germany in 1850, but reportedly that was caused by too much drink. The protrusion on the dome of the 4-meter reflector on Kitt Peak that Marc was using was a guide-rail for the lowest shutter segment. There was a warning sign, and there was an interlock such that the power of the dome rotation came off and its brakes came on if the door was opened, but the dome coasted briefly in its rotation at exactly the place to force the door shut.

Chapter 12

1. T. Gehrels and T. M. Teska, "The Wavelength Dependence of Polarization," Appl. Opt. **2**, 67 (1963).
2. M. D. Ross, "We Saw the World from the Edge of Space," *National Geographic Magazine* **120**, 670 (1961).
3. An early flight went up to about 20,000 feet when the balloon burst, but it was still a useful test and our equipment worked well.
4. The balloon program in India has used the cheaper hydrogen in hundreds of flights without accident.
5. T. Gehrels, "Ultraviolet Polarimetry Using High-Altitude Balloons," Appl. Opt. **6**, 231 (1967).
6. We had to take into account that the signals, traveling at the speed of 300,000 kilometers/second, would take about 80 minutes to travel one way between Earth and Saturn.

Chapter 13

1. J. Elliot and R. Kerr, *Rings: Discoveries from Galileo to Voyager* (MIT Press, Cambridge, MA, 1984) describes the occultations of Uranus and Neptune. Jim Elliot was the principal investigator on the Kuiper Airborne Observatory, the airplane named in honor of Gerard Kuiper, which carries a 91-centimeter pointable telescope.
2. The Tenerife accident is reported in *Aviation Accident Digest*, No. 23, Circular 153-AN/98, available from the International Civil Aviation Organization, 1000 Sherbrooke St. W., Suite 400, Montreal, Quebec, Canada H3A 2R2, for $10. J. M. Ramsden has evaluated the Spanish report in "Tenerife: the last analysis," Flight Int. **115**, 194 (1979).
3. N. Williams and G. Otis, *Terror at Tenerife* (Bible Voice, Van Nuys, CA, 1977).

Chapter 14

1. Surprisingly, some friends and colleagues who read this in the manuscript stage did not understand the word either: dollars!
2. V. Sakharov and U. Tosi, *High Treason* (Putnam, New York, 1980), p. 22.
3. *U.S.S.R. News Briefs, Human Rights*, published fortnightly by Das Land und die Welt e.V., Schwanthalerstr. 73, 8000 München 2, FRG.
4. A Letter to the Editor in *Physics Today*, October 1987, p. 152, presents a plea and some guidance for visits with refuseniks.

5. Four volumes of Andrei Sakharov's statements on public issues have been published in the United States: *Progress, Coexistence, and Intellectual Freedom* (Norton, New York, 1968); *Sakharov Speaks* (Knopf, New York, 1974); *My Country and the World* (Knopf, New York, 1975); *Alarm and Hope* (Knopf, New York, 1978). Other sources include: *On Sakharov* (Knopf, New York, 1982), which is based on a Festschrift compiled in Moscow and contains several articles written by Sakharov since 1978; *Collected Scientific Works* (Marcel Dekker, New York, 1982), which contains English translations of his scientific papers with commentary by Sakharov himself and by Western physicists. A letter by Sakharov on nuclear-arms issues in *Foreign Affairs*, June 1983, p. 1001.

6. L. Goldberg and T. Gehrels, "Reply to Bok," *Sky and Telescope* **63**, 445 (1982).

7. *The Fourth International Sakharov Hearing*, edited by S. Reznik (Overseas Pub. Interchange, London, 1983) (in Russian).

8. The bad treatment of Sakharov and his wife in Gorky are described in her book: E. Bonner, *Alone Together* (Knopf, New York, 1986).

9. *Unforgettable Fire. Pictures Drawn by Atomic Bomb Survivors* (Nippon Hoso Shuppan Kyokai, Tokyo, 1977).

10. The Boston Study Group, *The Price of Defense. A New Strategy for Military Spending* (Times Books, New York, 1979); The Boston Study Group, *Winding Down. The Price of Defense* (Freeman, San Francisco, 1982).

11. The Center for Defense Information, 1500 Massachusetts Ave., N.W., Washington, D.C. 20005. The center receives no funds from government or military contractors; it is financed by voluntary tax-deductible contributions.

12. The Eisenhower quote was transcribed from his televised presentation as we heard it on a NOVA television program in March 1987.

13. A. C. Clarke, "Star Wars and Star Peace," *Nineteenth Jawaharlal Nehru Memorial Lecture*, New Delhi, November 13, 1986 (India Offset Press, A-1 Mayapuri, New Delhi).

14. An example of a simple action to avoid what could be a major disaster: The Davis–Monthan Air Force Base lies in the desert south of Tucson; it has a single runway from which the controllers used to send, for no apparent reason, loaded KC-10 and KC-135 tanker planes taking off over the city and the University of Arizona campus, rather than over the desert. Failure of just one engine could cause a conflagration killing thousands. Once one of them seemed to just barely miss the basketball fieldhouse when there were 13,000 people in it. Letters to the proper places—the White House included—has stopped the practice.

15. I. Halperin, Department of Mathematics, University of Toronto, Toronto, Canada M55 1A1.

16. Amnesty International, 42 W. 88th St. No. 5R, New York, NY 10024. There is a national office in nearly each country.

17. Greenpeace, 2007 R Street., N.W., Washington, D.C. 20009. The Humane Farming Association (1550 California St. Suite 6, San Francisco, CA 94109) provides information on the shocking conditions for animals in large farms, and of dangers for human consumers, too, because of the use of pesticides and antibiotics; reading their material made me a vegetarian.

18. Common Cause, 2030 M Street N.W., Washington, D.C. 20009. The Center for Innovative Diplomacy (466 Green St., San Francisco, CA 94133) publishes the *Bulletin of Municipal Foreign Policy* and promotes nuclear free zones, sister cities in Nicaragua and the Soviet Union, and the conversion of local weapons plants to peacetime purposes through local city and county governments; this provides an alternative for trying to influence the US Congress.

19. The black list is reportedly kept in a computer at Fort Huachuca, Arizona. Our younger people were instructed not to cross the line, not to get arrested, as there is a fear of repercussion in case they later would apply for a federal position.

Chapter 15

1. His full name is Cornelis Adolf, as he was named after Cor.
2. E. A. Wood, *Science from Your Airplane Window* (Dover, New York, 1968) discusses some measurements one can make, and also the use of Polaroid glasses. She has written a children's version as well.
3. A favorite tale among flyers is of a new bombardier guiding his pilot over a target to be bombed: "Left a bit... Right a bit... Left a bit... Back a bit..."
4. "Science and the Citizen," *Scientific American*, May 1986, p. 70. The column contains an interview with George concerning the displacement and accretionary history of the Alexander terrane.

Chapter 16

1. T. Gehrels, "Les Asteroïdes. Études Fondamentales," l'Astronomie **98**, March 1984, p. 115, and April 1984 p. 159 (in French).
2. Pictures of the model of the yoke-mounted 72-inch are on the cover of *Sky and Telescope* **42**, March 1971, with a description on p. 170.
3. Mrs. Carolyn Shoemaker joined the search with the Palomar 18-inch Schmidt later. She has a special aptitude for discovering comets and near-Earth asteroids with a stereoscopic device that Gene designed to inspect each pair of photographic plates.
4. K. Serkowski, "Should We Search for Planets Around Stars?" Astron. Q. **1**, 5 (1977).

Chapter 17

1. R. Sather, "High School Teachers as Astronomers," *The Physics Teacher* **15**, 86 (1977), carries a description of the work by our high-school teachers.
2. *Physical Studies of Minor Planets*, edited by T. Gehrels, NASA SP-267 (1971), is actually not a part of the Space Science Series. It was a forerunner and a learning experience for the series. The following books have been published or are in preparation in the series (all are published by the University of Arizona Press, Tucson): *Planets, Stars and Nebulae, Studied with Photopolarimetry*, edited by T. Gehrels (1974).

   Jupiter, edited by T. Gehrels (1976); Russian translation published by Mir, Moscow.

   Planetary Satellites, edited by J. A. Burns (1977); Russian translation published by Mir, Moscow.

   Protostars and Planets, edited by T. Gehrels (1978); Russian translation published by Mir, Moscow.

   Asteroids, edited by T. Gehrels (1979).

   Comets, edited by L. L. Wilkening (1982).

   Satellites of Jupiter, edited by D. Morrison (1982); Russian translation published by Mir, Moscow.

   Venus, edited by D. M. Hunten, L. Colin, T. M. Donahue, and V. I. Moroz (1983).

   Saturn, edited by T. Gehrels and M. S. Matthews (1984).

   Planetary Rings, edited by R. Greenberg and A. Brahic (1984).

   Protostars and Planets II, edited by D. C. Black and M. S. Matthews (1985); being translated into Chinese.

Satellites, edited by J. A. Burns and M. S. Matthews (1986).

The Galaxy and the Solar System, edited by R. Smoluchowski, J. N. Bahcall, and M. S. Matthews (1986).

Mercury, edited by F. Vilas, C. R. Chapman, and M. S. Matthews (1988, in press).

Meteorites and the Early Solar System, edited by J. F. Kerridge and M. S. Matthews (1988, in press).

Origin and Evolution of Planetary and Satellite Atmosphere, edited by S. K. Atreya, J. B. Pollack, and M. S. Matthews (1988, in press).

Asteroids II, edited by R. P. Binzel, T. Gehrels, and M. S. Matthews (1989, in preparation).

Uranus, edited by J. T. Bergstralh, E. D. Miner, and M. S. Matthews (1989, in preparation).

Interior and Atmosphere of the Sun, edited by A. N. Cox, W. C. Livingston, and M. S. Matthews (1990, in preparation).

Mars, edited by H. M. Kieffer, B. M. Jakosky, C. W. Snyder, and M. S. Matthews (1990, in preparation).

The Sun in Time, edited by C. P. Sonett, M. S. Giampapa, and M. S. Matthews (1990, in preparation).

Comets in the Post-Halley Era, edited by P. L. Newburn, J. Rahe, and M. S. Matthews (1990, in preparation).

3. H. Shapley, *Through Rugged Ways to the Stars* (Scribner, New York, 1969).

Chapter 18

1. D. S. Evans and J. D. Mulholland, *Big and Bright. A History of the McDonald Observatory* (University of Texas Press, Austin, 1986).

2. Spacewatch Fund, Business Office, Space Sciences Building, University of Arizona, Tucson, Arizona 85721. Contributions are tax-exempt.

3. S. Cole, J. R. Cole, and G. A. Simon, "Chance and Consensus in Peer Review," Science **214**, 881 (1981).

4. E. J. Öpik, "About Dogma in Science, and Other Recollections of an Astronomer," Annu. Rev. Astron. Astrophys. **15**, 1 (1977).

5. S. C. Wolff, Mercury **17**, 29 (1988).

6. E. Stuhlinger, H. Alfvén, G. Arrhenius, R. Bourke, B. Doe, S. Dwornik, A. Friedlander, T. Gehrels, C. Guttmann, D. Strangway, and F. Whipple, *Comets and Asteroids. A Strategy for Exploration*, NASA TMX-64677 (1972). It was followed with more detail by F. Whipple, K. Atkins, G. Arrhenius, J. Brandt, B. Doe, T. Gehrels, N. Ness, T. Owen, S. I. Rasool, E. Stuhlinger, and J. Wasson, *The 1973 Report and Recommendations of the NASA Science Advisory Committee on Comets and Asteroids. A Program of Study*, NASA TMX-71917 (1973).

7. A simple system would be to have a single rank for the various people who work with us on the mountain, or as secretaries in town, and a second rank for their supervisors. We would pay them according to supply and demand, and capabilities and accomplishment.

8. This paragraph has been strongly endorsed by eight senior astronomers connected with the observatory or its Board of Directors. Two board members seemed neutral, and one disagreed, pointing out that the decision to have a Superdirector was made by the National Science Foundation, although the board could have objected, and that the closing of telescopes was unavoidable because of shortages in federal funding.

9. In his foreword to *Normandy to the Baltic* (Riverside, Cambridge, MA, 1948), Field Marshal B. L. Montgomery acknowledges the Allied team, General Eisenhower, and hospitality received in the liberated countries, but gives not a word about the wounded and the dead.

10. W. von Braun and F. I. Ordway, *History of Rocketry and Space Travel*, 3rd ed. (Crowell, New York, 1975), p. 108.

11. T. Gehrels, "Les Asteroïdes. Études Fondamentales." l'Astronomie **98**, March 1984, p. 115 and April 1984, p. 159, has the original version of my von Braun Lecture, translated into French.

12. A part of the area of the Dora concentration camp is preserved as an impressive memorial, near Nordhausen in East Germany. There are plans to open up also a section of the underground rocket factory as a memorial.

13. German underground activities are documented in W. Schumann and W. Bleyer, *Deutschland im zweiten Weltkrieg* (Pahl-Rugenstein, Köln, 1984), Vol. 5, p. 243 ff.

Chapter 19

1. T. Gehrels, "Possibilities for Astrophysics in Sri Lanka," in *Fundamental Studies and the Future of Science*, edited by C. Wickramasinghe (University College Cardiff Press, Cardiff, 1984), p. 377, has a proposal for the construction of an astronomical telescope in its Appendix.

2. A. C. Clarke, quoted in the epilogue of R. Beny and J. L. Opie, *Island Ceylon* (Viking, New York, 1971), p. 221.

3. A. C. Clarke, *Ascent to Orbit, A Scientific Autobiography* (Wiley, New York, 1984).

Chapter 20

1. D. K. Mishra, *Five Eminent Scientists* (Kalyani, Delhi, 1976), contains biographies of Raman, Saha, Bose, Bhabha, and Sarabhai.

2. C. V. Raman, quoted in W. A. Blanpied, "Pioneer scientists in pre-independence India," *Physics Today* **39**, May 1986, p. 36, an article on six Indian physicists including Saha, Bose, Raman, and Bhabha.

3. A. S. Eddington, quoted in S. Chandrasekhar, *Eddington* (Cambridge University Press, Cambridge, 1983), p. 17.

4. V. A. Sarabhai, *Science Policy and National Development* (MacMillan, Delhi, 1974).

5. P. L. Kapitza, quoted in the obituary by J. Rotblat, *Physics Today* **37**, September 1984, p. 95.

6. P. R. Pisharoty, in an unpublished article describing the Vikram Sarabhai Science Centre in Ahmedabad (December 1985).

7. J. Nehru, *Glimpses of World History* (John Day, London, 1942); a quote from this book is in Chapter 4.

8. This note was discovered in Mrs. Gandhi's own handwriting among her papers a few days after her death. There was no clue as to its date. (H. Y. Sharada Prasad, personal communication, 1978.)

9. *Time*, November 12, 1984.

10. P. Jayakar, "Reminiscences," transcript of an interview on Indian television, which she kindly sent me.

Chapter 21

1. Taken from a book of quotes, *Gandhi* (Voltas, Bombay, no date).
2. My nephew Marc Gehrels stayed in India six weeks, mostly in hostels and close to the people, and he strongly disagrees, saying that the people were constantly calling him and wanting to talk with him. I may have a bit of a colonial or upper-caste attitude not to react, and this may have a deterrent effect; the times I have to say "bug off" are rare.
3. Ahmedabad would seem to be the appropriate spelling, as the city was founded by Ahmed Shah in 1411. See K. L. Gillion, *Ahmedabad, a Study in Indian Urban History* (University of California Press, Los Angeles, 1968).
4. Mrinalini's sister Lakshmi, known as "Captain Lakshmi," fought the British during World War II in the Indian National Army under the command of Subbash Chandra Bose. She later became a medical doctor and active Marxist.
5. M. Sarabhai, "The Eight Nanikas: Heroines of the Classical Dance of India," Dance Perspectives **24**, 5 (1987).
6. M. Sarabhai, *This Alone Is True* (Hind, Delhi, 1977).

Chapter 22

1. B. Bok, quoted in *Lo Que Pasa*, January 24, 1983 (edited somewhat).
2. M. K. Gandhi, *An Autobiography, or The Story of My Experiments with Truth* (Penguin, Hammondsworth, UK, 1983); originally published in Gujarati as *Satya na Prayogo (Experiments in Truth)* (1927).
3. I should warn you that these ideas may not be acceptable to my colleagues. I worked them out in two papers and sent them to two different scientific journals, and they were returned as being of insufficient interest.

   The question is whether there is more to this than mere analogies of atomic properties and human characteristics, whether or not one can show that the properties of atoms propagated to determine the characteristics of people. This deserves further study, but it already now seems to me not to be a one-for-one correspondence. For instance, it may not be possible to prove rigorously that the dualism of atoms, in particles and waves, has propagated to the dualism of people, in body and thought. Similarly, it may not be possible to show that the fermions and bosons are the primary cause for the two genders generally encountered in nature. Atomic structure and life are too complex for that.

   However, the interaction of particles produces new particles; the mating of photons and electrons brings about new states that can give rise to other forms of energy. Holistically the connection seems inescapable. All the properties of atoms may be compared in their ensemble with all the characteristics of molecules, of crystals next, plants, animals, and finally people. The properties of the atoms are so phenomenal that there seems no problem for them to have produced the characteristics of people. The opposite must also be true: people must have obtained their characteristics from atoms for there was no other basic ingredient. The primordial, energetic atoms are the complete clue to all the characteristics of people and everything else in the universe. They are what everything is made of.

   These may be merely stabs in the dark at this point, but the ideas seem worthy of further pursuit. For me these ideas are the logical result of seeing the smallness of atoms and also the immensity of the observable universe, of seeing it from far outside that universe, as if we were gods ourselves. It is my latest liberation. I am thereby free to go a further step with the atomic analogies, namely to extrapolate the recycling of everything in the universe—the birth and death of photons and other subatomic particles, plants and people, of stars and stellar systems—to the universe itself. I make the simple heuristic argument that our universe must be a "closed system," that is, it must have enough mass to reverse the present expansion and thereby eventually contract to be regenerated again. This universe is a system fit enough to survive, having enough mass to cycle and regenerate itself. From that broad perspective I can imagine our universe among many others that have a diversity of masses, as the stars do. I then extrapolate the scales of time and space from those of the stars, taking account that massive stars recycle quickly, while the sun's cycle is slow enough for our type of life to have formed. Similarly, the mass of our universe apparently is not too large, or it would have collapsed gravitationally already. Nor is its mass too small, otherwise it would not have recycled; a previous collapse would explain how the present cycle of the universe began so hot, dense, and relatively uniform. [J. V. Narlikar and K. M. V. Apparao have modeled such a "bouncing cosmology" in Astrophys. Space Sci. **87**, 333 (1982).]

4. The volume of a person is about $5x10^4$ cubic centimeters and that of an atom is on the order of 10^{-23} cm^3. It follows that we have some $5x10^{27}$ atoms. By weight one obtains nearly the same result: 60 kilograms, say, divided by 10^{-23} grams.

5. F. H. Shu, *The Physical Universe, An Introduction to Astronomy* (University Science Books, Mill Valley, CA, 1982), gives an overview on the origin of atoms, as well as on human evolution.

6. W. Heisenberg, quoted in F. Capra, *The Tao of Physics. An Exploration of the Parallels Between Modern Physics and Eastern Mysticism* (Shambala, Boulder, 1975), p. 50; the book provides a basic introduction to atomic physics.

Chapter 23

1. A. Huxley, *The Art of Seeing* (Harper and Row, New York, 1942); *The Doors of Perception and Heaven and Hell* (Harper and Row, New York, 1954).

2. H. Mark, *The Space Station. A Personal Journey* (Duke University Press, Durham, NC, 1987). Mark describes his involvement with the Shuttle in this book.

Chapter 24

1. Numbers such as 72,000 are usually not to be taken literally, but rather to indicate that there is a multitude.

2. The exercises in India are not done in full sunshine, but rather in the early morning and evening. I do not know why there seems to be an avoidance of sunlight; there is no danger of skin cancer if one keeps moving, not exposing the same side longer than ten minutes or so.

3. M. D. Corbett, *Help Yourself to Better Sight* (Wilshire, North Hollywood, CA, 1949), which also includes relaxation exercises for better hearing; J. Jackson, *Seeing Yourself See, Eye Exercises for Total Vision* (Dutton, New York, 1975); G. Downing, *The Massage Book* (Random House, New York, 1972), is one of a series on holistic health. An introduction to eye care is: R. A. Kraskin, *How to Improve your Vision* (Wilshire, North Hollywood, CA, 1968), from which one learns that eye care and exercises should begin at a young age.

4. B. K. S. Iyengar, *Light on Yoga*, rev. ed. (Schocken, New York, 1977).

5. *Serendipity and the Three Princes*, edited by T. G. Remer (University of Oklahoma Press, Norman, OK, 1965).

Chapter 25

1. Our secretary, Shirley Marinus, as she was typing this asked the following questions, which others have also asked, and which I shall now try to answer.

Have you ever been in a situation where you asked for help from a higher power, as in a sickness or death in the family?

Last year when Neil called with the shocking news that their little daughter had just died, I flew out to be with them, near Washington, D.C., and flew back to Tucson the same evening. They had no need of help from a higher power, nor did I. But on the plane that night back to Tucson, the stewardess woke me because I was crying in my sleep; I told her why, and as she walked away she was crying too, while I meditated myself back to sleep. The point is that Neil, Ellen, and I—and Cor, too, apparently—would have felt odd to ask for help from any other than each other. From whom, after all? And from Whom would He ask for help?

Why is there such a rich diversity in forms of life—fish, flowers, babies, and the rest?

The atomic force is unimaginably resourceful and eternally seeks new forms; even within atoms "selection rules" give rise to a rich variety of structures.

Why do we have feelings, such as joy and sadness, rather than just going through life mechanically?

It may have something to do with the conditions at the beginning of the universe, when the great energies gave atoms esoteric properties.

Why is there a "male" and "female" in every living thing?

Probably because there are fermions and bosons, which are distinctly different, but interacting, among the elementary particles.

What was there before the beginning of the universe?

Probably another universe. But I do not really know. It will be a nice problem for our grandchildren!

2. Still, much is stated without explanation or reason; for instance, in yoga one is told not to eat spicy food, although such food seems healthy to me. In such situations I want to reason, question, and challenge. Descartes did have an influence on me, too.

3. Lin Yutang, *The Importance of Living* (Reynal and Hitchcock, New York, 1938).

4. F. Capra, *The Tao of Physics. An Exploration of the Parallels between Modern Physics and Eastern Mysticism* (Shambala, Boulder, 1975).

5. J. C. Smuts, *Holism and Evolution* (Greenwood, Westport, CT, 1973); originally published in 1926.

6. *Gandhi* (Voltas, Bombay, no date), a book of quotes. Gandhi's last utterance is described by his adopted daughter M. Gandhi in *The End of an Epoch* (Navajivan, Ahmedabad, no date).

Chapter 26

1. A. H. Ehrlich and P. R. Ehrlich, *Earth* (Franklin Watts, New York, 1987), p. 248.

2. F. J. Dyson, *Weapons and Hope* (Harper and Row, New York, 1984) carries as a byline to its title: "Hope is the capacity to live with danger without being overwhelmed by it; hope is the will to struggle against obstacles even when they appear insuperable."

3. World Watch Institute, *State of the World* (W. W. Norton and Co., New York, annually). The institute has an international staff and board of directors. In 1986 the report included chapters, for example, on deficits, ecology, water efficiency, range lands, energy supply, electric power, nuclear power, banishing tobacco, children, Africa's decline, and national security.

4. M. Sarabhai, *Creations* (Mapin, Ahmedabad, 1986; distributed by University of Washington Press, Seattle), p. 78.

5. R. Tagore, *Collected Poems and Plays* (MacMillan, London, 1937), p. 16.

6. K. Clark, *Civilization, a Personal View* (Harper and Row, New York, 1969), p. 4.

Index

C

D

E

F

G

H

I

J

K

L

M

N

O

P

Q

R

S

T

U

V

W

X

Y

Z